New Experimental Modalities in the Control of Neoplasia

NATO ASI Series

Advanced Science Institutes Series

A series presenting the results of activities sponsored by the NATO Science Committee, which aims at the dissemination of advanced scientific and technological knowledge, with a view to strengthening links between scientific communities.

The series is published by an international board of publishers in conjunction with the NATO Scientific Affairs Division

A	Life Sciences	Plenum Publishing Corporation
B	Physics	New York and London
C	Mathematical and Physical Sciences	D. Reidel Publishing Company Dordrecht, Boston, and Lancaster
D	Behavioral and Social Sciences	Martinus Nijhoff Publishers
E	Engineering and Materials Sciences	The Hague, Boston, and Lancaster
F	Computer and Systems Sciences	Springer-Verlag
G	Ecological Sciences	Berlin, Heidelberg, New York, and Tokyo

Recent Volumes in this Series

Volume 112—Human Apolipoprotein Mutants: Impact on Atherosclerosis and Longevity
edited by C. R. Sirtori, A. V. Nichols, and G. Franceschini

Volume 113—Targeting of Drugs with Synthetic Systems
edited by Gregory Gregoriadis, Judith Senior, and George Poste

Volume 114—Cardiorespiratory and Cardiosomatic Psychophysiology
edited by P. Grossman, K. H. Janssen, and D. Vaitl

Volume 115—Mechanisms of Secondary Brain Damage
edited by A. Baethmann, K. G. Go, and A. Unterberg

Volume 116—Enzymes of Lipid Metabolism II
edited by Louis Freysz, Henri Dreyfus, Raphaël Massarelli, and Shimon Gatt

Volume 117—Iron, Siderophores, and Plant Diseases
edited by T. R. Swinburne

Volume 118—Somites in Developing Embryos
edited by Ruth Bellairs, Donald A. Ede, and James W. Lash

Volume 119—Auditory Frequency Selectivity
edited by Brian C. J. Moore and Roy D. Patterson

Volume 120—New Experimental Modalities in the Control of Neoplasia
edited by Prakash Chandra

Series A: Life Sciences

NATO Advanced Study Institute on New Experimental Modalities in the Control of Neoplasia (1985: Maratea, Italy)

New Experimental Modalities in the Control of Neoplasia

Edited by
Prakash Chandra
University of Frankfurt
Frankfurt am Main, Federal Republic of Germany

RC270.8
N39
1985

Plenum Press
New York and London
Published in cooperation with NATO Scientific Affairs Division

Proceedings of a NATO Advanced Study Institute on
New Experimental Modalities in the Control of Neoplasia,
held September 23-October 3, 1985,
in Maratea, Italy

Library of Congress Cataloging in Publication Data

NATO Advanced Study Institute on New Experimental Modalities in the Control
of Neoplasia (1985: Maratea, Italy)
  New experimental modalities in the control of neoplasia.

  (NATO ASI series. Series A, Life sciences; v. 120)
  "Proceedings of a NATO Advanced Study Institute on New Experimental Mo-
dalities in the Control of Neoplasia, held September 23-October 3, 1985, in
Maratea, Italy"—T.p. verso.
  "Published in cooperation with NATO Scientific Affairs Division."
  Includes bibliographies and index.
  1. Cancer—Treatment—Evaluation—Congresses. 2. Cancer—Chemotherapy
—Evaluation—Congresses. 3. Immunotherapy—Evaluation—Congresses. I.
Chandra, Prakash. II. North Atlantic Treaty Organization. Scientific Affairs Divi-
sion. III. Title. IV. Series. [DNLM: 1. Neoplasms—drug therapy—congresses. 2.
Neoplasms—immunology—congresses. QZ 267 N279n 1985]
RC270.8.N39   1985   616.99′2061   86-22709
ISBN 0-306-42464-9

© 1986 Plenum Press, New York
A Division of Plenum Publishing Corporation
233 Spring Street, New York, N.Y. 10013

All rights reserved. No part of this book may be reproduced, stored in a retrieval system,
or transmitted in any form or by any means, electronic, mechanical, photocopying,
microfilming, recording, or otherwise, without written permission from the Publisher

Printed in the United States of America

Preface

Acquisition of new knowledge about the biological and biochemical nature of neoplastic cells has led to the design and development of several experimental approaches in the treatment of cancer. These approaches emerge from the recent work in tumor virology, e.g. the control of vital cellular genes by viral regulatory signals; the implication of monoclonal antibodies as a vehicle for the targeted drug delivery and selective destruction of tumor cells; immunologic advances in the recognition of some specific events during metastatic growth; the role of biological response modifiers in modifying or reversing malignant growth; and biochemical advances, such as the role of gene amplification in drug resistance and the approaches to the reversal of resistance in drug refractory cancer, the role of membranes in designing useful strategies, the identification of new enzymic targets in some types of cancer cells, and the characterization of metabolically active forms of cytostatic compounds. These are the important issues which were addressed in an international NATO Advanced Study Institute (ASI) attended by scientists from all over the world representing a wide spectrum of scientific disciplines.

Although the acquired immune deficiency syndrome (AIDS) is not a malignant disease, but the etiological involvement of a retrovirus belonging to the family of human T-cell lymphotropic viruses which cause leukemia and lymphomas in man, justifies its inclusion in the agenda of this Study Institute. Scientific presentations from Drs. Gallo, Wong-Staal, Sarin, Sarngadharan and Zagury document the most recent facts about the molecular events in the pathology of AIDS which offer unique strategies for its control, as discussed by our group in this volume.

It is hoped that the state of the art presented in this volume will stimulate new ideas leading to additional approaches in the treatment of neoplastic and related diseases.

July, 1986
Frankfurt

Prakash Chandra

Acknowledgments

I am especially indebted to Gianni Bonadonna (Milan), Robert C. Gallo (Bethesda), Michel Feldman (Rehovot), John E. Ultmann (Chicago) and Harald zur Hausen (Heidelberg), who served as members of the scientific committee of this Advanced Study Institute. Their specific suggestions and general encouragement provided invaluable assistance in designing the concept of this Study Institute. I am grateful to the session chairpersons, Allan Goldstein (Washington), J.R. Hobbs (London), Robert W. Baldwin (Nottingham), John E. Ultmann (Chicago), J. Bertino (Yale), M. Epstein (Oxford), Eva Klein (Stockholm), Hilary Koprowski (Philadelphia), Flossie Wong-Staal, Dominique Stehlin (Lille) and Isaac P. Witz (Tel Aviv), for organizing stimulating discussions.

I wish to express my appreciation to the NATO Scientific Affairs Division, the support of which made this Institute possible. Additional financial assistance was provided by Hoechst AG (Frankfurt), E.I. du Pont de Nemours (Wilmington) Biotech Research Laboratries (Rockville, Maryland) and Litton Bionetics, Inc. (Kensington, Maryland).

The help of my coworkers, Angelika Chandra, Ilhan Demirhan, Thomas Gerber, Maria Dzwonkowski and Knuth Krug has been invaluable in the organization and management of this Institute.

Last, but not least, I want to thank the contributors and the publishers, especially Miss M. Carter, for making the publication of this volume possible. I would also like to acknowledge the assistance of Ruth Träger for her competent preparation of the meeting proceedings into manuscript form.

July, 1986　　　　　　　　　　　　　　　　　　　　　Prakash Chandra

Frankfurt

Contents

1. A Developmental Biologist's View of Cancer 1
   F. Gros, M. Fiszman, and D. Montarras

2. Oncogenes in Development, Neoplasia and Evolution. . 15
   F. Anders, A. Anders, M. Schartl, T. Gronau,
   W. Luke, C.-R. Schmidt, and A. Barnekow

3. Acquisition of Metastatic Properties via Somatic
   Cell Fusion: Implications for Tumor
   Progression in vivo. 41
   P. De Baetselier, E. Roos, H. Verschueren,
   S. Verhagen, D. Dekegel, L. Brys, and
   M. Feldman

4. Oncogenes and Tyrosine Kinase Activities as a
   Function of the Metastatic Phenotype 57
   L. Eisenbach and M. Feldman

5. Immune Reactivities During the Precancer and
   Early Cancer Periods: Novel Approaches
   for Immunomodulation 71
   I.P. Witz, L. Agassy-Cahalon, B. Fish,
   Y. Lidor, Y. Ovadia, H. Pinkas, M. Ran,
   M. Schickler, N. Smorodinsky, B. Sredni,
   and I. Yron

6. Immunomodulation of Tumor Metastases 81
   L. Eisenbach, S. Katzav, and M. Feldman

7. Monoclonal Antibody Targeting of Anti-Cancer
   Agents 91
   R.W. Baldwin

8. Treatment of Human B Cell Lymphoma with
   Monoclonal Anti-Idiotype Antibodies 101
   E.M. Rankin and A. Hekman

9. Human Monoclonal Antibodies Directed Against
   HuMTV Antigens form Human Breast Cancer
   Cells 123
   I. Keydar, I. Tsarfaty, N. Smorodinsky,
   E. Sahar, Y. Schoenfeld, and S. Chaitchik

10. Recent Advances in Thymic Hormone Research 137
    M.B. Sztein and A.L. Goldstein

11. T-Cell Growth Factor (Interleukin -2) 165
    S.K. Arya and M.G. Sarngadharan

12. Drug Resistance: New Approaches to Treatment . . . 183
    J.R. Bertino, S. Srimatkandada, M.D. Carman,
    M. Jastreboff, L. Mehlman, W.D. Medina,
    E. Mini, B.A. Moroson, A.R. Cashmore,
    and S.K. Dube

13. Mechanisms of Membrane-Mediated Cytotoxicity
    by Adriamycin 195
    T. Grace, Y.H. Ehrlich, and T.R. Tritton

14. Terminal Transferase and Adenosine Deaminase
    Activities in Human Neoplasia: Their
    Role in Modulating Cancer Treatment 203
    P.S. Sarin, A. Thornton, and D. Sun

15. Inhibitors of Terminal Transerase: A New
    Strategy for the Treatment of
    Human Leukemia 213
    R. McCaffrey, A. Ahrens, R. Bell,
    R. Duff, H. Roppe, A. Lillquist, and
    Z. Spiegelman

16. DNA-Protein Crosslinking of Platinum
    Coordination Complex in Living Cells:
    Implication to Evaluate the Cytotoxic
    Effects of Chemotherapeutic Agents 223
    L.S. Hnilica, R. Olinski, Z.M. Banjar,
    W.N. Schmidt, and R.C. Briggs

17. Antitumor Activity, Pharmacology and Clinical
    Trials of Elliptinium (NSC 264-137) 235
    A.K. Larsen and C. Paoletti

18. Therapeutic Efficacy of Oxazaphosphorines
    by Immunomodulation 243
    R. Voegeli, J. Pohl, T. Reissmann,
    J. Stekar, and P. Hilgard

19. The Role of Cellular Glutathione in Protecting
    Mammalian Cells from X Radiation and
    Chemotherapy Agents *in vitro* 249
    D.C. Shrieve

20. The Role of Papilloma Viruses in Human Cancer . . . 255
    H. Zur Hausen and E.-M. de Villiers

21. Anti-Viral Vaccine Control of EB Virus-
    Associated Cancers 263
    M.A. Epstein

22. Bovine Leukosis Virus as a Model for Human
    Retroviruses 279
    C. Bruck, R. Kettmann, D. Portetelle,
    D. Couez, and A. Burny

23. Human T-Lymphotropic Retroviruses and their
    Role in Human Diseases 287
    R.C. Gallo and F. Wong-Staal

24. Molecular Biology of the Human T-Lymphotropic
      Retroviruses 293
    F. Wong-Staal

25. Antiviral Approaches in the Treatment of
      Acquired Immune Deficiency Syndrome. 303
    P. Chandra, A. Chandra, I. Demirhan,
    and T. Gerber

26. Inhibition of HTLV-III Replication in
      Cell Cultures. 329
    P.S. Sarin, D. Sun, A. Thornton,
    and Y. Taguchi

27. Proteins Encoded by the Human T-Lymphotropic
      Virus Type III/Lymphadenopathy Associated
      Virus (HTLV-III/LAV) Genes 343
    M.G. Sarngadharan, F. diMarzo Veronese,
    and R.C. Gallo

28. Cytopathogenic Mechanisms which Lead to
      Cell Death of HTLV-III Infected T. Cells . . . 357
    D. Zagury, R. Cheynier, J. Bernard,
    R. Leonard, M. Feldman, P. Sarin,
    and R.C. Gallo

29. Clinical Aspects of Thymic Factors in the
      Treatment of Immunodeficiency Diseases
      and Neoplasia: Achievements and
      Failures 363
    J.R. Hobbs

30. An Overview of the Current Understanding and
      Management of the Non-Hodgkin's Lymphomas . . 377
    E.R. Gaynor and J.E. Ultmann

Participants 393

Index 399

A DEVELOPMENTAL BIOLOGIST'S VIEW ON CANCER

Francois Gros, Marc Fiszman and Didier Montarras

Department of Molecular Biology
Institut Pasteur
Paris, France

INTRODUCTION

It makes very little doubt that if the causes for cancer are legion (cancer being a multifactorial disease) they must nonetheless involve a common genetic denominator. The very nature of this "substratum" has escaped for some time until the relatively recent discovery that oncogenic sequences of retroviruses do in fact originate from cellular oncogenic genes, designated as proto-oncogenes, cellular oncogenes or c-oncs for short (1,2).

That there is a genetic origin of cancer has been an "article of faith" for almost half of the past century. The question was quite clearly raised by T. Boveri, as early as 1914, and the Mendelian analyses have given shape to this belief.

But it is much later that the genetic determinants of carcinogenesis were identified in various experimental systems including fish, insects and plants. Examination of human pedigrees has given rise to the concept of "cancer genes" whose hereditary abnormalities predispose to certain kinds of neoplasms.

But is is mainly the discovery of the vertical transmission of leukaemias in inbred mice which has emphasized the role of retroviruses leading to the belief that retroviral oncogenes reside within the germ lines of all species (3,4).

In 1970, following Bader's observation concerning the effect of 5-Bu on the transformation or multiplication of RNA viruses (5), Temin, Mizutani (6) and Baltimore (7) discovered the RNA-dependent DNA polymerase, supporting strongly the hypothesis concerning the existence of DNA proviruses integrated in cellular genomes and thereby explaining the retrovirus cycle.

Yet the phylogenetic origin of these viruses remained obscure until the discovery by D. Stehelin et al. in 1976 (8) and Spector et al. in 1978 (9) that DNAs from birds and mammals contain sequences that are related to the oncogenic gene of the RSV, the well known src gene. Since this period, two main strategies have been utilized to unravel the existence of other cellular oncogenes and to provide for the first time a strong explanatory basis to the origin of tissue specific cancers. These involves: insertional mutagenesis, and transfer of transforming genes to in vitro cultivated cells by means of transfection.

We shall not analyze this type of result in detail inasmuch as they will probably be covered by the following speakers. We would rather like to illustrate some of the most significant aspects regarding the study of cellular and viral oncogenes in relation to developmental biology.

CELLULAR ONCOGENES

The present view is that many etiologic factors of cancer would act on a restricted subset of proto-oncogenes that exhibit a latent potential for transforming cells and contributing to their malignant phenotype (1,2).

At present, about two dozens of distinct oncogenic substances, transduced by retroviruses, have been described. In table 1 we have described some of their chracteristics, concerning the type of tumorigenicity and the nature of the encoded products both in terms of cellular locations and functions. The following comments can be made.

i) Certain oncogenes: e.g. src, fps/fes, fgr, yes, ros, abl encode an entity with a tyrosine-specific protein kinase activity (10). These proteins are closely related over a continous stretch of about 260 amino acids which defines the catalytic domain of the phosphotyrosine kinase (PTK) family. The substrates of these phosphotransferase activities are still actively investigated. Several proteins containing phospho-tyrosine, have been detected in RSV transformed cells, and cells transformed by other viruses whose oncogenes encode PTKs. It should be noticed that these proteins are not phosphorylated by the product of the cellular src gene. Three of the proteins, p36, p81 and vinculin (see for review 11) are found associated with the internal layer of the plasma membrane and the cytoskeleton where they could be direct targets for the viral PTKs, many of which are membrane associated.

- It has been quite well documented that this kinase activity plays an important and presumably direct role in the in vitro transforming activity of some oncogenic sequences.

- Klarlund and co-workers have recently reported that vanadium salts act as potent and quite specific inhibitors of the protein phosphatase which removes phosphate groups from phosphotyrosine both in vitro and inside the cell. Sodium vanadate was shown to induce transformation of a non-tumorigenic derivative of normal kidney cells, the NRK-1 strain, while increasing about 40fold the steady state level of phosphotyrosine in the cell (12).

- In a RSV mutant, $RSV-SF_2$ - in which a lysine residue implicated in the binding of ATP to $pp60^{v-src}$ is replaced at position 295 - one observes the failure of the mutant protein to become phosphorylated at the serine or tyrosine residues, concomitantly with the loss of PTK activity. In parallel, the mutation totally abolishes the capacity for inducing transformation and producing tumors (13).

The phosphorylation of tyrosine within the p36 or other targets is, at any rate, largely implicated in the cellular response to mitogenic factors. Furthermore a PTK activity has been identified as component of various growth factor receptors or polypeptide hormone receptors, including EGF receptor, PDGF receptor, insulin like growth factor receptor and insulin receptor (10,14).

Table 1. Oncogenes transduced by retroviruses

Viral Oncogenes	Tumorigenicity	Proposed Biochemical function	Protein Products Location
V-src	Sarcoma	PK (tyr)	Plasma membrane
V-fps/V-fes	Sarcoma	PK (tyr)	Plasma membrane?
V-yes	Sarcoma	PK (tyr)	?
V-ros	Sarcoma	PK (tyr)	?
V-fgr	Sarcoma	PK (tyr)	?
V-abl	B-cell lymphoma	PK (tyr)	Plasma membrane
V-ski	Sarcoma		Nucleus
V-myc	Carcinoma-Sarcoma and myelocytoma	Binds DNA	Nucleus
V-myb	Myeloblastic Leukemia	Binds DNA	Nucleus
V-fos	Sarcoma	?	Nucleus
V-erbB	Eurythroleukemia and Sarcoma	Truncated EGF-R PK (tyr)	Plasma membrane
V-fms	Sarcoma	CSF-1 Receptor PK (tyr)	Plasma membrane
V-sis	Sarcoma	PDGF-2/B	Secreted. Cytoplasm?
V-ras	Sarcoma and Erythroleukemia	Regulates adenylate cyclase	Plasma membrane
V-mos	Sarcoma	PK (thr and ser)?	Cytoplasm
V-mil/raf		PK (thr and ser)?	Cytoplasm
V-ets	Erythroblastic Leukemia	?	Nucleus (fused with V-myb)
V-erbA			Cytoplasm
V-rel	Lymphatic Leukemia	?	?

ii) More recently the role of oncogenes in the alteration of growth control was more clearly examplified: Several proteins encoded by viral oncogenes were shown to represent abnormal versions of cell surface growth factor receptors and one of these proteins corresponds to a portion of a growth factor.

In 1983, Waterfield et al. (15) have shown that v-sis is encoding a protein very closely related to a naturally occuring growth factor, namely PDGF.

In the same vein was the finding by Downward et al. (16) and Ullrich et al. (17) that v-erbB, an AEV oncogene, is coding for a truncated EGF receptor. In this protein, the normal EGF binding domain is lost while the protein kinase activity is conserved. This incidently provides a good explanatory model for what happens during the c-onc to v-onc conversion, since a deregulated receptor, constitutively expressed in its activated state, is formed irrespective of the presence of its cognate factor.

A novel locus which corresponds to the neu oncogene (18) or the c-erb B-2 gene (19) has been described recently in vertebrates. This gene whose product encodes a transmembrane protein highly related to

EGF-R has been repeatedly associated with tumor formation and transforms cells in vitro.

Recently, the product of the c-fms proto-oncogene was identified by Sherr and his associates (20) as a 170 Kd glycoprotein with associated tyrosine kinase activity. This glycoprotein was expressed by mature cat macrophages derived from peritoneal inflammatory exsudates and from the spleen. An antibody directed against a v-fms-encoded polypeptide synthesized by transformed bacteria, precipitated the feline c-fms product and cross-reacted with the 165 Kd glycoprotein from feline and mouse macrophages. The murine c-fms gene product was capable of specifically binding CSF-1, a macrophage growth factor.

iii) Another family of genes, the ras gene family, has been characterized, whose products might also interact with the cell machinery via phosphorylation.

Three classes of ras genes have been described (21): two classes c-Ha ras and c-Ki ras correspond to the transduced oncogenes of Harvey and Kirsten sarcoma viruses. The third class N-ras was isolated from the DNA of a human tumor as a transforming gene.

Mammalian and viral versions of the ras genes encode products of the same size (21 Kd) localized on the inner face of the plasma membrane and with very similar activities including binding and hydrolysis of GTP.

The discovery of highly conserved ras genes in yeast (22,23) has led to the finding that the products of these genes stimulate adenylate cyclase (24). This does not seem to be the case for the mammalian ras gene products (25). However, results obtained in yeast suggest the possibility of a control of protein phosphorylation at a distance.

Besides this, structural data also suggest that other oncogenes such as v-mil/raf and v-mos might induce cell transformation via phosphorylation of serine and threonine residues (10,21).

iv) Aside from oncogenes whose products are localized on cell plasma membrane, or which exert a growth factor like activity, another class comprises genes coding for proteins with a nuclear localization. Four retroviral oncogene products respectively encoded by v-fos, v-ski, v-myb and v-myc are found in the nucleus of the transformed cells (21). Two of them, the products of v-myb and v-myc are found in the nuclear matrix and have the ability to bind double stranded DNA (21). No enzymatic activity has been atributed yet to these viral oncogene products or to their cellular counterparts.

In brief, the main oncogenic classes can thus far be distinguished based upon the overall functions of their products: One is coding for proteins with protein kinase activities located mostly at the tyrosine residue, another is coding for cellular growth factors or their corresponding receptors, a third one consists of the ras gene product family, and a fourth one corresponds to various nuclear proteins whose functions are still unknown.

ACTIVATION OF CELLULAR ONCOGENES

There exist, grossly speaking, two mechanisms whereby cellular oncogenes can be "activated" so as to trigger a malignant transformation.

In certain oncogenic sequences, the prototype of which is the c-ras gene, activation seems related to point mutations, in some bladder neoplasms, for instance, a single amino acid substitution of the c-ras protein (26,27).

In most other cases, however, one is rather dealing with alterations in control elements of the c-onc allele that render it insensitive to normal exogenous signals or cause an increase in its rate of expression. Archetypes of this group are c-myc or c-abl genes.

Such alterations in gene controlling elements might be the result of chromosomal abnormalities some of which often accompany the neoplastic state. These abnormalities can be of primary nature, including such events as primary translocations, deletions, inversions, or they can be regarded as secondary, in which case a sequence of chromosomal rearrangements might culminate into the uncontrolled expression of the onc genes located in the vicinity or at the site of rearrangements involving such phenomena as secondary translocations, gene amplification, aneuploidy, etc.

A well documented example of oncogenic activation consecutive to a primary translocation relates to that involving the c-myc gene. As shown by several authors (28), close to 100 % of the Burkitt lymphoma (whether the EB virus be detected or not) display some chromosomal translocations, and the same seems to apply to murine plasmacytomas.

The c-myc oncogene normally behaves as a Mendelian allele. It normally maps on chromosome Nr. 8 (at the level of the band 9.24), but in the malignant situation associated with lymphoma it undergoes transposition to one of the following chromosomes: 14, 22 or 2 (which are coding for the lambda or kappa chains of immunoglobulins). Translocation places it in the 5' position of a gene coding for the constant IgG domain, at distances varying between 10 and 50 Kb of the gene.

As discussed by Robertson (29) the normal c-myc gene contains 3 exons, 2 of which (2 and 3) correspond to the coding region (whereas exon 1 is thought to have a regulatory function). It is lost upon conversion of c-myc to its retroviral counterpart, while exons 2 and 3 are unaltered. According to Leder et al. (30) exon 1 would encompass a binding site for a transcriptional repressor. This repressor being nothing but the product of the c-myc gene itself. In the normal allelic state, c-myc would undergo some type of autogenous repression which would maintain its transcription rate at a low level. In plasmocytoma-associated translocations, the first exon is always lost, in good accordance with Leder's scheme, but this is not always so in the rearrangements accompanying Burkitt lymphomas. The difficulty can be matched if one assumes that, at receiving chromosomal site, the constitutive (derepressed) activity of the IgG locus would cause cis-activation of exons 2 and 3. Alternatively it has been advanced that this region of frequent somatic mutations would result in some mutations of exon 1, thus altering its regulatory function. Yet, more careful examination by Croce et al. (29) has led the authors to conclude that, in the lymphoma-associated translocation, exon 1 remains, in most cases, free of any mutation. A possible way to account for these discrepancies would be to postulate some long range modifications in some c-myc upstream controlling element (enhancer?) during the conversion of the normal lymphoblastoid cells into lymphoma. This does not rule out the possibility that, when it functions as a normal allele, in relation to the cell cycle for instance, the c-myc gene would be modulated by an exon 1 dependent control.

Another well documented example illustrating c-onc activation following a translocation event is that afforded by human chronic myelogenic leukemia (28,31). Here a translocation of the c-abl gene from chromosome 9 in 3' position relative to the bcr sequence of chromosome 22 is observed.

Aside from point mutations or typical DNA rearrangements, other chromosomal events might cause activation, the most frequent of which is amplification. A good example is that of n-myc (a c-myc related DNA-sequence which is amplified 20 to 140fold in some human neuroblastomas, 32). Amplification and consequent increased expression of n-myc may represent a critical step toward malignancy.

COOPERATION

About two years ago, R. Weinberg and his colleagues have underlined a very salient aspect of cancerogenesis in its relationship with the activation of oncogenic sequences. This is the phenomenon of "cooperation" whereby "activation" of more than a single oncogenic sequence is a prerequisite for malignant transformation. More precisely, oncogenic sequences would fall into two operationally distinct and complementary categories. One group would include those oncogenes, such as ras, whose activation primarily results in cell transformation, the other one, exemplified by myc, being rather related with cell immortalization. Induction of a neoplastic state in cultivated cell lines as well as in vivo tumorigenesis would require cooperation of two onc genes belonging to each of these categories, neither of which being active by itself.

Other observations, however, indicate that our present conception of the cooperation mechanisms is probably too simplistic since it has been shown that mutated c-ras allele alone can induce tumorigenic growth of embryonic rodent cells (33).

According to Bernstein and Weinberg (34), a third category of oncogenic sequences could be envisaged, functionally speaking. The one that would be involved in imparting the tumor with the capacity to form metastases. We shall hear much more about this subject from Professor Feldman's presentation.

PHYLOGENY

Proto-oncogenic sequences are present in a very large spectrum of eukaryotic cells from yeast to man.

In yeast, only one type of c-onc gene has been clearly identified so far, which belongs to the ras gene family.

On the other hand, studies performed on the fruit fly Drosophila melanogaster have revealed the presence of an impressive number of c-onc sequences, indlucing c-src (35-37), c-abl (37), c-erbB (or EGF-R, 38), c-ras (39,40) and c-myb (41), and other Drosophila proto-oncogenes are in the process of identification. Nucleotide and amino acid sequence comparisons argue in favor of the fact that cellular oncogenes (particularly those of the PTK family such as src, abl, EGF-R) may have evolved from one or a small number of archetypes existing prior to the divergence of the Annelid Arthropod and Echinoderm chordate superphyla.

ONC GENE PRODUCTS AND CELL DIFFERENTIATION

Why did "mother Nature" maintain over such a large evolutionary scale (probably 1 billion years or more) genomic sequences of the c-onc type in the chromosomes of eukaryotic cells? It is hard to escape the view that these genes in their normal allelic state must fulfill functions which are vital for metazoan metabolism. Two hypotheses, which do not appear mutually exclusive, particularly to developmental biologists, have been proposed.

One, based on what we have learnt from retroviral oncogenes, is that c-onc genes might be involved in the control of normal cell division. The second is that these genes would be implicated in differentiation processes.

For, one of the most challenging aspects to people concerned with somatic cell differentiation is the subtle balance existing between (mitogenic) factors ensuring the process of mitotic recycling and those which precisely cause the cell to withdraw from the mitotic cycle and to express its tissue specific program.

It is with the prospect of understanding better the mechanisms operating during the choice between these two alternatives, that many investigators have undertaken to analyze how retroviral oncogenic expression reflects on cellular differentiation.

A lot of work has been devoted to the mechanisms whereby transforming viruses can cause pleiotropic blockade of somatic cell differentiation in cultivated cell lines. In these studies the RSV, which has probably been the most widely used transforming virus, appeared to prevent further differentiation of distinct cell types such as myoblasts (42), melanoblasts (43) and chondroblasts (44). Similar results were also obtained in the case of erythroblasts transformed by AEV (45).

Two mechanisms by which v-onc genes prevent differentiation of these various cell types still await clarification. One possibility is that the inhibition of cell differentiation would be due to the rephasing of the cells into an active replication cycle by mitogenic factors released upon transformation.

Along this line conflicting results have been obtained by Falcone et al. (46) and ourselves (D. Montarras, unpublished observation) in the case of RSV transformed avian myoblasts. It was shown in our laboratory that RVS ts NY68-dependent inhibition of myoblast differentiation, observed at 35 °C under permissive condition for transformation, could be totally released upon mitomycin C (a potent inhibitor of DNA synthesis) treatment of the cells. This result is in disagreement with the results of Falcone et al. who observed no differentiation of RSV transformed myoblasts under conditions similar to ours. We do not explain this difference except if we consider that chicken myoblasts behave differently from the quail myoblasts that we have used in our experiments.

As we have seen, the effect of viral oncogenes on the expression of tissue specific genetic programs most often consists in a pleiotropic inhibition of these programs. The situation seems to be totally different in the case of c-onc genes. Neurogenesis, for instance, examplifies an interesting case.

Neural tissues were shown to contain high level of c-src gene product (47,48) at the time of neuronal differentiation in chicken embryos. More recently, other authors found that neurons and astrocytes derived from the rat central nervous system express high level of pp60 c-src protein (49). Similar conclusions have been drawn from studies performed in the fruit fly Drosophila melanogaster (36).

Put together, all these results indicate that the product of pp60^{c-src} specifically accumulates in non-proliferating differentiating neural cells. In this respect several observations have already suggested that changes in the rates of c-onc gene expression is a tissue specific and stage-specific developmental event (50-52). However, it is still difficult to distinguish between the genes whose expression would be activated as a consequence of development and those whose activity would direct development.

It was recently shown that PC12 phaeochromocytoma cells, that differentiate into neurons upon treatment with nerve growth factor, also do so when infected by RSV (53). This result indicates that the product of pp60^{v-src} may have an inductive effect on differentiation. However, it does not constitute an argument to conclude that the product of pp60^{c-src} would also do so.

PROTO-ONCOGENES AND THE ECONOMY OF THE NORMAL CELL

One of the most challenging questions as regards present developmental biology concerns the remarkable conservation of oncogenic sequences within normal eukaryotic cells over a very large phylogenetic scale, a situation suggesting that c-onc genes must be endowed with a very basic function in the normal control of cell division.

Many facts emerging recently from the literature have led to the current belief that cellular oncogenes are probably involved in the transduction of exogenous signals - such as those actually conveyed by growth factors in such a way as to ensure a proper response from the target cells.

For instance, it has been shown in varous laboratories (Campisi et al., 54; Kelly et al., 55) that the c-myc product is involved at the onset of the cell cycle, at which phase the cell is progressing from the quiescent G0 state towards the competence for starting a new cycle of replication, usually refered to as priming for growth competence. In accordance with this view is the observation that PDGF, a potent fibroblast mitogenic factor, as well as ConA and LPS, two strong lymphocytic mitogens, all of which can prime for growth competence, trigger marked activation of c-myc transcription in the target cells within 1 - 2 hours following their addition. The burst of c-myc transcription is transient and there is a return to the background level before the onset of the S phase.

This factor dependent stimulation of c-myc gene expression seems to be lost during malignant transformation and, as shown by Campisi et al. (54) using transformed fibroblast lines, transcription of this oncogenic sequence becomes constitutive, i.e. can now proceed in the absence of any exogenous factor.

The view that oncogenic sequences could take part in a cascade of transducing events, starting from exogenous signals culminating in balanced cellular growth, finds support in some recent observations showing that there is a sequential program of protooncogene expression induced by various growth factors, that this program is probably a common program of gene control serving a variety of cell types during their differentiation.

For instance Kruijer et al. (56), as well as Müller et al. (57) observe that growth factors such as PDGF and FGF or TPA cause immediate expression of the c-<u>fos</u> gene in normal mouse fibroblasts; 10 min later c-<u>fos</u> mRNA synthesis is turned off and there is a marked increase in the cellular level of c-<u>myc</u> mRNA. This situation is in good accordance with the view that c-<u>onc</u> gene products intervene within a <u>cascade</u> of events linking the chemical exogenous signal to a cell proliferation response. Moreover different factors, each of which interacts with a particular membrane associated receptor in different cell types, seem to induce very similar changes in the transcription of cellular genes. This common program of protooncogene expression is illustrated by the works of M.E. Greenberg et al. (58): not only addition of serum, or better of PDGF and FGF to quiescent 3T3 cells induces sequential transcription of c-<u>fos</u>, β-Actin and c-<u>myc</u>, confirming previous data, but the NGF dependent differentiation of a pheochromocytoma cell line, PC12, is accompanied by an enhanced transcription of the same genes with kinetics that directly parallel the changes first observed in PDGF induced fibroblasts. Thus induction of c-<u>fos</u>, c-<u>myc</u> and actin could be regarded as a general nuclear response to growth factors involving different receptors and different cell types.

Besides these observations, the idea also emerged that protooncogene expression may be associated with both specific differentiation pathways and normal cell growth (59). This idea is supported by the recent finding of Thompson et al. (60). These authors determined that an increase of c-<u>myb</u> expression correlates with cell proliferation, with no apparent tissue specificity except in thymus. This strongly suggests that the c-<u>myb</u> gene may have a specific role in the differentiation of thymic cells and a more general role in the regulation of proliferation of other cell types.

As regards the biochemical mechanisms involved, we still know relatively little. Berridge and Irvine's model (61), according to which growth factor dependent stimulation of mitotic response is channelled through two major events: a) a transient intracellular increase in pH and b) a transient increase in cytoplasmic Ca^{++} ions, has been at the focus of many interesting studies and discussions. In this context is is significant to recall that different cellular protooncogenes are coding for products whose interplay fit in relatively well with a chain of transducing events involving, in addition to growth factors, a membrane receptor, receptor bound protein kinases, phosphotidyl-inositol, diacylglycerol, proteinase C and the subsequent changes in pH and Ca^{++} level. Accordingly c-<u>sis</u>, c-<u>erbB</u>, c-<u>src</u> and c-<u>ras</u> gene products whould correspond to some of these defined intermediates.

These considerations, although fragmentary, show nonetheless that oncogenic sequences are establishing a strong point of convergence between two preoccupations: the mechanisms of malignant transformation and the more general phenomenon of signal transduction whose mediation is responsible for the subtle balance between proliferation and differentiation of eukaryotic cells.

References

1. J. M. Bishop, Cellular oncogenes and retroviruses, Ann. Rev. Biochem. 52:301 (1983).
2. H. E. Varmus, The molecular genetics of cellular oncogenes, Ann. Rev. Genet. 18:553 (1984).
3. R. J. Huebner and G. J. Todaro, Oncogenes of RNA tumor viruses as determinants of cancer, Proc. Natl. Acad. Sci. 64:1087 (1969).
4. G. J. Todaro and R. J. Huebner, The viral oncogene hypothesis: New evidence, Proc. Natl. Acad. Sci. 69:1009 (1972).
5. J. P. Bader, The requirement for DNA synthesis in the growth of Rous Sarcoma and Rous associated Viruses, Virology 26:253 (1965).
6. H. M. Temin and S. Mizutani, RNA directed DNA polymerase in virions of Rous sarcoma virus, Nature 226:1211 (1970).
7. D. Baltimore, RNA dependent DNA polymerase in virions of RNA tumor viruses, Nature 226:1209 (1970).
8. D. Stehelin, H. E. Varmus, J. M. Bishop, and P. K. Vogt, DNA related to the transforming gene(s) of avian sarcoma viruses is present in normal avian DNA, Nature 260:170 (1976).
9. D. H. Spector, H. E. Varmus, and J. M. Bishop, Nucleotide sequences related to the transforming gene of avian sarcoma virus are present in the DNA of uninfected vertebrates, Proc. Natl. Acad. Sci. 75:4102 (1978).
10. T. Hunter and J. A. Cooper, Protein tyrosine kinases, Ann. Rev. Biochem. 54:897 (1985).
11. B. M. Sefton, The viral tyrosine protein kinases, Curr. Top. Microbiol. Immunol. 129:39 (1986).
12. J. K. Klarlund, Transformation of cells by an inhibitor of phosphatase acting on phosphotyrosine in proteins, Cell 41:707 (1985).
13. M. A. Snyder, J. M. Bishop, J. P. McGrath, and A. D. Levinson, A mutation of the ATP binding site of $pp60^{v-src}$ abolishes kinase activity, transformation and tumorigenicity, J. Mol. Cell. Biol. 5:1772 (1985).
14. C. H. Heldin and B. Westermark, Growth factors: mechanism of action and relation to oncogenes, Cell 37:9 (1984).
15. M. D. Waterfield, T. Scrace, N. Whittle, P. Stroobant, A. Johnson, A. Wasteson, B. Westermark, C. H. Heldin, J.S. Huang, and T.F. Deuel, Platelet-derived growth factor is structurally related to the putative transforming protein $p28^{sis}$ of simian sarcoma virus, Nature 304:35 (1983).
16. J. Downward, Y. Yarden, E. Mayes, G. Scrace, N. Totty, P. Stockwell, A. Ullrich, J. Schlessing, and M. D. Waterfield, Close similarity of epidermal growth factor receptor and v-erb-B oncogene protein sequences, Nature 307:521 (1984).
17. A. Ullrich, L. Coussens, J. S. Hayflick, T. J. Dull, A. Gray, A.W. Tam, J. Lee, Y. Yarden, T. A. Libermann, J. Schlessinger, J. Downward, E.L.V. Mayes, N. Whittle, M. D. Waterfield, and P.H. Seeburg, Human epidermal growth factor receptor cDNA sequenceΔ and aberrant expression of the amplified gene in A431 epidermoid carcinoma cells, Nature 309:418 (1984).
18. C. I. Bargmann, Mien-Chie Hung, and R. A. Weinberg, The neu oncogene encodes an epidermal growth factor receptor-related protein, Nature 319:226 (1986).
19. T. Yamamoto, S. Ikawa, T. Akiyama, K. Semba, N. Nomura, N. Miyajima, T. Saito, and K. Toyoshima, Similarity of protein encoded by the human c-erb B-2 gene to epidermal growth factor receptor, Nature 319:230 (1986).
20. C. J. Sherr, C. W. Rettenmier, R. Sacca, M. Roussel, A.T. Look, and E.R. Stanley, The c-fms proto-oncogene product is related to the receptor of the mononuclear phagocyte growth factor CSF-1, Cell 41:665 (1985).

21. J. M. Bishop, Viral oncogenes, Cell 42:23 (1985).
22. R. Dhar, A. Nieto, R. Koller, D. Defoe-Jones, and E.M. Scolnick, Nucleotide sequence of two rasH related genes isolated from the yeast Saccharomyces cerevisiae, Nucleic Acids Res. 12:3611 (1984)
23. S. Powers, S. Kataoka, O. Fasano, M. Goldfarb, J. Strathern, J.B. Broach, and M. Wigler, Genes in S. cerevisiae encoding proteins with domains homologous to the mammalian ras proteins, Cell 36:607 (1984).
24. T. Toda, I. Uno, T. Ishikawa, S. Powers, T. Kataoka, D. Broek, S. Cameron, J. Broach, K. Matsumoto, and M. Wigler, In yeast ras proteins are controlling elements of adenylate cyclase, Cell 40:27 (1985).
25. S. K. Beckner, S. Hattori, and T. Shih, The ras oncogene product p21 is not a regulatory component of adenylate cyclase, Nature 317:71 (1985).
26. C.J. Tabin, S.M. Bradley, C.I. Bargmann, R. Weinberg, A.G. Papageorge, E.M. Scolnick, R. Dhar, D.R. Lowy, and E.H. Chang, Mechanism of activation of a human oncogene, Nature 300:143(1982)
27. E. Premkumar-Reddy, R.K. Reynolds, E. Santos, and M. Barbacid, A point mutation is responsible for the acquisition of transforming properties by T24 human bladder carcinoma oncogene, Nature 300:149 (1982).
28. G. Klein and E. Klein, Evolution of tumours and the impact of molecular oncology, Nature 315:190 (1985).
29. M. Robertson, Message of myc in context, Nature 309:585 (1984).
30. P. Leder, J. Battey, G. Lenoir, C. Moulding, W. Murphy, H. Potter, T. Stewart, and R. Taub, Translocations among antibody genes in human cancer, Science 222:765 (1983).
31. J.B. Konopka, S.M. Watanabe, J.W. Singer, S.J. Collins, and O.N. Witte, Cell lines and clinical isolates derived from Ph^1-positive chronic myelogenous leukemia patients express c-abl proteins with a common structural alteration, Proc. Natl. Acad. Sci. 82:1810 (1985).
32. M. Schwab, J. Ellison, M. Busch, W. Rosenau, H.E. Varmus, and J.M. Bishop, Enhanced expression of the human gene N-myc consequent to amplification of DNA may contribute to malignant progression of neuroblastoma, Proc. Natl. Acad. Sci. 81:4940 (1984).
33. D.A. Spandidos and N.M. Wilkie, Malignant transformation of early passage rodent cells by a single mutated human oncogene, Nature 310:469 (1984).
34. S.C. Bernstein and R.A. Weinberg, Expression of the metastatic phenotype in cells transfected with human metastatic tumor DNA, Proc. Natl. Acad. Sci. 82:1726 (1985).
35. M.A. Simon, T.B. Kornberg, and J.M. Bishop, Three loci related to the src oncogene and tyrosine specific protein kinase activity, Nature 302:837 (1983).
36. M.A. Simon, B. Drees, T. Kornberg, and J.M. Bishop, The nucleotide sequence and the tissue specific expression of Drosophila c-src, Cell 42:831 (1985).
37. F.M. Hoffmann, L.D. Fresco, H. Hoffmann-Falk, and B.Z. Shilo, Nucleotide sequences of the Drosophila src and abl homologs: conservation and variability in the src family oncogenes, Cell 35:393 (1983).
38. E. Lineh, L. Glazer, D. Segal, J. Schlessinger, and B.Z. Shilo, The Drosophila EGF Receptor gene homog: conservation of both hormone binding and kinase domains, Cell 40:599 (1985).
39. F.S. Neuman-Silberberg, E. Schejter, F.M. Hoffman, and B. Shilo, The Drosophila ras oncogenes: structure and nucleotide sequence, Cell 37:1027 (1984).
40. B. Mozer, R. Marlor, S. Parkhurst, and V. Corces, Characterization and developmental expression of a Drosophila ras oncogene, J.

Mol. Cell. Biol. 5:885 (1985).
41. A.L. Katzen, T.B. Kornberg, and J.M. Bishop, Isolation of the proto-oncogene c-myb from D. melanogaster, Cell 41:449 (1985).
42. M.Y. Fishman and P. Fuchs, Temperature sensitive expression and differentiation in transformed myoblasts, Nature 254:429 (1975).
43. D. Boettiger, K. Roby, J. Brumbaugh, J. Bielh, and H. Holtzer, Transformation of chicken embryo retinal melanoblasts by a temperature sensitive mutant of Rous Sarcoma Virus, Cell 11:881 (1977).
44. M. Pacifici, D. Boettiger, K. Roby, and H. Holtzer, Transformation of chondroblasts by Rous Sarcoma Virus and synthesis of sulphated proteoglycan matrix, Cell 11:891 (1977).
45. H. Weintraub, H. Beug, M. Groudine, and T. Graf, Temperature sensitive changes in the structure of globin chromatin in lines of red cell precursor transformed by ts AEV, Cell 28:931 (1982).
46. G. Falcone, D. Boettiger, S. Alema, and F. Tato, Role of cell division in differentiation of myoblasts infected with a temperature sensitive mutant of Rous sarcoma virus, EMBO Journal 3:1327 (1984).
47. L.K. Sorge, B.T. Levy, and P.F. Maness, $pp60^{c-src}$ is developmentally regulated in the neural retina, Cell 36:249 (1984).
48. D.W. Fults, A.C. Towle, J.M. Lauder, and P.F. Maness, $pp60^{src}$ in the developing cerebellum, J. Mol. Cell. Biol. 5:27 (1985).
49. J.S. Brugge, P.C. Cotton, A.E. Queral, J.N. Barrett, D. Nonner, and R.W. Keane, Neurons express high level of a structurally modified activated form of $pp60^{c-src}$, Nature 316:554 (1985).
50. T.J. Gonda, D.K. Sheiness, and J.M. Bishop, Transcripts from the cellular homologues of retroviral oncogenes: Distribution among chicken tissues, J. Mol. Cell. Biol. 2:617 (1982).
51. R. Müller, D.J. Salmon, J.M. Tremblay, M.J. Cline, and I.M. Verma, Differential expression of cellular oncogenes during pre- and post-natal development of the mouse, Nature 299:640 (1982).
52. R. Müller, J.M. Tremblay, E.D. Adamson, and I.M. Verma, Tissue and cell type specific expression of two human c-onc genes, Nature 304:454 (1983).
53. S. Alema, P. Casalbore, E. Agostini, and F. Tato, Differentiation of PC12 phaeochromocytoma cells induced by v-src oncogene, Nature 316:557 (1985).
54. J. Campisi, H.E. Gray, A.B. Pardee, M. Dean, and G.E. Sonenshein, Cell cycle control of c-myc but not c-ras expression is lost following chemical transformation, Cell 36:241 (1984).
55. K. Kelly, B.H. Cochran, C.D. Stiles, and P. Leder, Cell specific regulation of the c-myc gene by lymphocyte mitogens and platelet derived growth factor, Cell 35:603 (1983).
56. W. Kruijer, J.A. Cooper, T. Hunter, and I.M. Verma, Platelet-derived growth factor induces rapid but transient expression of the c-fos gene and protein, Nature 312:711 (1984).
57. R. Müller, R. Bravo, J. Burckhardt, and T. Curran, Induction of c-fos gene and protein by growth factor precedes activation of c-myc, Nature 312:716 (1984).
58. M.E. Greenberg, L.A. Greene, and E.B. Ziff, Evidence for a common program of proto-oncogene expression induced by diverse growth factors, in: "Current communications in Molecular Biology, Eukaryotic Transcription, The Role of cis- and trans-acting elements in initiation", Y. Gluzman, ed., Cold Spring Harbor Laboratory, p. 161 (1985).
59. R. Müller, T. Curran, D. Müller, and L. Gilbert, Induction of c-fos during myelomonocytic differentiation and macrophage proliferation, Nature 314:546 (1985).
60. C.B. Thompson, P.B. Challoner, P.E. neiman, and M. Groudine, Expression of the c-myb proto-oncogene during cellular

proliferation, *Nature* 319:374 (1986).
61. M. J. Berridge and R. F. Irvine, Inositol triphosphate, a novel second messenger in cellular signal transduction, *Nature* 312:315 1984).

ONCOGENES IN DEVELOPMENT, NEOPLASIA, AND EVOLUTION[*]

Fritz Anders[1], Annerose Anders[1], Manfred Schartl[1,3], Thomas Gronau[1], Wolfgang Lüke[1], Carl-Rudolf Schmidt[1], and Angelika Barnekow[2]

[1] Genetisches Institut der Justus-Liebig-Universität Giessen, D-6300 Giessen, Bundesrepublik Deutschland
[2] Institut für Virologie (FB Humanmedizin), Justus-Liebig-Universität Giessen, Bundesrepublik Deutschland
[3] Max-Planck-Institut für Biochemie, D-8033 Martinsried Bundesrepublik Deutschland

INTRODUCTION: HISTORY OF THE ONCOGENE CONCEPT

The concept of genes that code for neoplastic transformation, called "oncogenes" today, originates from two sources: from virology and from animal genetics. The virological source can be traced back to the year 1911 when Peyton Rous discovered the virus that causes sarcoma in chickens. It took, however, about sixty years until evidence was brought about that the cancer determinants located in the genome of this virus and of related viruses (retroviruses) are genes (Huebner and Todaro, 1969; Bentvelzen, 1972). The source that originates from animal genetics can be traced back to the year 1928 when Myron Gordon, Georg Häussler, and Curt Kosswig indepently discovered that the F_1 hybrids between certain domesticated ornamental breeds of the Central American fish species Xiphophorus maculatus (platyfish) and Xiphophorus helleri (swordtail) spontaneously develop melanoma that is inherited in the hybrid generations like the phenotype of any normal Mendelian gene located in the genome of the fish. Both, Rous' sarcoma virus (RSV) and Xiphophorus fish represent up to date highly suitable models for research on oncogenes.

---

[*] This work has been generously supported for many years by the President of the Justus-Liebig-Universität Giessen, by the Deutsche Forschungsgemeinschaft (SFB 103, Marburg; SFB 47, Giessen; Schwerpunkt "Onkozytogenetik"), and by the Bundesminister für Forschung und Technologie. It contains parts of the PhD theses of Thomas Gronau and Wolfgang Lüke. We are grateful to Dr. Chandra and to Dr. Gallo to be invited to be contributors to this volume. Thanks are due to Mrs. Kristine Krüger and Mrs. Sybill Lenz for their help in the preparation of the manuscript.

While many investigators focussed their attention on the retroviral oncogenes, symbolized as v-oncs, we focussed on the animal cellular oncogenes, symbolized as c-oncs.

We started our research on oncogenes in Xiphophorus in 1957 with systematic crossings between populations, races and species, and with mutagenesis studies in purebred and hybrid fish. Subsequently we found that melanoma and a large variety of other neoplasms developing spontaneously or after treatment with carcinogens can be assigned to particular chromosomal genes (F. Anders et al., 1961; F. Anders and Klinke, 1966). These genes are "oncogenes" by definition. They are present in several copies in all specimens of all wild populations of all species of Xiphophorus. In the wild populations they are under stringent control exerted by regulatory genes that are organized as regulatory gene-systems. The genetic make-up of the regulatory gene-system is highly different between the species of the genus Xiphophorus, less different between the races of a certain species, but still different between populations of a certain race. Certain regulatory gene-systems are completely linked to the particular oncogene of a particular chromosome, others are distributed over several chromosomes. In the first case, the oncogene never mediates neoplastic transformation. In the latter case, however, following interpopulational, interracial, or interspecific hybridization that may replace chromosomes containing regulatory genes by homologous chromosomes lacking them, the particular oncogene becomes deregulated in the zygote and codes for neoplastic transformation as soon as the sequence of its differential activity in the early life of the fish requires regulation. Any damage of the regulatory gene-system in the germ line cells induced by physical or chemical agents showed the same effect, i.e. "spontaneous" development of neoplasia in the early life of the fish, even in the early embryo. If, in contrast, the regulatory gene-system becomes damaged in a somatic cell of the fish, then, within a given time after the treatment, the oncogene becomes released from control and codes for neoplastic transformation. Thus, the common primary events in the causation of neoplasia in Xiphophorus are changes in the regulatory gene-system controlling the oncogenes rather than changes of the oncogenes themselves (F. Anders, 1967; A. Anders et al., 1973; Ahuja and F. Anders, 1977; A. Anders and F. Anders, 1978; F. Anders, 1981; F. Anders et al., 1984a; 1985).

New ideas in the development of the cellular oncogene concept were introduced when Stehelin, Varmus, and Bishop (1976) published that the transforming gene of RSV contains nucleotide sequences that are also present in the noninfected cells of chickens, and when Spector et al. (1978) found these sequences in the noninfected genome of humans, cattle and even in the genome of salmon. Subsequently we also looked for these sequences and identified them in the genome of Xiphophorus.

Based on a fifty years' history, the present article deals with the recent development of the cellular oncogene concept elaborated by means of the Xiphophorus fish tumour system.

THE ANIMAL SYSTEM

Xiphophorus, including platyfish and swordtail (Fig. 1), lives in genetically isolated populations in brooks, rivers, lakes, ponds, and pools in Central America and has evolved into innumerable genotypically and phenotypically distinguishable groups. Based on certain morphological, biochemical, ecological, and ethological characteristics, 17 of these groups have been listed as species (Rosen, 1979; Radda,

1980; Meyer et al., 1985). All individuals of this genus, however, can be hybridized in the laboratory without difficulty, and all hybrids are fertile. This finding and the results on the conformity of genome organization (Schwab, 1980), the low degree of enzyme polymorphism (Scholl and Anders, 1973), and the normal chromosome pairing during meiosis in the hybrids (Kollinger and Siegmund, 1981) led us to the conclusion that the taxonomic differences between these groups of Xiphophorus are not at the species level but at the level of elementary local populations as well as ecological and geographical races (F. Anders and Schartl, 1984). The ease of hybridizing animals from the different taxonomic groups of Xiphophorus listed as species, therefore, is no curiosity in the animal kingdom, but is comparable to the frequently occurring hybridization between animals of different populations or local races from other teleosts as well as amphibia, lizards, birds, and mammals (Scholl and Anders, 1973), including humans (F. Anders, 1981).

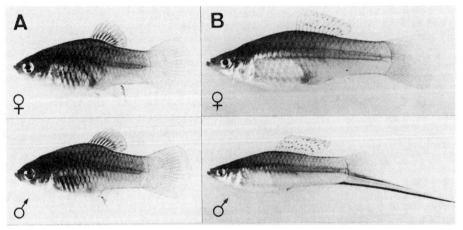

Fig. 1. A Xiphophorus maculatus from a population of the Rio Usumacinta and B Xiphophorus helleri from Rio Lancetilla (up to about 8 cm in length).

ONCOGENES IN DEVELOPMENT

Cellular counterparts of viral oncogenes as normal constituents of the genome

14 retroviral oncogenes (v-oncs) were at our disposal[*]. The cellular counterparts (c-oncs) of 12 of them have been detected in DNA from X. helleri and X. maculatus by Southern blot hybridization with v-onc probes (Table 1). The oncogenes c-ros and c-mos could not be identified in Xiphophorus. Since Xiphophorus is capable of developing all histotypes of neoplasms known of mammals (Schwab et al., 1978a,b), c-ros and c-mos are probably not essentials for neoplastic transformation.

From c-src, c-myc, and c-yes, and recently from c-sis and c-ras (Raulf and Schartl, 1985), and from c-erb B (Schartl et al., 1985a) we identified the transcripts by Northern blot hybridization. Furthermore, the product of c-src, the $pp60^{c-src}$-kinase, was identified by immunoprecipitation with antisera from RSV tumor-bearing rabbits, and phosphorylation of tyrosine exerted by the kinase was proven (Barnekow et al., 1982). In conclusion, these results suggest to us that the

---

[*]Viral onc probes were gifts from R. C. Gallo and K. Toyoshima.

cellular counterparts of the viral oncogenes found in Xiphophorus are normal constituents of the genome of this fish genus and operate in the normal animals (Barnekow et al., 1982; Schartl and Barnekow, 1984; Schartl et al., 1982, 1985b; Schartl and Barnekow 1982; Raulf and Schartl, 1985; F. Anders, 1983; F. Anders et al., 1984a,b; 1985).

Since most of our knowledge about oncogenes in Xiphophorus comes from studies on c-src, this oncogene in particular, and its product, will be mainly considered in the following paragraphs.

Table 1. Homologues of retroviral oncogenes detected in Xiphophorus

c-onc	Prototypic virus
c-erb A, B	Avian erythroblastosis virus
c-src	Rous sarcoma virus
c-myc	Avian myelocytomatosis virus MC 29
c-yes	Yamaguichi-73 sarcoma virus
c-abl	Abelson murine leukemia virus
c-fgr	Gardner-Rasheed feline sarcoma virus
c-fes	Gardner-Arnstein feline sarcoma virus
c-myb	Avian myeloblastosis virus
c-ras H	Harvey murine sarcoma virus
c-ras K	Kirsten murine sarcoma virus
c-sis	Simian sarcoma virus
c-fos	FBJ osteosarcoma virus
not detected were:	
c-ros	UR II sarcoma virus
c-mos	Moloney murine sarcoma virus

Organ-specificity of c-src activity

Considerable differences in kinase activity were detected in the different tissues of Xiphophorus (Fig. 2). Very low activity was always found in muscle. Skin, Liver, and testes showed always low activity as compared to the spleen. In brain tissue the kinase activity was always high.

Kinase activity was also measured throughout the lifetime of the fish, e.g. in entire neonates, and in the organs, for the most part exemplarily in the brain of 6-week-old, 12-week-old, 1-year-old, and 2-year-old fish. No significant changes of c-src activity were detected (Barnekow et al., 1982).

Since we obtained always the same results in the different organs throughout the lifetime of both X. helleri and X. maculatus, we conclude that the degree of c-src activity is organ-specific. Moreover, our results suggest to us that c-src governs common cell functions which, however, require a much higher activity in the brain than in any other organ (Barnekow et al., 1982).

High c-src activity was also found in the brain or the nervous cell system, respectively, of chicken, frog, shark, lamprey, lancelet, cockroach, crab, cuttle fish, and sea anemone (Schartl and Barnekow, 1982).

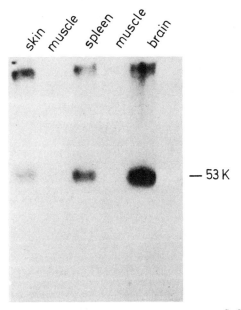

Fig. 2. Demonstration of organ-specificity of pp60^{c-src}-associated protein kinase activity in extracts of different fish tissues of X. helleri (from Barnekow et al., 1982, modified).

Trosko (1986), investigating gap junctions, found also high c-src activity in the nerve cell systems of different animals, and suggested that c-src might operate in cell-cell communication.

c-src activity in embryos

pp60^{c-src}-kinase activity in Xiphophorus is detectable from the very outset of cleavage and strongly increases during early organogenesis (Fig. 3). While organogenesis morphologically culminates, kinase activity apparently becomes choked, decreases during the growth phase before and after birth to a lower level, and thereafter (not shown in Fig. 3) remains constant at a basic level throughout the life of the fish (Schartl and Barnekow, 1984). The same sequences of differential c-src activity were confirmed at the level of mRNA by Northern blot analysis. Differential activity at this level, although in different sequences, was also observed for c-sis and c-ras (Raulf and Schartl, 1985).

Sequences of pp60^{c-src}-kinase activity similar to those of Xiphophorus were found in embryos of chicken (Fig. 3) and frogs (Schartl and Barnekow, 1984). Slamon and Cline (1984), investigating the mRNA of several oncogenes in mice embryos, found that c-src, c-erb, and c-sis (not c-myc and not c-ras) show the highest activities during a period which is comparable to that of organogenesis in Xiphophorus. Bishop (1985), investigating mRNA of oncogenes in Drosophila, mentioned results of M. Simon (unpublished, and Simon et al., 1983) that show abundance of transcripts of c-src, c-abl, and a not further characterized c-src- and c-fps-related oncogene very nearly in development (c-src, for instance, in 2-to-12-hour-old embryos).

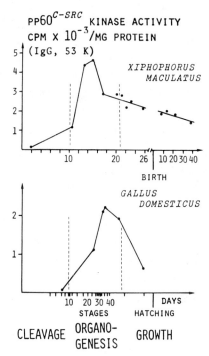

Fig. 3. pp60^{c-src} activity during cleavage, organogenesis and growth in Xiphophorus and chicken (data from Schartl and Barnekow, 1984).

Our observations together with those of other investigators obtained by means of other systems suggest to us that many of the oncogenes exert important normal functions as developmental genes. These functions are related to cell differentiation in organogenesis rather than to cell proliferation. When the phase of organogenesis progresses to the growth phase the oncogenes become choked but keep still functioning. Organ specificity of pp60^{c-src}-kinase activity in adults, in particular the high activity that was always found in the brain, indicates that the oncogenes may even change their functions in the course of development from embryogenesis to maturity.

c-src activity in carcinogen-treated animals

The purebred descendants of the wild populations of Xiphophorus proved to be highly insensitive to carcinogens such as X rays (A. Anders et al., 1971; 1973; Pursglove et al., 1971), N-methyl-N-nitrosourea (Schwab et al., 1978a,b; Schartl et al., 1985; C. R. Schmidt et al.,

1986), N-ethyl-N-nitrosurea, diethylnitrosamine, dimethyl sulfoxide, benzo(a)pyrene (C. R. Schmidt et al., 1986), and tumor promoters such as 12-O-tetradecanoylphorbol-13-acetate (Schwab, 1982; C. R. Schmidt et al., 1982) as well as to potential promoters such as saccharin, cyclamate and phenobartibal (C. R. Schmidt et al., 1986). In a recent broad-scale experiment about fifteen thousand purebred individuals survived treatment with X rays and N-methyl-N-nitrosurea, but none developed neoplasia.

Measurements of pp60^{c-src} kinase activity in specimens of the treated animals showed no changes as compared to the nontreated controls. As will be shown later, nontumorous organs of tumorous fish, too, show no changes in c-src activity (Schartl et al., 1985).

Conclusions

Xiphophorus is abundantly equipped with c-oncs. These genes are apparently normal constituents of the genome. Studies on c-src and on some other c-oncs of Xiphophorus as well as data reported in the literature suggest to us that many of these genes are active from the outset of cleavage all over the lifetime of an animal, with very high activities during organogenesis, and different activities in different organs. It appears that many of them operate differently at different times and for different purposes during the course of development. They are stringently regulated since even powerful carcinogens are not capable to change their activity and to induce neoplasia in the fish.

ONCOGENES IN NEOPLASIA

c-src activity in particular neoplasms

Based on the fact that neoplasia of Xiphophorus was discovered in hybrids (see introduction), and that we failed to induce neoplasia in the purebred wild fish (see previous chapter), we crossed specimens of different geographical and ecological provenance and treated the F_1, F_2, and BC generations with carcinogens such as X rays and N-methyl-N-nitrosourea (MNU). The followig result was obtained: out of 10,195 hybrids, 805 (7.9 %) individuals developed neurogenic, epithelial and mesenchymal neoplasms. Many animals, mostly the elder ones, developed several different kinds of neoplasms. Neoplasms could also be induced by tumor promoters such as 12-O-tetradecanoylphorbol-13-acetate (TPA) and methyltestosterone (C. R. Schmidt et al., 1986). In certain hybrid combinations tumors developed even spontaneously and were inherited according to the Mendelian prediction.

As shown in Table 2, pp60^{c-src} kinase activity is elevated in the neoplasms. The elevation ranges from 2-fold up to 50-fold compared to the highest activities found in any of the normal organs. The factors correspond to the degree of malignancy. No change of c-src-activity was found in the nontumorous organs of the fish carrying the induced tumors as is demonstrated by means of the brain. In contrast, the fish that developed the hereditary tumors showed an elevation of pp60^{c-src}-kinase activity in both tumorous and nontumorous tissues. The factor by which kinase activity is elevated in these tumors parallels the factor of elevation in the brain. Moreover, the degree of elevation in tumor and brain corresponds to the degree of malignancy of the tumor. Animals developing malignant hereditary tumors showed a higher elevation of pp60^{c-src} kinase in both the tumor and the brain than those developing benign hereditary tumors. We took advantage of this observation in several analytical experiments on tumor formation discussed in the chapter "Oncogenes in the causation of neoplasia".

Oncogene amplification in the neoplasms

Several observations suggest to us that the oncogene activity in the tumors corresponds to oncogene amplification. In melanomas, for instance, in which kinase activity was elevated by the factor 30 as compared to the nontumorous tissue (see Table 2), factors up to 50 were estimated for c-src gene amplification (Fig. 4) (Gronau, 1986). c-scr amplification increases in parallel with the age and with the growth of the tumors. c-yes, the only oncogene so far tested at this level, is also amplified in the tumors. About 70 % of the transformed cells in cell cultures derived from tumors of embryos exhibit double minute chromosomes that, according to Schimke (1984), might be interpreted as carriers of amplified genes. It appears that oncogene amplification, like elevation of oncogene activity, might represent a common phenomenon in the tumors of Xiphophorus.

As discussed by several authors in the context of transposition of a) copia elements in Drosophila (Flavell, 1984; Shiba and Saigo, 1983), b) Ty plasmids in yeast (Garfinkel et al., 1985; Mellor et al., 1985), c) A-type particles in mice (Kuff et al., 1983; and others), and d) the generation of pseudogenes and alu-elements in humans (Sharp, 1983), RNA-dependent DNA polymerase might be related to these processes. These discussions encouraged us to search in Xiphophorus for cellular sequences homologous to the viral pol gene that codes for RNA-dependent DNA polymerase. Furthermore, we were encouraged to search for RNA-dependent DNA polymerase activity in normal and neoplastic tissues.

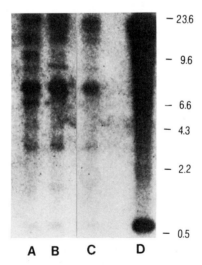

Fig. 4. Hind III restriction fragments of the c-src oncogene in Xiphophorus. Southern blot, 10 µg each A brain tissue, B skin, C muscle of the non-tumorous littermate of D, D MNU-induced melanoma. - Note the signals at 25.0 and 0.5 Kb which are 30 to 50 times amplified in D as compared to A, B, and C. Amplification is in parallel with the elevation of $pp60^{c-src}$ kinase activity; see Table 2.

Table 2. pp60^{c-src} kinase activity in _Xiphophorus_ hybrids bearing tumors of different etiology (data from Schartl et al., 1985)

Tumor	Etiology	Remarks	Factor by which the kinase activity is elevated	
			tumor	brain
	carcinogen-induced			
Melanomas	X ray, adult	invasive, malignant	5[a]	no
Squamous cell carcinoma	X ray, adult	invasive	2[a]	no
Epithelioma	X ray, adult	benign	2[a]	no
Fibrosarcoma	ENU**, adult	invasive, malignant	13[c]	no
Fibrosarcoma	MNU*, adult	malignant	10[c]	1.4
Fibrosarcoma	MNU, adult	invasive, malignant	10[c]	n.t.
Retinoblastoma	MNU, adult	invasive, highly malignant	50[b]	no
Retinoblastoma	MNU, adult	progressive growth	3[b]	no
Melanoma	MNU, adult	progressive growth	3[b]	no
Melanoma	MNU, embr.	invasive	8[a]	no
Rhabdomyosarcoma	MNU, embr.	invasive	10[a]	no
Rhabdomyosarcoma	MNU, embr.		6[c]	n.t.
	MNU, adult	highly malignant, invasive	50[c]	no
	promoter-induced			
Mesenchymal tumor	MNU + Testosterone	exophytic, slow growing	7[c]	n.t.
Melanoma, amelanotic	Testosterone	highly malignant	30[a]	no
	hereditary			
Melanoma (n = 15)	spontaneous	benign	2-3[a]	1.5-2
Melanoma (n = 28)	spontaneous	malignant	4-8[a]	2-3
	unknown			
Rhabdomyosarcoma	spontaneous	invasive	20[c]	n.t.

For comparison nontumorous organs were used:
[a] skin (including epidermis, stratum compactum, dermis, vasculary system); [b] eye (whole bulbus); [c] muscle (including muscle fibres, surrounding connective tissue, vascular system); *N-ethyl-N-nitrosourea; **N-methyl-N-nitrosourea

By Southern blot hybridization of Xiphophorus DNA with a pol probe of Mouse Mammary Tumor Virus (MMTV), sequences probably related to a cellular pol gene (c-pol) were detected (Lu, 1986). When several template primers were tested unter poly(rC)-conditions (Fig. 5, at left), poly(rC)p(dG)12-18 which is an appropriate substrate for RNA-dependent DNA polymerase is preferentially transcribed in the melanoma as well as in the muscle which served as the control. Polymerase activity in the hereditary melanoma is 3 to 4 times higher than in the muscle. The supernatant of a cell culture derived from an early embryo which developed an extreme malignant hereditary whole body melanoma (A. Anders and F. Anders, 1978) exhibited mRNA-dependent DNA polymerase activity that showed a preference for the template primer poly(rA)-p(dT)12-18 (Fig. 5, at left, lower curves). It appears that the melanomas of the adults and the whole body melanomas of the embryos express different RNA-dependent DNA polymerases. Since the template primer poly(rC)p(dG) 12-18, besides being a substrate for RNA-dependent DNA polymerase, might also serve in a small extent as a substrate for DNA-dependent DNA polymerase γ, we examined muscle (as the control), malignant melanoma, and highly malignant melanoma under poly(rCm)-conditions with the template primer poly(rCm)p(dG)12-18, which is more specific for RNA-dependent DNA polymerases (Gerard et al., 1980) (Fig. 5, at right). The factors by which the activity of the enzyme is elevated in the tissue tested under these conditions are in the ratio 1 to 2 to 5, which is in parallel with $pp60^{c-src}$ kinase activity measured in the hereditary tumors of different degree of malignancy.

Conclusions

Regardless of any future findings that might bring or might not bring the elevation of oncogene activity in causal relation to gene amplification and RNA-dependent DNA polymerase activity, these parameters incontestably parallel the appearance of neoplasia in Xiphophorus.

ONCOGENES IN THE CAUSATION OF NEOPLASIA

Genetic analysis of the regulation of oncogene activity in the formation of neoplasia

Our analytical work on neoplasia is mainly based on the development of crossing-conditioned hereditary melanoma in Xiphophorus as outlined in Fig. 6 (see F. Anders, 1967; F. Anders et al., 1985; 1986).

Crosses of a spotted wild platyfish female from Rio Jamapa (Fig.6A) with a non-spotted wild swordtail male from Rio Lancetilla (Fig.6B) result in melanoma developing F_1 hybrids (Fig. 6C). These hybrids develop uniformly in all individuals with melanoma consisting mainly of well-differentiated transformed pigment cells which are morphologically similar to those of the spots of the parental platyfish. The transformed pigment cells occur only in those compartments of the body where the platyfish parent exhibits the spots, i.e. in the skin of the dorsal fin and in the skin of the posterior part of the body. In older F_1 animals, the compartment-specific melanomas combine to form a large benign melanoma. Backcrosses of the F_1 hybrids with the swordtail as the recurrent parent result in offspring (BC_1) exhibiting three types of segregants: 25 % of the BC_1 (Fig. 6D) develop benign melanoma like that of the F_1, 25 % (Fig. 6E) develop malignant melanoma consisting mainly of incompletely differentiated transformed cells which invade other tissues (except for brain, gonads, intestine) and eventually kill

---

* Commercial DNA from Salmon showed also c-pol signals.

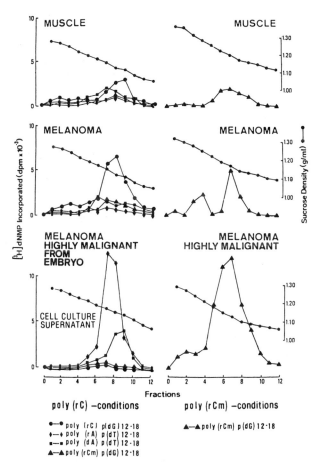

Fig. 5. RNA-dependent DNA polymerase activity, measured after centrifugation of the microsomal pellets in a sucrose gradient (20 % - 60 %) to equilibrium. Left: The enzyme activity was tested with several template primers under poly(rC)-conditions. Right: The enzyme activity was tested with poly(rCm)p(dG)12-18 under poly(rCm)-conditions (from Lüke, 1985; Lüke et al., 1986).

the fish, whereas 50 % (Fig. 6F,G) develop neither spots nor melanomas. Further backcrosses (not shown in Fig. 6) of the BC_1 fish carrying benign melanoma with the swordtail result in exactly the same segregation. The same applies for further backcrosses of this kind. At present we breed a BC_{25}. Backcrosses of the fish carrying the malignant melanoma with the swordtail show a different result: 50 % of the BC segregants develop malignant melanoma, whereas the remaining 50 % are melanoma free; benign melanomas do not occur. Whenever melanomas occur in these crossing experiments, the develop in both the compartment of the dorsal fin and the compartment of the posterior part of the body. Experiments which were discussed earlier (A. Anders et al., 1973) indicated that the inheritance of the spots and the melanomas are X chromosome-linked. If a terminal Giemsa band of the X chromosome is deleted hemizygously in the male of the platyfish and in the hybrids or homozygously in the female of the platyfish, neither spots nor melanomas develop (Ahuja, 1979; Ahuja et al., 1979).

The spontaneous occurence of melanomas in the hybrids induced by this crossing experiment (and by a large variety of similar experiments) may be explained by the assumption of four prominent genetic components which are contributed by X. maculatus to the hybrid genome.

1. The Mendelian oncogene Tu. This gene is apparently responsible for the process of neoplastic transformation. The restrictions of its transforming activity to pigment cells, to the dorsal fin, and to the posterior part of the body, as well as its capacity to mediate spots, benign melanomas and malignant melanomas come from the other genetic components.

2. The melanophore-specific regulatory gene R-Mel. This gene is a member of a series of regulatory genes that control the oncogene Tu specifically in the tissues of mesenchymal (R-Mes), epithelial (R-Epi) and neurogenic (R-Nerv and R-Mel) origin (see Fig. 7). In the experiment outlined in Fig. 6 it was impaired (see R-Mel') permitting the Mendelian oncogene Tu to mediate neoplastic transformation in certain pigment cell precursors. Gene transfer experiments showed co-transfer of Tu and R-Mel indicating a very close linkage between both genes (J. Vielkind et al., 1982).

3. The dorsal fin-specific and posterior part-specific regulatory genes R-Df and R-Pp. These genes are members of a multigene family consisting of at least 14 compartment-specific regulatory genes (R-Co in total; see Fig. 7) that correspond to the same number of compartments of the body of the fish (mouth, eye, dorsal fin, anterior part, posterior part, tail fin, peritoneum, meninx primitiva, etc.). They were identified by mutagenesis studies (A. Anders et al., 1973; A. Anders and F. Anders, 1978). In their active state they suppress Tu specifically in the respective compartments. In the present experiment the regulatory genes specific to the compartments of the dorsal fin (Df) and of the posterior part of the body (Pp) are impaired (R-Df' and R-Pp', see Fig. 6) thus permitting neoplastic transformation specific to these parts of the body. - Many of these compartments clearly correspond to the segmental organization of the fish anatomy (Peter, 1986). Restriction enzyme studies (Eco RI, and probes of fushi tarazu and antennapedia from Drosophila[*]) showed that the inheritance of certain homeo box-containing fragments is in parallel with the inheritance of the compartment-specific regulatory genes (Paulsen, 1986).

4. The differentiation gene Diff. This gene can be recognized by the 1:1

---

[*] The probes were gifts from W. Gehring.

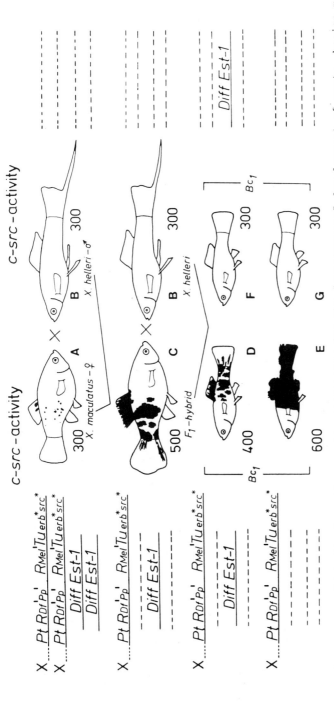

Fig. 6. Crossing scheme which displays the genetic conditions for the "spontaneous" development of spots, benign melanoma and malignant melanoma. ——, chromosomes of X. maculatus; ----, chromosomes of X. helleri; Tu, tumor gene; erb and src, critical copies of the cellular homologues of the retroviral oncogenes v-erb and v-src; R-Mel, impaired regulatory gene specific to pigment cells but nonspecific to the compartments; R-Pp' and R-Df', impaired regulatory genes controlling Tu in the compartments of the posterior part of the body (Pp) and the dorsal fin (Df); Diff, regulatory gene controlling differentiation of neoplastically transformed cells; Pt, pterinophore locus; Est-1, locus for esterase-1 (a marker gene for Diff) of X. maculatus; c-src-activity, pp60 c-src kinase activity expressed as counts per minute/milligram protein; note basic and enhanced activity, and correlation between src expression and Tu expression (from F. Anders et al., 1984 a,b, modified and actualized).

segregation of benign melanoma bearing and malignant melanoma bearing fish. The segregation indicates furthermore that Diff is not linked to Tu. If Diff is lacking in the system, the majority of the melanoma cells remains poorly differentiated thus forming malignant melanoma. If, however, Diff is present in single dosage in the system, the majority of the melanoma cells becomes rather well differentiated thus building up a benign melanoma. If, finally, Diff is present in double dosage, the transformed pigment cells become terminally differentiated thus forming the spots (U. Vielkind, 1976). The Diff activity is closely related to the appearance and disappearance of certain modified tRNAs (A. Anders et al., 1985b).

Xiphophorus helleri also contains the Tu oncogene and presumably the linked regulatory genes, which, however, are not mutated and thus are fully active. Tu of the swordtail and its linked regulatory genes, therefore, are not detectable with the methods used in this experiment. Furthermore, since the linked regulatory genes act only in the cis position, this "oncogene-regulatory-gene complex", contributed by the swordtail, does not significantly influence the expression of the platyfish-derived Tu copy in the hybrids. No Diff nor any other nonlinked regulatory gene was found in the swordtail that might correspond to the Tu-nonlinked regulatory genes of the platyfish. As indicated by the unexpected high Tu expression in F_1 (the Tu:Diff relation is 1:1 like in the platyfish female), X. helleri contributes "intensifier genes" to the genome (F. Anders, 1967). These genes are not taken into consideration in this context.

The outcome of this crossing experiment suggests the following interpretation: In the wild X. maculatus three out of about twenty important regulatory genes linked to the X-chromosomal oncogene Tu (R-Mel; R-Df and R-Pp) are impaired. This impairment, however, is compensated by the powerful autosomal regulatory gene Diff and some other nonlinked regulatory genes that have evolved in the wild population. The linked regulatory genes act in cis position only, while the nonlinked ones, such as Diff, act in the trans position. X. helleri contains an intact regulatory gene system that is entirely linked to the same chromosome where the oncogene Tu is located (not shown in Fig. 6). In the course of the crossings and backcrossings the autosome of the platyfish carrying regulatory genes are replaced stepwise by the homologous chromosomes of the swordtail lacking these genes. Thus the hybrids carrying the critical Tu of the platyfish but lacking the non-linked regulatory genes develop neoplasia. The neoplasms are melanomas because the pigment cell-specific R-Mel is impaired. The melanomas are restricted to the dorsal fin and the posterior part of the body because the compartment-specific R-Df and R-Pp are impaired. If there is still one copy of Diff present in the hybrid genome, the melanomas are benign. If Diff is lacking the melanomas are malignant.

The correctness of this interpretation was confirmed by successive backcrossings of the melanoma bearing hybrids with the platyfish, i.e. by the stepwise reintroduction of Diff and other platyfish-specific regulatory genes into the genome. The resulting animals are not capable to develop melanoma although they contain the "critical" Tu. As a reminescence to their ancestors that suffered from melanoma they may exhibit some small spots at the dorsal fin and at the posterior part of the body (for crossing procedures and photographs see F. Anders et al. 1984b).

Genetically and molecularly defined oncogenes in melanoma formation

Data on the relation between Mendelian and molecularly defined oncogenes came from serological studies on the already described purebred and hybrid fish shown in Fig. 6. Both, tumorous and nontumorous fish show a tissue-specific $pp60^{c-src}$ kinase activity (see chapter "Oncogenes in development"). $pp60^{c-src}$ activity in both brain and melanoma of these fishes is elevated (see data on hereditary melanomas in Tab. 2), and varies genotype-specifically in both brain and melanoma in the same direction (Barnekow et al., 1982; Schartl et al., 1982). Hence, we could determine $pp60^{c-src}$ kinase activity in brain tissue extracts of the crossing-conditioned tumorous fish and relate the activity observed to the expression of the critical Tu ascertained by the development of melanoma. It appears unlikely that the differences in kinase activity measured in the fish of different Tu genotypes are due to epiphenomena of the melanoma. The results reflect the actual genetic activity of the c-src oncogene in the nontumorous brain tissue of the crossing-conditioned tumorous and nontumorous fish.

As indicated in Fig. 6, the purebred platyfish female carrying the critical "Tu-regulatory-gene-complex" in both of its X chromosomes, as well as the purebred swordtail and the BC hybrids lacking this special complex, display the same low activity of c-src kinase in the brain. We interpret this activity to be the basic expression of c-src. In contrast, the hybrids which contain the melanoma mediating Tu show an increase in c-src activity, with the malignant melanoma bearing BC hybrids diplaying the highest activities in the brain (and in the tumor).

To accomplish these experiments, we crossed two melanoma bearing backcross hybrids. Compared with the parental BC-hybrids, the resulting offspring carrying 2 Tu-copies show an oncogene dosage effect in both melanoma formation and $pp60^{c-src}$ kinase activity (Schartl et al., 1982; F. Anders, 1983; F. Anders et al., 1984). On the other hand, a deletion of the terminal Giemsa band of the platyfish-derived X chromosome in BC-hybrids (deletion involving the R-Co R-Mel Tu fragment but excluding the pterinophore locus Pt-Dr-Ar which is a marker for red coloration; see Fig. 7) results in a loss of both the capability of melanoma formation and the elevation of $pp60^{c-src}$ kinase activity in the brain.

In conclusion, the expression of the c-src oncogene, specified by the $pp60^{c-src}$ kinase activity in the brain, parallels the phenotypic expression of the critical Tu oncogene, specified by the development of the melanoma. The parameters are interrelated with respect to a) the dosage of the critical Tu-carrying X chromosome (homozygous, hemizygous, lacking) and b) the dosage of the Diff-carrying autosome (also homozygous, hemizygous, lacking).

"Critical" copies of the molecularly defined oncogenes

More information about the parallel inheritance of Tu and c-src comes from Eco RI restriction enzyme analysis by means of total genomic Fish DNA and v-src probes (Southern blot) (Gronau, 1986). First of all it is important to note that the Eco RI restriction enzyme patterns of c-src are species-, race-, and population-specific. The same applies for c-erb. Genotype-specificity was also confirmed by Hind III digestion (Gronau, 1986; Zechel, 1986). The oncogenes that are highly conserved during evolution differ from species to species and even within the species from population to population. Furthermore, the Eco RI restriction enzyme patterns of c-src of the different population of X. maculatus are female- and male-specific indicating that the c-src copies located on the X and Y chromosome are also different.

The analysis was performed with females of the inbred X. maculatus from Rio Jamapa, and males of the inbred X. helleri from Rio Lancetilla,

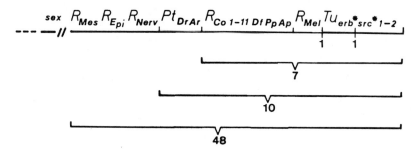

Fig. 7. Preliminary map of the sex chromosomes (X and Y) of X. maculatus from Rio Jamapa (and partly of X. variatus and X. xiphidium) based on a) 11 X-Y crossovers, b) 19 deletions, c) 14 duplications, d) 4 translocations, e) 14 compartment-specific mutants, f) Giemsa banding studies, and g) restriction enzyme analysis of the Tu erb src oncogene complex. The brackets indicate the regions within which the indicated number of structural changes occurred; one translocation separated Tu from all linked regulatory genes, another translocation showed that src* 1-2 is located distally to erb*. sex, sex determining region; R-Mes, R-Epi, R-Nerv, sets of regulatory genes controlling Tu in mesenchymal and epithelial tissues, and in the nervous cell system; Pt, pterinophore locus including the compartment-specific regulatory genes Dr (dorsal red) and Ar (anal red); R-Co, region containing at least 14 compartment-specific regulatory genes such as R-Df (dorsal fin-specific), R-Pp (posterior part-specific), R-Ap (anterior part-specific); R-Mel, melanophore specific regulatory gene; erb*, cellular homolog of the transforming gene of the avian erythroblastosis virus; src* 1-2, 2 copies of the cellular homolog of the transforming gene of Rous' sarcoma virus. The unit including R-Co, R-Mel, and Tu corresponds to the specific "color genes" known as Sd (spotted dorsal), Sp (spotted), etc. in the literature. Additional regulatory genes for Tu are distributed throughout other chromosomes, e.g. Diff (nonlinked to Tu), see text (A. Anders and F. Anders, 1978, actualized).

their F_1 hybrids (Fig. 8,I), and with melanoma free and malignant melanoma developing BC_{12} hybrids (Fig. 8,II). The fish indicated by the capital letters in Fig. 8 correspond to the fish indicated by the same letters in Fig. 6.

Depending upon the selective backcrossings of the hybrids with the swordtail, the nontumorous BC_{12} segregants probably contain only swordtail chromosomes whereas the tumorous segregants have retained the critical Tu-carrying X chromosome which is the only platyfish-chromosome in the swordtail genome of these hybrids. The analysis was accomplished by studying BC_7 animals having lost the above mentioned terminal Giemsa band that involves the critical R-Co R-Mel Tu segment of the platyfish-specific X chromosome.

All animals used for restriction enzyme analysis were subadults. In this developmental stage melanoma development is at the very beginning and does not seriously modify the bands of the restriction enzyme patterns by oncogene amplification (see paragraph "Oncogene amplification in the neoplasms" in the chapter "Oncogenes in neoplasia").

The following results were obtained. The Eco RI restriction enzyme pattern of c-src of the X. maculatus females (containing the critical Tu) differs considerably from that of the X. helleri males and from that of the hybrids (Fig. 8; some prominent bands are indicated by full arrows and full arrow heads if they are present, and by empty arrows and empty arrow heads if they are lacking). As was shown by application of particular conditions for the separation of the smaller or the larger fragments, respectively, the enzyme pattern of c-src of the X. maculatus females comprises in total 6 bands, and that of the X. helleri males 9 bands. 4 of the bands correspond to both the platyfish and the swordtail. 2 bands are platyfish-specific, and 5 are swordtail-specific. The pattern of the F_1 hybrids (containing the critical Tu) represents a combination of the species-non-specific and the species-specific bands, comprising altogether 11 bands.

The Eco RI restriction enzyme pattern of the melanoma free BC_{12} segregants (lacking the critical Tu) is identical to that of X. helleri. The same applies for the nontumorous backcross animals containing the Tu deletion-chromosome inherited from the platyfish. Differing from these genotypes, the Eco RI restriction enzyme pattern of the melanoma bearing BC_{12} segregants that contain the critical Tu-oncogene inherited from the platyfish, shows some platyfish-specific bands of the c-src-oncogene. The most striking band (see the lower full arrow) represents a 5.0 Kb fragment that can clearly be assigned to the terminal Giemsa-band of the X Chromosome of X. maculatus where the genetic information for neoplastic transformation, i.e. the critical Mendelian oncogene Tu is located. Following the deletion of the Giemsa band (deletion of Tu) this 5.0 Kb fragment disappears. The same applies for a 15.0 Kb fragment (see the upper arrow) which in this run of electrophoresis is incompletely separated from an adjacent band but was clearly identified in other runs on gels for the separation of the larger fragments. We consider these platyfish-specific bands as copies or parts of copies of the "critical" c-src, and symbolize these critical copies as c-src*.

The Eco RI restriction enzyme pattern of c-src allows some more conclusions. The 15 Kb band representing a critical c-src copy (c-src*) and the largest band (24.0 Kb) in the run show homologies with both the 5' and the 3' end of the viral src-probe (Pst I fragments) indicating that they contain a complete c-src copy each; they are distinguished by sequences of their flanking regions. Future analyses with other restriction enzymes will show whether these c-src copies are identical or different. The other fragments show homologies with only one end of the viral probe indicating that they are parts of an unknown number of copies, one of which (the 5.0 Kb band) is part of a src* copy. The haploid genome of the melanoma developing Xiphophorus hybrid (Fig. 8E, Fig. 6E), therefore, contains at least 4 copies of c-src comprising at least 3 different forms of organization. The genetically and cytogenetically determined Tu-oncogene contains at least two src* copies (c-src* 1-2, in Fig. 7) that might be involved in melanoma formation.

The same kind of studies was performed with the c-erb oncogene, and in principle, the same results were obtained. In this case a 2.8 Kb fragment of c-erb could be assigned to a critical copy of this oncogene (erb*) that is located in the chromosome region of the critical Tu. As is shown by a translocation, the erb* is proximally located to src* (see Fig. 7).

Conclusions

The X chromosome of X. maculatus from Rio Jamapa contains a terminally located gene complex involved in melanoma formation. Certain Copies of c-src and c-erb appear as the "critical" copies (c-src* 1-2 and c-erb ; Fig. 7) that might be involved in neoplastic transformation. They are probably constituents of this gene complex.

ONCOGENES IN EVOLUTION

Neoplasia was found in all taxonomic groups of eumetazoa (Dawe and Harshbarger, 1969; Dawe et al., 1981; Krieg, 1973; Kraybill et al., 1977) and was even detected in fossils of prehistoric animals, such as sauria, and in fossil mammals including humans (Kaiser, 1981). Oncogenes, therefore, are expected to be ubiquitously distributed in the animal kingdom (F. Anders, 1968; F. Anders et al., 1981).

After c-src was specified in non-infected chicken, as well as in humans, cattle, salmon, and Xiphophorus (see introduction), we started a systematic study on the taxonomic distribution of this oncogene, partly by Southern blot hybridization with v-src probes and partly by identification of pp60^{c-src}.

First of all, specimens of 8 species of the genus Xiphophorus, thereafter specimens of 5 genera taxonomically related to Xiphophorus, and of 8 taxonomically non-related groups of bony fish were investigated. Since all fish specimens contained c-src, we extended our search for this oncogene to several prominent groups of organisms. Our results together with those of Spector et al. (1978) obtained with humans, calf, and salmon, and those of Shilo and Weinberg (1981) obtained with Drosophila are listed in Table 3.

c-src was not detected in unicellular eukaryotes including the protists Euglena, Cryptomonas, Chlorogenium, Paramecium, Tetrahymena as well as the colony forming Volvox. The same applies to Trichoplax, which is regarded to represent an intermediate form between the prozozoan and metazoan organization. More refined methods will show whether there are still some src-related sequences in the src-negative organisms. To detect such sequences could be of very high value for developing ideas about how an oncogene such as c-src might have evolved from the very outset of living beings on the earth.

The ubiquity of c-src in metazoa from humans down to the sponges which are known to have evolved in the proterozoikum 1.5×10^9 years ago, and the lack of c-ros and c-mos in Xiphophorus (see Tab. 1) as well as the presence of ras-, mos-, and abl-related DNA fragments in yeast, slime molds and other filamentous fungi (Prakash and Seligy, 1985) gave rise to a preliminary study on the taxonomic distribution of 6 additional c-oncs.

Although this study (Fig. 9) is still incomplete and needs improvement, it gives rise to the idea that the different oncogenes tested have probably evolved in parallel with the evolution of the large taxonomic groups: c-ras - heterotrophic Eukaryotes, c-src - Metazoa, c-abl - Bilateria, c-sis - Chordata, c-yes - Gnathostomata, c-ros - Tetrapoda, c-mos - Mammalia. Hence, the number of different oncogenes increased probably in parallel with the complexity of the genome, of the differentiation and of the development in the course of evolution in the animal kingdom (Tab. 4). Mammals contain probably the highest number of oncogenes. They have more oncogene species that may become

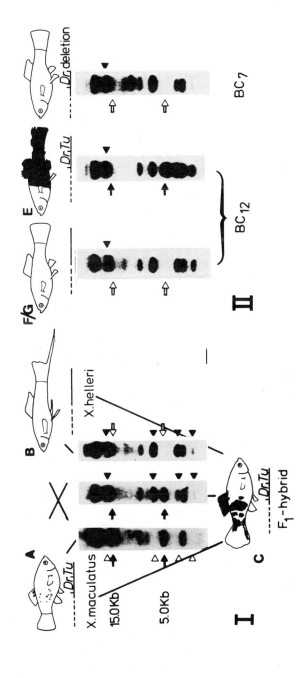

Fig. 8. c-src Eco RI restriction fragments. Some prominent bands are indicated by full arrows and full arrow heads. Prominent bands that are lacking are indicated by empty arrows and empty arrow heads. Note the arrows corresponding to the 5.0 and 15.0 Kb fragments: they are platyfish-specific, are present in melanoma-developing fish, and are lacking in swordtails and non-melanomous hybrids. - Dr (dorsal red) is a compartment-specific pterinophore gene (see Fig. Pt Dr Ar) causing a reddish coloration. The deletion genotype has lost the terminal Giemsa band involving the R-Co R-Mel Tu fragment (see Fig. 7). The fish indicated by the capital letters correspond to those indicated by the same letters in Fig. 6 (data from Gonau, 1986).

33

Tab. 3. c-src in Eukaryotes

Mammals	Bony fish	Bony fish	Cartilag. fish	Mollusc
Humans	Poeciliidae	Flat fish	Shark	Cuttle fish
Calf	Xiphophorus	Sea robin	Jawless fish	Coelenterates
Rat	Girardinus	Mackerel	Lamprey	Sea anemone
Mouse	Poecilia	Roach	Acrania	Sponges
Birds	Belonesox	Gudgeon	Amphioxus	Marine
Chicken	Heterandria	Salmon	Insects	sponge
Quail	Xenotoca	Codfish	Cockroach	Freshwater
Frogs		Cichlid	Drosophila	sponge
Xenopus				

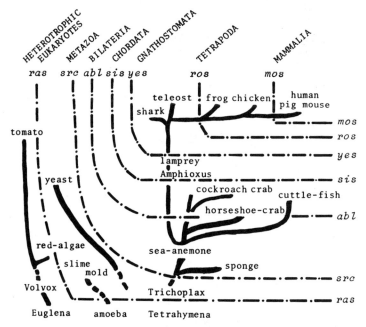

Fig. 9. Preliminary study on the distribution of cellular oncogenes. Mammals contain all oncogenes tested; c-mos seems to be restricted to mammals; c-ros appears first in frogs, c-yes in sharks, c-sis in Amphioxus, c-abl in crabs, c-src in sponges, c-ras in yeast (data from Schartl; Anders et al., 1985a).

Table 4. Increase of the number of oncogenes in parallel to complexity of different species.

Yeast	ras						
Sponges	ras	src					
Crabs	ras	src	abl				
Amphioxus	ras	src	abl	sis			
Sharks	ras	src	abl	sis	yes		
Frogs	ras	src	abl	sis	yes	ros	
Mammals	ras	src	abl	sis	yes	ros	mos

deregulated as compared to the lower animals and therefore, may run the highest risk to become susceptible to neoplasia, or to become sensitive to carcinogenes, respectively.

GENERAL CONCLUSIONS

Many facts about the structure and the activity of particular oncogenes have been elaborated by many research groups but little became known about the significance of these genes in development, neoplasia and evolution.

First of all it appears that the oncogenes are indispensable developmental genes that are probably used for normogenesis and for the maintenance of the differentiated state of all heterotrophic eukaryotes including yeast as well s humans. During the history of life on the earth they were, therefore, submitted to a concerted evolution involving a) the development from the zygote to the germ cell-producing individuum in the succeeding generations, and b) the phylogeny from heterotrophic microbes to the highly complex multicellular animals. Keeping in mind both development and phylogeny it becomes evident that the oncogenes on the one hand appear as highly conserved genetic elements with tissue and developmental stage specificity while, on the other hand, they appear as evolving multigene families with the properties of multiplicity, close linkage and overlapping functions. Even in the different wild populations of a given species of Xiphophorus they occur in different numbers of copies, different forms of organization, different functions, and are located on different chromosomes. This altogether and the new combinations of the genome in the course of interpopulational hybridization that favors microevolution, may result in the de novo origin of selectable phenotypic variations that finally may lead to higher taxonomic groups. The probably existing parallel between the number of oncogenes and the taxonomic groups of animals as demonstrated in Figure 9 and Table 4 has to be considered in this sense. A stepwise addition of oncogenes in the course of evolution is not conceivable.

Because the oncogenes are responsible for important developmental functions for different purposes at different times, they require stringent regulation. Regulatory gene systems that consist of several multi-gene families acting on different levels of biological organization were submitted to the concerted evolution mentioned above, and evolved oncogene-specifity. Impairment of a regulatory gene system results, therefore, in the loss of control over the oncogene directed cell reproduction and cell differentiation but not in the loss of the ability of the cell to grow, to devide and to differentiate. Thus, the oncogenes of an adult may become reactivated to exert their early embryo-specific functions which, as an extension of the cellular development in the embryo appears as neoplastic transformation.

REFERANCES

Ahuja, M.R., 1979, On the nature of genetic change as an underlying cause for the origin of neoplasms, in: Antiviral Mechanisms in the Control of Neoplasia", P. Chandra, ed., Plenum Press, New York, London.

Ahuja, M.R., and Anders, F., 1977, Cancer as a problem of gene regulation, in: "Recent Advances in cancer Research: Cellbiology, Molecular Biology, and Tumor Virology, 1" R. C. Gallo, ed., CRC Press, Cleveland, Ohio.

Ahuja, M.R., Lepper, K., and Anders, F., 1979, Sexchromosome aberrations involving loss and translocation of tumor-inducing loci in Xiphophorus, Experientia, 35:28.

Anders, A., and Anders, F., 1978, Etiology of cancer as studied in the platyfish-swordtail system, Biochim. Biophys. Acta, 516:61.

Anders, A., Anders, F., and Klinke, K., 1973, Regulation of gene expression in the Gordon-Kosswig melanoma system. I. The distribution of the controlling genes in the genome of the xiphophorine fish, Platypoecilus maculatus and Platypoecilus variatus. II. The arrangement of chromatophore determining loci and regulating elements in the sexchromosomes of xiphophorine fish, Platypoecilus maculatus and Platypoecilus variatus, in: "Genetics and Mutagenesis of Fish", J.H. Schröder, ed., Springer-Verlag, Heidelberg.

Anders, A., Anders, F., Lüke, W., Henze, M., Schartl, M., and Schmidt, C.-R., 1985a, Onkogene in Evolution, Entwicklung u. Tumorbildung, in: Sonderforschungsbereich 103, Zellenergetik und Zelldifferenzierung, Ergebnisbericht 1983-1985, Philipps-Universität Marburg.

Anders, A., Anders, F., and Pursglove, D.L., 1971, X ray-induced mutations of the genetically-determined melanoma system of xiphophorine fish, Experientia, 27:931.

Anders, A., Dess, G., Nishimura, S., and Kersten, H., 1985b, A molecular approach to the study of malignancy and benignancy in melanoma of Xiphophorus, in: "Pigment Cell 1985, Biological, Molecular and Clinical Aspects of Pigmentation", J. Bagnara, S. N. Klaus, E. Paul, and M. Schartl, eds., University of Tokyo Press, Tokyo.

Anders, F., 1967, Tumour formation in platyfish-swordtail hybrids as a problem of gene regulation, Experientia, 23:1.

Anders, F., 1968, Genetische Faktoren bei der Entstehung von Neoplasmen, Zbl. Vet. Med., 15:29.

Anders, F., 1981, Erb- und Umweltfaktoren im Ursachengefüge des neoplastischen Wachstums nach Studien an Xiphophorus, Klin. Wochenschr., 59:943.

Anders, F., 1983, The biology of an oncogene, based upon studies on neoplasia in Xiphophorus, in: "Haematology and Blood Transfusion Vol. 28, Modern Trends in Human Leukemia V", R. Neth, R. C. Gallo, M. F. Greaves, M. A. S. Moore, and K. Winkler, eds., Springer-Verlag, Berlin, Heidelberg.

Anders, F., Gronau, T., Schartl, M., Barnekow, A., Jaenel-Dess, G., and Anders, A., 1986, Cellular oncogenes as ubiquitous genomic constituents in the animal kingdom and as fundamentals in melanoma formation, in: "Proceedings of the First International Conference on Skin Melanoma", N. Cascinelli, ed., in press.

Anders, F., and Klinke, K., 1966, Über Gen-Dosis, Gen-Dosiseffekt und Gen-Dosiskompensation, Verh. Dtsch. Zool. Anz., 30 (Suppl.):391.

Anders, F., and Schartl, M., 1984, Wertung von Parametern für die taxonomische Klassifizierung im Genus Xiphophorus (Teleostei: Poeciliidae), Verh. Dtsch. Ges. Zool. Anz., 77:254.

Anders, F., Schartl, M., and Barnekow, A., 1984b, Xiphophorus as an in vivo model for studies on oncogenes, in: "Use of Small Fish Species in Carcinogenicity Testing", K. L. Hoover, ed., Natl. Cancer Inst. Monogr. 65, National Cancer Institute, Bethesda, Maryland.

Anders, F., Schartl, M., Barnekow, A., and Anders, A., 1984a, Xiphophorus as an in vivo model for studies on normal and defective control of oncogenes, Adv. Cancer Res., 42:191.

Anders, F., Schartl, M., Barnekow, A., Lüke, W., Jaenel-Dess, G., and Anders, A., 1985, The genes that carcinogens act upon, in: "Modern Trends in Human Leukemia, IV", R. Neth, R. C. Gallo, M. F. Greaves, and G. Janka, eds., Springer-Verlag, Berlin, Heidelberg, New York.

Anders, F., Schartl, M., and Scholl, E., 1981, Evaluation of environmental and hereditary factors in carcinogenesis, based on studies in Xiphophorus, in: "Phyletic Approaches to Cancer", C. D. Dawe, J. C. Harshbarger, S. Kondo, T. Sugimura, and S. Takayama, eds., Japan Scientific Societies Press, Tokyo.

Anders, F., Vester, F., Klinke, K., and Schumacher, H., 1961, Genetisch bedingte Tumoren und der Gehalt an freien Aminosäuren bei lebendgebärenden Zahnkarpfen (Poeciliidae), Experientia, 17:549.

Barnekow, A., and Schartl, M., 1984, Cellular src gene product detected in the freshwater sponge Spongilla lacustris, Mol. Cell. Biol., 4:1179.

Barnekow, A., Schartl, M., Anders, F., and Bauer, H., 1982, Identification of a fish protein associated with kinase activity and related to the Rous sarcoma virus transforming protein, Cancer Res., 42:4222.

Bentvelzen, P., 1972, Hereditary infections with mammary tumor viruses in mice, in: "RNA Viruses and Host Genome in Oncogenesis", P. Emmelot and P. Bentvelzen, eds., North-Holland Publ., Amsterdam.

Bishop, J. M., 1985, Retroviruses and cancer genes, in: "Genetics, Cell Differentiation, and Cancer", P. A. Marks, ed., Academic Press, Inc., New York.

Dawe, C.D., and Harshbarger, J.C., 1969, "Neoplasms and Related Disorders of Invertebrate and Lower Vertebrate Animals", Natl. Cancer Inst. Monogr. 31, Bethesda.

Dawe, C.D., Harshbarger, J.C., Kondo, S., Sugimura, T., and Takayama, S., 1981, "Phyletic Approaches to Cancer", Japan Scientific Societies Press, Tokyo.

Flavell, N., 1984, Role of reverse transcription in the generation of extra-chromosomal copia mobile genetic elements, Nature, 310:514.

Garfinkel, D.J., Boeke, J.D., and Fink, G.R., 1985, Ty element transposition: Reverse transcriptase and virus-like particles, Cell, 42:507.

Gerard, G., Loewenstein, P., and Green, M., 1980, Characterization of a DNA-polymerase activity in cultured human melanoma cells that copies poly(rCm)p(dG), J. Biol. Chem., 255:1015.

Gordon, M., 1928, Pigment inheritance in the Mexican killifish: Interaction of factors in Platypoecilus maculatus, J. Hered., 19:253.

Gronau, T., 1986, Onkogene und Onkogenamplifikation bei der Tumorbildung von Xiphophorus, Thesis, Univ. Giessen.

Häussler, G., 1928, Über Melanombildungen bei Bastarden von Xiphophorus helleri und Platypoecilus maculatus var. Rubra, Klin. Wochenschr. 7:1561.

Huebner, R.J., and Todaro, G.J., 1969, Oncogenes of RNA tumor viruses as determinants of cancer, Proc. Natl. Acad. Sci. USA, 64:1087.

Kaiser, H.E., 1981, Phylogeny and paleopathology of animal and human neoplasms, in: "Neoplasms - Comparative Pathology of Growth in Animals, Plants and Man", H. E. Kaiser, ed., Williams and Wilkins Baltimore, London.

Kollinger, G., and Siegmund, E., 1981, Meiosis of species and interspecific hybrids of Xiphophorus, Verh. Dtsch. Zool. Ges., 1981:206.

Kosswig, C., 1928, Über Kreuzungen zwischen den Teleostiern Xiphophorus helleri und Platypoecilus maculatus, Z. indukt. Abstammungs- und Vererbungslehre, 47:150.

Kraybill, H.F., Dawe, C.J., Harshbarger, J.C., and Tardiff, R.G., 1977, "Aquatic Pollutants and Biological Effects with Emphasis on Neoplasia", Ann. N. Y. Acad. Sci., 298.

Krieg, K., 1973, Zur Bedeutung wirbelloser Tierspezies für die Geschwulstforschung, Biol. Zentralbl., 92:617.

Kuff, E.L., Feenstra, A., Lueders, K., Smith, L., Hawley, R., Hozumi, N., and Shulman, M., 1983; Intracisternal A-particle genes as movable elements in the mouse genome, Proc. Natl. Acad. Sci. USA, 80:1992.

Lu, K., 1986, Isolation and sequencing of the pol-related gene from Xiphophorus, Thesis, Univ. Giessen.

Lüke, W., 1985, RNA-abhängige DNA-Polymerase-Aktivität im Xiphophorus Tumor System, Thesis, Univ. Giessen.

Lüke, W., and Anders, F., 1986, RNA-dependent DNA polymerase activity in normal and neoplastic tissues of Xiphophorus, Biochim. Biophys. Acta, submitted.

Mellor, J., Malim, M.H., Gull, K., Tuite, M.F., McCready, S., Dibbayawan, T., Kingsman, S.M., and Kingsman, A.J., 1985, Reverse transcriptase activity and Ty RNA are associated with virus-like particles in yeast, Nature, 318:583.

Meyer, M.K., Wischnath, L., and Foerster, W., 1985, "Lebendgebärende Zierfische, Arten der Welt", Mergus-Verlag, Melle.

Paulsen, N., 1986, Homoeoboxen in Xiphophorus, Thesis, Univ. Giessen.

Peter, R.U., 1986, Vergleichend-anatomische Untersuchungen über die extrakutanen Pigmentierungsmuster der Poeciliiden (Pisces, Teleostei), Thesis, Univ. Giessen.

Prakash, K., and Seligy, V.L., 1985, Oncogene related sequences in fungi: Linkage of some to actin, Biochem. Biophys. Res. Commun.,133:293.

Pursglove, D.L., Anders, A., Döll, G., and Anders, F., 1971, Effects of X-irradiation on the genetically-determined melanoma system of xiphophorine fish, Experientia, 27:695.

Radda, A.C., 1980, Synopsis der Gattung Xiphophorus HECKEL, Aquaria,27:39

Raulf, F., and Schartl, M., 1985, Differential expression of protooncogenes during embryogenesis of Xiphophorus, Europ. J. Cell Biol. (Suppl.), 39:27.

Rosen, D.E., 1979, Fishes from the uplands and intermontane basins of Guatemala: revisionary studies and comparative geography, Bull. Am. Mus. Natl. Hist., 162:267.

Rous, P., 1911, A sarcoma transmitted by an agent separable from the tumor cell, Proc. N. York Path. Soc., XI:8.

Schartl, A., Schartl, M., and Anders, F., 1982, Promotion and regression of neoplasia by testosterone-promoted cell differentiation in Xiphophorus and Girardinus, in: "Cocarcinogenesis and Biological Effects of Tumor Promoters, Carcinogenesis - a Comprehensive Survey, 7", E. Hecker, N.E. Fusenig, W. Kunz, F. Marks, and H. W. Thielmann, eds., Raven Press, New York.

Schartl, M., and Barnekow, A., 1982, The expression in eukaryotes of a tyrosine kinase which is reactive with $pp60^{c-src}$ antibodies, Differentiation, 23:108.

Schartl, M., and Barnekow, A., 1984, Differential expression of the cellular src gene during vertebrate development, Dev. Biol., 105:415.

Schartl, M., Barnekow, A., Bauer, H., and Anders, F., 1982, Correlation of inheritance and expression between a tumor gene and the cellular homolog of the Rous Sarcoma Virus-transforming gene in Xiphophorus, Cancer Res., 42:4222.

Schartl, M. Mäueler, W., Raulf, F., and Barnekow, A., 1985a, Differential expression of cellular oncogenes in normal and neoplastic tissues of vertebrates, Europ. J. Cell Biol. (Suppl.), 39:30.

Schartl, M., Schmidt, C.-R., Anders, A., and Barnekow, A., 1985b, Elevated expression of the cellular src gene in tumors of differing etiologies in Xiphophorus, Int. J. Cancer, 36:199.

Schmidt, C.-R., Herbert, H., and Anders, A., 1986, Selective tumor formation in sensitive tester strains of Xiphophorus indicating initiating and/or promoting activities of carcinogens, in preparation.

Schimke, R.T., 1984, Gene amplification, drug resistance, and cancer, Cancer Res., 44:1735.

Scholl, A., and Anders, F., 1973, Electrophoretic variation of enzyme proteins in platyfish and swordtail (Poeciliidae; Teleostei), Archiv für Genetik, 46:121.

Schwab, M., 1980, Genome organization in Xiphophorus (Poeciliidae; Teleostei), Mol. Gen. Genet., 188:410.

Schwab, M., 1982, How can altered differentiation induced by 12-O-tetradecanoylphorbol-13-acetate be related to tumor promotion, in: "Cocarcinogenesis and Biological Effects of Tumor Promoters, Carcinogenesis - a Comprehensive Survey, 7", E. Hecker, N. E. Fusenig, W. Kunz, F. Marks, and H. W. Thielmann, eds., Raven Press, New York.

Schwab, M., Abdo, S., Ahuja, M.R., Kollinger, G., Anders, A., Anders, F., and Frese, K., 1978a, Genetics of susceptibility in the platyfish/swordtail tumor system to develop fibrosarcoma and rhabdomyosarcoma following treatment with N-methyl-N-nitrosourea (MNU), Z. Krebsforsch., 91:301.

Schwab, M., Haas, J., Abdo, S., Ahuja, M.R., Kollinger, G., Anders, A., and Anders, F., 1978b, Genetic basis of susceptibility for development of neoplasms following treatment with N-methyl-N-nitrosourea (MNU) or X-rays in the platyfish-swordtail system, Experientia, 34:780.

Schwab, M., Kollinger, G., Haas, J., Ahuja, M.R., Abdo, S., Anders, A., and Anders, F., 1979, Genetic basis of susceptibility for neuroblastoma following treatment with N-methyl-N-nitrosourea and X-rays in Xiphophorus, Cancer Res., 39:519.

Sharp, P., 1983, Conversion of RNA to DNA in mammals: Alu-like elements and pseudogenes, Nature, 301:471.

Shiba, S., and Saigo, K., 1983, Retrovirus-like particles containing RNA homologous to the transposable element copia in Drosophila melanogaster, Nature, 302:119.

Shilo, B.Z., and Weinberg, R.A., 1981, DNA sequences homologous to vertebrate oncogenes are conserved in Drosophila melanogaster, Proc. Natl. Acad. Sci. USA, 78:6789.

Simon, M.A., Kornberg, T.B., and Bishop, J.M., 1983, Three loci related to the src oncogene and tyrosine-specific protein kinase activity in Drosophila, Nature, 302:837.

Slamon, D., and Cline, M.J., 1984, Expression of cellular oncogenes during embryonic and fetal development of mouse, Proc. Natl. Acad. Sci. USA, 81:7141.

Spector, D.H., Varmus, H.E., and Bishop, J.M., 1978, Nucleotide sequences related to the transforming gene of avian sarcoma virus are present in DNA of uninfected vertebrates, Proc. Natl. Acad. Sci. USA, 75:4102.

Stehelin, D., Varmus, H.E., and Bishop, J.M., 1976, DNA related to the transforming gene(s) of avian sarcoma viruses is present in normal avian DNA, Nature, 260:170.

Trosko, J.E., 1986, personal communication.
Vielkind, J., Haas-Andela, H., Vielkind, U., and Anders, F., 1982, The introduction of a specific pigment cell type by total genomic DNA injected into the neural crest region of fish embryos of the genus Xiphophorus, Mol. Gen. Genet., 185:379.
Vielkind, U., 1976, Genetic control of cell differentiation in platyfish-swordtail melanomas, J. Exp. Zool., 196:197.
Zechel, Ch., 1986, Subklonierung und Sequenzierung von c-erb von Xiphophorus, Thesis, Univ. Giessen.

ACQUISITION OF METASTATIC PROPERTIES VIA SOMATIC CELL FUSION: IMPLICATIONS FOR TUMOR PROGRESSION IN VIVO

Patrick De Baetselier[1], Ed Roos[2], Hendrik Verschueren[3], Steven Verhaegen[1], Daniel Dekegel[3], Lea Brys[1] and Michael Feldman[4]

Instituut voor Moleculaire Biologie, Vrije Universiteit Brussel, Belgium[1]; Division of Cell Biology, The Netherlands Cancer Institute, Amsterdam, The Netherlands[2] Pasteur Instituut van Brabant, Brussels, Belgium[3] and Department of Cell Biology, Weizmann Institute of Science Rehovot, Israel[4]

SOMATIC CELL HYBRIDIZATION AND TUMOR PROGRESSION

Somatic hybridization of normal somatic cells with neoplastic cells has been widely adopted as a tool to identify chromosomes or genes involved in the suppression or expression of malignancy. Originally, the neoplastic trait appeared to be dominant (1,2) but now experimental results seem to indicate that fusion of transformed cells with normal cells results in hybrids that are, initially at least, non tumorigenic (3,4). Tumorigenic lines, when they arise, are thought to do so by chromosomal segregation and the loss of specific chromosomes (5,6). This suppression of in vivo growth capacity of tumorigenic lines has often been referred to as "suppression of malignancy" and recently it has been proposed that a group of recessive genes, the socalled "tumor suppressors" or "anti-oncogenes" are implicated in this process (7).

There is a considerable body of literature indicating that fusion between cells does occur in experimentally transplanted tumors (8,9) and furthermore there is some strong circumstantial evidence that such a phenomenon also takes place in naturally occuring neoplasms (10). The key question then emerges as to whether hybrids of this nature have acquired an increased malignant phenotype. Indeed, since it is now generally accepted that the trait of malignancy is suppressed in tumor-normal cell hybrids made in vitro, it is then appropriate to ask whether there is evidence that some of the malignant segregants might be significantly more malignant than the parental tumor cells once they arise in vivo. This question is difficult to answer since in most investigations on the malignant status of tumor-host cell hybrids and their segregants "malignancy" was evaluated by assessing local tumor growth namely "tumorigenicity". The tumorigenic potential of a given tumor cell population is, however, not a complete reflection of the malignant status since many authors (11) have stressed that invasive and metastatic capacity are the traits which characterize the truly malignant potential of tumor cells. Quite recently, from results obtained in a few model systems, it was suggested that fusion of neoplastic cells with certain types of host cells may result in hybrids which are not suppressed and even express a higher malignant status as reflected by invasiveness and metastatic capacity. The first intriguing observation in this direction was made by Goldenberg et al. (12) who reported a human tumor which, upon transplantation into hamster cheek pouch, metastasized. Cytogenetic analysis revealed that the metastases were generated by tumor-host cell hybrids. These workers proposed that such in vivo hybridization in xenogeneic and possibly isogeneic systems was a possible mechanism by which tumor cells could progress to more advanced states of malignancy. Such a notion has been tested in a few experimental model systems both in vivo and in vitro (13-16), and it has been proposed that fusion of neoplastic cells with certain types of host cells may actually enhance the metastatic potential of the resultant hybrids by conferring the normal cell's special properties on these hybrids.

We will herein summarize our experimental evidence indicating that tumor cells can acquire or express invasive and metastatic properties following somatic cell hybridization with normal somatic cells both in vivo and in vitro. The implications of these observations for the progression of both clinical and experimental tumors, the evolution of tumor heterogeneity in vivo and other aspects of tumor biology will be discussed in the following chapters.

NONMETASTATIC LYMPHOID TUMOR CELLS ACQUIRE METASTATIC PROPERTIES FOLLOWING CELL FUSION WITH NORMAL LYMPHO-RETICULAR CELLS IN VITRO

In our initial studies we selected as an experimental tumor model nonmetastatic B-cell tumors such as NSI and SP2/O that are currently used as universal fusers for the immortalization of B lymphocytes and we addressed the question of whether fusions between nonmetastatic plasmocytomas and normal B lymphocytes would result in hybrids manifesting metastatic properties. We found that all B-cell hybridomas, irrespective of the class and antigenic specificity of their secreted immunoglobulins did generate metastases following subcutaneous or intraveneous inoculation. Hereby two types of metastatic patterns were observed, namely hybridomas that generated metastases in spleen and liver and hybridomas that generated metastases solely in the liver (13). Unlike B-cell hybridomas, a NSI derived hybridoma of non B-cell origin, for which a putative macrophage origin was suggested by morphological and functional criteria, produced metastases in the lungs, whereas the liver and spleen were not infiltrated (17). These experiments demonstrated that nonmetastatic B-cell tumors can acquire metastatic properties following somatic cell hybridization with normal lympho-reticular cells such as B lymphocytes or macrophage-like cells. Yet it appears that the organ-specificity of the acquired metastatic competence is determined by the normal somatic parental cell.

Subsequently, we have tested whether other neoplastic cells, such as thymoma cells, could also acquire metastatic properties through somatic hybridization with normal somatic cells. To test whether, in principle, somatic fusion could confer metastatic properties on non-metastatic T-cell tumors, a model system was selected in which somatic hybridization could readily be achieved in culture and in which the normal non-neoplastic partner was "an invasive cell" and thus possessed some of the dessiminating properties of a metastatic cell. Thus, we investigated the outcome of hybridization of the nonmetastatic AKR BW5147 T-cell lymphoma with T-cells that had been activated for five days in mixed lymphocyte cultures (MLC). Such MLC activated T lymphocytes exhibit properties considered to be characteristic for metastatic tumor cells such as invasiveness in cultures of endothelial cells and hepatocytes

(18) or the degradation of extracellular matrix components (19). When tested for their capacity to develop organ-specific metastases, we found that these cells manifested, upon i.v. inoculation, extensive metastasis formation in spleen, liver, kidneys, and ovaries (20). The inherent metastatic capacity of the T-cell hybridomas, as compared to BW cells, was further confirmed in an in vitro hepatocyte invasion test system (18). In such an in vitro invasion model adhesion of BW to the hepatocytes was minimal and adherent BW cells did not infiltrate the monolayer. On the other hand, hybrid T-cells adhered to the hepatocytes and rapidly infiltrated the cultures. Such in vitro observations indicate that hybrid T-cells have acquired invasive properties that enable these to diffusely infiltrate liver tissue. Interestingly the pattern of metastatic development in the liver of T-cell hybrids as compared to liver-seeking B-cell hybridomas was strikingly different. As shown in Fig. 1, the T-cell hybrids manifested a diffuse pattern of development, as compared to distinct nodular growth of liver metastases produced by SP2/0 B-cell hybridomas. Such differences in metastatic growth patterns may reflect differences in organ-tumor interaction. In fact we found that liver specific metastatic B-cell hybridomas do not adhere to monolayers of hepatocytes, nor do they infiltrate them (17). Thus, diffuse liver infiltration observed with T-cell hybridomas might be due to their capacity to specifically adhere to and infiltrate hepatocytes, while nodular liver growth by B-cell hybridomas might reflect a local organ-specific growth effect. In this context it is worth mentioning that metastatic B-cell hybridomas as compared to metastatic T-cell hybridomas were found to be relatively less malignant probably due to their differential capacity to infiltrate liver tissue.

Testing different sets of T-cell hybridomas we found a close association between high invasiveness in vitro and metastatic potential in vivo. Indeed, all of 29 T-cell hybridomas generated from activated T-cells were highly invasive in hepatocyte cultures and 6 representative hybrids and clones thereof were found to be highly metastatic (20). Other fusion experiments between BW and unstimulated T-cells yielded a number of non-invasive T-cell hybrids that were non-metastatic. Hence invasiveness and metastatic potential of T-cell hybridomas appear to be a property conferred by the normal, activated T-cell fusion partner. Such notion is strengthened by experiments demonstrating that BW tumor cells acquire similar invasive and metastatic properties through somatic cell hybridization with an IL-2 dependent continuous T-cell line (CTL-D). This cell line, which can be considered as activated since it expresses IL-2 receptors, was found to be invasive in in vitro hepatocyte cultures. T-cell hybridomas derived from fusions between CTL-D and BW were found to be invasive in vitro and metastatic in vivo. Interestingly a distinct organotropism was observed with certain of these hybridomas such as 82-2-14 T-cell hybrids which metastasized mainly or exclusively to the kidneys while the majority of the other hybrids tested, metastasized extensively to liver and spleen and to a lesser extent to the kidneys (Fig. 2). Apparently in vitro invasiveness in hepatocyte cultures is related to metastatic capacity in vivo, yet it is not unequivocally associated with the capacity to metastasize extensively to the liver.

In conclusion, the above described experiments demonstrate that nonmetastatic B and T-cell tumors acquire invasive and metastatic properties through in vitro somatic hybridization with lympho-reticular cells and as such B and T-cell hybrids can be considered as more malignant (summarized in table 1). These results appear to disagree with the repeated finding that metastatic capacity behaves as a recessive trait in normal x tumor cell hybrids (21,22). A possible explanation for this discrepancy might reside in the fact that in those cases quite distinct cell types were fused, namely lymphocytes with either mammary

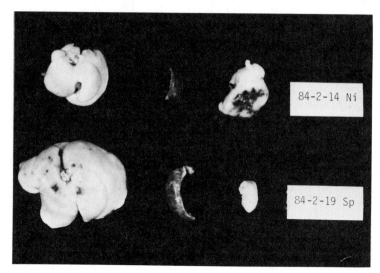

Fig. 1. Metastatic burden and pattern in the livers of mice given metastatic B-cell hybridoma cells (11H11 Li,Li) or T-cell hybridoma cells (84-2-19 SP,SP). Note the nodular tumor growth in the liver by B-cell hybrids and the diffuse tumor growth in the liver by T-cell hybrids.

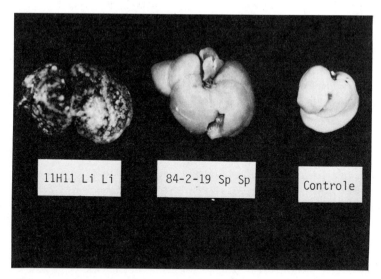

Fig. 2. Metastatic burden and pattern in the organs of mice given metastatic T-cell hybridoma cells (84-2-14, Ni; 84-2-19,SP). 84-2-14, Ni cells seek exclusively the kidneys (upper, from left to right: liver, spleen, kidneys). 84-2-19, SP cells seek the liver and spleen (lower, from left to right: liver, spleen, kidneys).

TABLE 1. LYMPHOID TUMOR CELLS ACQUIRE METASTATIC PROPERTIES FOLLOWING CELL FUSION WITH LYMPHO-RETICULAR CELLS

Tumor cells	origin	invasive-ness[a]	metastatic capacity[b] organotropism	pattern
B-cell tumors				
NSI	plasmacytoma	NT	-	-
Hy1	NSI x B-cell hybrid (IgG1)	NT	spleen mainly	nodular
Hy2	NSI x B-cell hybrid (IgG1)	NT	liver mainly	nodular
Hy8	NSI x B-cell hybrid (IgM)	NT	liver	nodular
Hy2	NSI x B-cell hybrid (IgM)	NT	spleen, liver	nodular
Hy38	NSI x macrophage hybrid	NT	lungs	nodular
SP2/0	myeloma	-(0.0)	-	-
11H11	SP2/0 x B-cell hybrid (IgM)	-(0.0)	spleen, liver	nodular
T-cell tumors				
BW	thymoma	-(0.0)	-	-
TAM2,D2	BW x T-cell hybrid (MLC activated T-cell)	+(1.5)	spleen, liver, kidneys	diffuse
TAM4,A6	BW x T-cell hybrid (MLC activated T-cell)	+(1.3)	spleen, liver, kidneys	diffuse
TAM4,D1	BW x T-cell hybrid (MLC activated T-cell)	+(0.9)	spleen, liver, kidneys	diffuse
TAS5,C4	BW x T-cell hybrid (non-activated T-cell)	-(0.1)	-	-
TAS5,C6	BW x T-cell hybrid (non-activated T-cell)	-(0.1)	-	-
CTL-D	IL-2 dependent T-cell line	+(0.5)	-	-
84-2-14	BW x CTL-D hybrid	+(0.6)	kidneys	diffuse
84-2-19	BW x CTL-D hybrid	+(0.8)	liver, spleen	diffuse
84-2-21	BW x CTL-D hybrid	+(0.9)	liver	diffuse

a. Invasive capacity was assessed in an *in vitro* hepatocyte invasion assay. Invasive capacity in hepatocyte cultures was estimated by counting the number of infiltrating cells per hepatocyte nuclei (i.e., the infiltration index). An arbitrary upper limit for non-invasiveness was set at an infiltration index of 0.1 (NT = not tested)

b. The metastatic potential of the tumor cells was assessed by inoculating 2×10^6 cells i.v. to syngeneic mice. The mice were killed when they looked moribund and examined macroscopically for evidence of metastatic growth in different organs

carcinoma (21) or melanoma (22) cells, which may have led to an inappropriate combination of properties. We have fused quite comparable cell types, namely T-lymphoma cells with T lymphocytes and B-lymphoma with B lymphocytes, so that this problem was avoided. A complete phenotypic compatibility between the two cellular partners is, however, not a general prerequisite for the manifestation of a metastatic potential since a NSI plasmacytoma x macrophage hybridoma manifested a hihgly metastatic organ specific growth and furthermore other investigations have recently demonstrated that the nonmetastatic Eb T-cell lymphoma acquires metastatic capability through cell fusion with bone-marrow derived macrophages (16). Our results are also in contrast to the repeatedly observed recessive nature of tumorigenicity in somatic cell hybrids (3,4). Since segregation of only one normal chromosome appears sufficient for tumorigenicity to reappear (4), it is not excluded that all the hybrids tested in vivo so far have lost this chromosome and karyotyping will have to be performed to test this possibility. Alternatively it could be speculated that repressors of tumorigenicity and metastatic potential are present in most normal cell types, but that their activity is suppressed in certain leucocytes depending on their phenotype and/or state of activation.

NONMETASTATIC LYMPHOID TUMOR CELLS ACQUIRE METASTATIC PROPERTIES FOLLOWING CELL FUSION WITH NORMAL LYMPHORETICULAR CELLS IN VIVO

The fact that fusion of neoplastic cells with certain types of host cells in vitro may result in hybrids which are not suppressed but rather express a higher metastatic capacity prompted us to assess whether such a phenomenon might represent a mechanism for tumor progression in vivo. To monitor the possible events of tumor-host cell fusion in vivo and subsequent generation of metastatic variants, a tumor-host model had to be selected in which both the tumor cells and host somatic cells would readily express identifiable, distinguishable markers. Furthermore, to increase the probability of fusion in vivo, the tumor cells had to be highly fusogenic in vitro. In view of the results obtained with the metastatic BW x T and BW x CTL-D hybridoma cells, we chose as tumor model the AKR derived T-cell tumor BW 5147. As recipients for BW tumor inoculation we selected allogeneic CBA and semi-allogeneic (CBA x AKR)F1 recipients since the CBA strain shares with AKR the same H-2 haplotype but differs in the expression of minor surface alloantigens (i.e. CBA: Thy 1.2 and Lyt 1.1, AKR: Thy 1.1 and Lyt 1.2).

Thus, CBA and (CBA x AKR)F1 animals were inoculated i.v. with BW 5147 lymphosarcoma cells and one month after inoculation two animals developed modular growth in the liver. Cells derived from these liver nodules (termed CBA derived BW-Li cells or (CBA x AKR)F1 derived BW-O-Li cells) were grown in culture and re-inoculated i.v. to either CBA or (CBA x AKR)F1 recipients. The animals were killed 2 weeks after tumor inoculation and autopsy revealed diffuse tumor growth in spleen, liver, kidneys and ovaries in 100 % of the animals. In particular, the liver was heavily infiltrated in a diffuse manner, distinct from compact nodular growth. Furthermore, these BW liver variants appeared to have acquired inherent invasive properties enabling them to infiltrate in vitro monolayers of hepatocytes. Thus inoculation of an AKR T-cell lymphosarcoma to CBA or (CBA x AKR)F1 animals has led within one transplantation cycle to the generation of a tumor cell which manifests the capacity to invade different organs. Finally, these BW liver variants were new genetic variants since BW cells were 8-azaguanine drug-resistant tumor cells which died in HAT-medium while BW-Li and BW-O-Li cells were 8-azaguanine-sensitive and HAT-resistant.

We next addressed the question of whether the BW liver variants were real metastatic variants derived from the BW tumor cell. Flow cytofluorographic analysis of cells stained with monoclonal anti-mouse Thy 1 antibodies revealed that the BW liver variants expressed the Thy 1 marker of the parental BW lymphoma (i.e. Thy 1.1) and the Thy 1 marker of the CBA host (i.e. Thy 1.2), unlike the BW cells which stained completely negative for the membrane expression of Thy 1.2. These results imply that the BW liver variants originated from the BW lymphoma and had acquired membrane components of the CBA host such as Thy 1.2. Additional staining experiments with monoclonal reagents (i.c. anti-H-2K^k, anti-H-2D^k, anti-Lyt 1, anti-Lyt 2) revealed that BW liver variants differed from the parental BW cells in the membrane expression of two additional membrane markers, namely Lyt 1 and H-2K alloantigens. Interestingly, the Lyt 1 alloantigen appeared to have an AKR origin (Lyt 1.2) and not a CBA origin (Lyt 1.1). Thus BW-Li liver variants express one host-derived alloantigen (Thy 1.2) and two BW tumor derived alloantigens (Thy 1.1 and Lyt 1.2). The expression of a mature T-cell marker (i.e. Lyt 1) on BW-Li variants raised the question of whether these cells would exhibit T-cell functions such as lymphokine secretion. In fact, BW liver variants but not BW cells, were found to be inducible to secrete low, yet reproducible levels of IL-2 after stimulation with ConA. Hence the transition from the non-metastatic phenotype of BW cells to the high-metastatic phenotype of BW-Li and BW-O-Li cells was associated with the acquisition or expression of certain differentiation antigens (i.e. Thy 1.2, Lyt 1.2, H-2^k) and functional characteristics (i.e. IL-2 production). Some of the newly expressed differentiation antigens were unequivocally of host origin (i.e. Thy 1.2). Thus, on the basis of phenotypic, functional and drug-sensitive characteristics (summarized in table 2) we proposed that BW-Li and BW-O-Li cells have originated from an in vivo fusion between BW cells and host CBA or (CBA x AKR)F1 T-cells (14).

TABLE 2. CHARACTERISTICS OF BW CELLS AND METASTATIC BW VARIANTS GENERATED THROUGH IN VIVO CELL FUSIONS

Characteristics		BW	BW-Li	BW-O-Li
Invasiveness		−	+	+
Metastatic capacity		−	+	+
AZA resistance		+	−	−
HAT resistance		−	+	+
Membrane phenotype:	Thy 1.2	−	+	+
	Thy 1.1	+	+	+
	Lyt 1.2	−	+	+
	Lyt 2	−	−	−
	H-2D^k	−	+	+
	H-2K^k	−	−	+
IL-2 secretion		−	+	+

a. Tumor cells derived from a metastatic liver lesion of a CBA mice inoculated i.v. with BW cells.

b. Tumor cells derived from a metastatic liver lesion of a (AKR x CBA)F1 mice inoculated i.v. with BW cells.

Similar results, indicating that the in vivo formation of tumor-host cell hybrids could lead to the emergence of more malignant tumor cells as defined by metastatic ability were reported by Kerbel and colleagues (15). Using a nonmetastatic variant (MDW-D4) of the highly metastatic MDAY-D2 DBA/2 (H-2^d) mouse tumor, these authors demonstrated that such variants became metastatic in vivo subsequent to a cellular change, such as extinction of recessive lectin or drug sensitivity or acquisition of a higher chromosome number. Supportive evidence for a spontaneous cell fusion in vivo between the MDW4 tumor cells and host cells was provided by the observation that growth of MDW4 tumor cells in either $(H-2^k \times H-2^d)F1$ mice or $(H-2^k) \rightarrow H-2^d$ bone marrow radiation chimeras led to the appearance of metastatic MDW4 derived cells which express $H-2^k$ antigens. Thus in this experimental tumor model, tumor progression might have arisen as a consequence of tumor-host cell fusion in vivo.

Using another murine tumor model, the DBA/2 Eb/Esb lymphoma suggestive evidence was provided recently by Larizza et al. (16) that the highly metastatic variant Esb of the T-cell lymphoma Eb was derived from spontaneous fusion with a host macrophage. Esb cells represent a spontaneous high metastatic variant which arose during successive i.p. transplantation of the low-metastatic Eb T-lymphoma and shifts from Eb to Esb have been thereafter observed several times (23). Hybridization of a spontaneously selected drug marked variant of the Eb line (Eb TGR) with syngeneic bone marrow derived macrophages resulted in the generation of highly metastatic cell lines (16). These results from the in vitro hybridization experiments between Eb cells and macrophages suggest the possibility that somatic cell hybridization followed by chromosomal segregation may also be the cause of spontaneous in vivo shifts from low to high metastatic capacity in the Eb/Esb system. Supportive evidence for this was provided by a detailed cell surface marker analysis of Eb versus Esb cells. Indeed Esb cells were found to express several cell surface markers that are quite unusual in a T-cell derived lymphoma, namely class II histocompatibility antigens and the macrophage differentiation antigen Mac.1. Like Fc-receptors which are expressed on Esb but not on Eb cells, these antigens could well be traits of a host macrophage introduced into the Eb T lymphoma through cell fusion. In conclusion, experimental evidence provided by three different murine lymphoid tumors (BW 5147, MDW-D4 and Eb) suggests that lymphoid tumors may fuse in vivo with host cells. Additional events following this fusion process, such as generation of segregants by chromosome loss or host-mediated selection, may lead to the emergence of hybrids with an increased malignant phenotype as defined by invasive and metastatic characteristics.

Finally, although the evidence for spontaneous in vivo hybridization of tumor and host cell is not so compelling in human cancers as in experimental tumor systems, indirect evidence for fusion events occuring in human tumors is worth mentioning. Presumptive evidence for tumor-host fusion in human malignancies has been mainly inferred from cytological and karyological data such as high incidence of multinucleated cells, hyperploid chromosome constitution and premature chromosome condensation (PCC). PCC is a well recognized consequence of spontaneous or artificially induced cellular fusion since it is induced when two cells are fused while one of them is in mitosis and the other in interphase. PCC has been reported to occur in various human malignancies such as bladder carcinoma (10), acute myeloblastic leukemia (24), acute myelomonoblastic leukemia (24), and breast carcinoma (25). Provided further evidence for somatic mating will emerge from the study of human tumors, then in vivo fusion could be considered as a natural phenomenon, not a laboratory artefact created by repeated transplantation of tumor cells.

POSSIBLE MECHANISMS IMPLICATED IN THE GENERATION OF METASTATIC VARIANTS THROUGH SOMATIC CELL FUSION

Most of the hybrid systems, investigated so far show that fusion either in vivo or in vitro of nonmetastatic tumors with lympho-reticular cells or bone marrow derived normal cells can lead to the generation of metastatic tumor variants (13-17). We tend to suggest that the "circulating" capacity of the normal parental partners such as lymphocytes or macrophages might represent molecular properties which determine the acquisition of metastatic competence via cell fusion. Indeed lympho-reticular cells share with disseminating tumor cells a number of properties relevant to the metastatic process such as: the capacity to migrate via the blood circulation, to survive while migrating, to penetrate blood vessels, to invade different tissues and to home to defined regions of the lymphoid system. Hence, fusion of lympho-reticular cells with neoplastic cells may confer such properties on the hybrid tumor cells, resulting in the manifestation of metastatic competence. In fact, our results obtained with the invasive, metastatic T-cell hybrids, suggest that invasiveness is a T-cell property, dominantly expressed in the hybrids. Thus, particular properties of normal cells, when introduced into a tumor cell, may apparently give rise to a metastatic behaviour.

Alternatively, it has been proposed that somatic cell fusion is a source of somatic rearrangement leading to metastatic variants (26). Conceivably, extensive chromosome rearrangements and segregational mechanisms in tumor-host hybrids may represent mechanisms causing altered oncogene activity. Activation of new oncogenes by transposition or amplification as well as the amplification of previously activated oncogenes could influence the phenotypic character of tumors resulting in transitions from low to high malignancy. Both mechanisms, namely acquisition of an invasive machinery from a normal host cell or induction of so called "metastogenes", may be implicated during somatic cell fusion and are not mutually exclusive. In fact, in the context of the BW tumor model it might be premature to state that the in vivo generated BW liver variants have acquired their invasive potential from a normal activated host T-cell. Indeed, these variants express the Lyt 1.2 alloantigen which belongs genotypically to BW tumor cells but is not expressed on BW cells, indicating that these lymphoma cells are arrested in a stage of differentiation. Cell fusion of BW cells with a normal somatic cell might lead to large genotypic changes resulting in the activation of silent differentiation genes as manifested by Lyt 1.2 expression and IL-2 secretion. The resulting hybrid would then express the phenotype of a more differentiated and/or activated T-cell in which invasiveness is constitutively expressed. Such an assumption would lead to the prediction that BW tumor cells possess a silent, repressed invasive phenotype and recent experimental evidence points towards this possibility.

To test whether differentiation events could be induced in BW cells resulting in the expression of invasive and metastatic capabilities, BW cells were inoculated in a hematopoietic inducing microenvironment such as the spleen. This organ was selected since in the mouse the spleen has been reported to provide an appropriate microenvironment for the generation of differentiation or growth inductive signals both for normal and malignant lympho-reticular cells. Indeed, B lymphocytes and macrophages require the spleen for an optimal muturation to fully immunocompetent cells (27,28). Furthermore, the capacity of the spleen in producing a "microenvironmental growth milieu" resulting in enhanced tumorigenicity and peripheral growth has been reported for certain murine leukemias (29) (BCL1) and B-cell hybridomas (30). Finally, with regards to metastatic spread we have reported that spleen and liver B-cell hybridomas require an inductive signal emitted by the spleen in order to generate liver metastases (13). Taken together, these data indicate that the spleen plays a growth controlling role in lymphoid tumor development, or that certain lymphoid tumor cells may be induced in the spleen to undergo a differentiation event necessary for their spread to other organs. This last aspect was substantiated with the BW tumor model since intra-splenic inoculation of BW cells to either semi-allogeneic (AKR x Balb/c)F1 or syngeneic (AKR) recipients resulted in metastatic development in spleen and liver. Cells derived from these metastatic lesions behaved as real metastatic variants (termed BW-O-IS), generating liver and spleen metastases in 100 % of the recipients. These variants did not result from a cell fusion process with host cells as evidenced by the presence of the recessive drug resistance marker to 8-azaguanine and the absence of host membrane markers (i.e. Thy 1.2 and $H-2^d$ of Balb/c origin). Furthermore, these variants expressed syngeneic membrane antigens, that are absent on BW cells such as $H-2^k$ and Lyt 1.2 alloantigens. Thus intra-splenic inoculation of BW cells has resulted in the induction of metastatic variants (table 3) that resemble phenotypically the BW liver variants generated by cell fusion in vivo (table 2).

TABLE 3. CHARACTERISTICS OF BW CELLS AND SPLEEN INDUCED METASTATIC BW VARIANTS

Characteristics		BW	BW-IS-Li (AKR)	BW-IS-Li (AKR x BALB/C)
Invasiveness		−	+	+
Metastatic capacity		−	+	+
AZA resistance		+	+	+
HAT resistance		−	−	−
Membrane phenotype:	Thy 1.2	−	−	−
	Thy 1.1	+	+	+
	Lyt 1.2	−	+	+
	$Lyt 2$	−	−	−
	$H-2^k$	−	+	+
	$H-2^d$	−	−	−

a. Tumor cells derived from a metastatic liver lesion of a AKR mice inoculated intra-splenically with BW cells

b. Tumor cells derived from a metastatic liver lesion of a (AKR x BALB/C) F1 mice inoculated intra-splenically with BW cells

Recently, a model was proposed in which signals from the tumor microenvironment could activate genetic programs within tumor cell subpopulations causing phenotypic changes similar to those seen in normal tissue differentiation (31). This model suggests that [I] multiple phenotypes can become expressed by one common genotype and [II] that microenvironmental factors have a regulatory role on tumor cell phenotypes and expression of heterogeneity. Obviously, the spleen may provide an ideal microenvironment for the generation of differentiation related inductive signals in particular when dealing with tumors of lymphoid origin.

In conclusion, different mechanisms might be implicated in the expression of metastatic properties by BW cells following somatic hybridization with host T-cells. As shown in Fig. 3 cell fusion of BW cells with activated T lymphocytes could result in the acquisition of invasive properties of the normal somatic cell partner resulting in

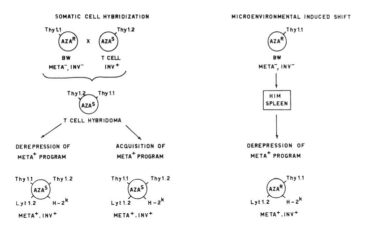

Fig. 3. Generation of metastatic BW variants through either somatic cell hybridization or microenvironmental induced shifts.

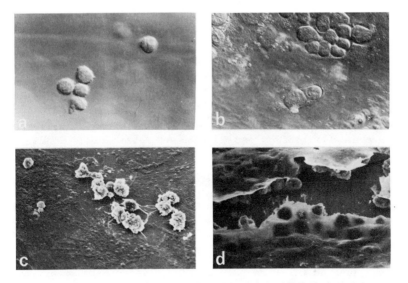

Fig. 4. Interaction of BW or BW-O-Li cells with 10T1/2 monolayers.

a and b, differential interference micrographs: (a) micrograph focused on BW cells lying on top of a 10T1/2 monolayer, (b) micrograph of invasive BW-O-Li cells lying in the same focal plane as the invaded 10T1/2 monolayer.

c and d, scanning electron micrographs: (c) BW cells lying on top of a 10T1/2 monolayer, (d) BW-O-Li cells have invaded the monolayer. The damaged 10T1/2 layer reveals the presence of underlying lymphoma cells.

invasive T-cell hybridomas that are highly metastatic. Alternatively cell fusion might have led to the induction of a repressed invasive phenotype. This last possibility cannot be excluded since BW cells possess metastasis-associated genetic programs that can be activated in a hematopietic-inducing microenvironment (HIM).

SOMATIC CELL HYBRIDIZATION AS A TOOL TO ANALYSE FACTORS INVOLVED IN TUMOR CELL METASTASIS

Besides representing a possible mechanism through which tumor cells may acquire metastatic properties in vivo, somatic cell hybridization may provide us with a new useful method to analyse cellular and molecular properties which control tumor cell metastasis. Indeed the use of somatic hybridization to engineer sets of malignant cells is an attractive tool for metastasis research for the following reasons: First, a large number of closely related but distinct cell lines of a defined genetic background can be generated. Phenomena such as different target organ specificity by related T-cell hybridomas (i.e. 84-2-14 versus 84-2-19 hybrid cells, Fig. 2) may be based on different membrane properties. Clearly, such closely related sets of metastatic hybrids could be adopted to screen for membrane molecules that control the metastatic pattern of tumor cells. Secondly, whereas some hybridomas are quite stable in chromosomal constitution, others show an extensive chromosome segregation. Thus, by comparing many different cell lines, or clones of one cell line obtained soon after fusion, it should be possible to focus on properties closely associated with invasive potential. We have used such sets of related T-cell hybridomas to develop an in vitro model permitting detailed studies on malignant cell implantation as will be illustrated below.

As indicated above (table 1) a close relationship was observed with T-cell hybrids between invasiveness in vitro using a hepatocyte monolayer invasion assay and metastatic potential in vivo. Prior to organ tissue infiltration blood-borne malignant cells have to interact with endothelial cells which are lining the microcirculation. Invasion of endothelial cells will in fact determine whether metastatic cells will successfully implant in the microcirculation at selected sites, extravasate in the tissues and develop as metastatic colonies. We have developed a monolayer invasion system which may mimick to a certain extent initial steps in the process of extravasation such as interactions with endothelial cells. This invasion assay is based on a continuous, cloned line of fibroblastic mouse embryo cells (10T1/2) which are similar in morphology to cultured endothelial cells. Non-metastatic BW cells did not invade this 10T1/2 monolayer and stayed as rounded lymphoma cells on top of the confluent 10T1/2 cells as visualized by differential interference contrast microscopy (Fig. 4a) or scanning electron microscopy (Fig. 4c). Metastatic T-cell hybrids or activated T lymphocytes such as CTL-D on the contrary invaded extensively the 10T1/2 monolayer (Fig. 4b,d). A quantitative estimation of invading tumor cells was readily obtained by counting the number of underlying flattened, invasive cells in phase contrast microscopy. Using this invasion system we have compared different T-cell hybrids and a consistent correlation was apparent between the 10T1/2 invasion index, the hepatocyte infiltration index and in vivo metastatic potential (Table 4). Using this relatively easy invasion assay and matched pairs of invasive and non-invasive hybrids a model system is available to assess the effect of extrinsic factors, i. e. anti-invasive and chemotherapeutic drugs, antibodies, physical parameters on invasive properties of tumor cells.

TABLE 4. INVASION OF T-CELL HYBRIDS IN FIBROBLAST AND HEPATOCYTE CULTURES

Tumor cells (origin)	Metastatic capacity[a]	Invasiveness 10T1/2 cultures[b]	Hepatocyte cultures[c]
BW (Thymoma)	-	-(<2)	-(<0.1)
TAS-5 (BW x T-cell hybrid, in vitro)	-	-(<2)	-(<0.1)
TCM-6 (BW x T-cell hybrid, in vitro)	+	+(3-3.7)	+(1.5)
BW-O-Li (BW x T-cell hybrid, in vivo)	+	+(3.8-4.2)	+(1.44)
CTL-D (IL-2 dependent T-cell line)	-	+(3.8-4.2)	+(0.61)
84-2-19 (BW x CTL-D hybrid, in vitro)	+	+(3.8-3.9)	+(1.67)
84-2-14 (BW x CTL-D hybrid, in vitro)	+	+(3.3-3.7)	+(1.77)

a. The metastatic potential of the tumor cells was assessed by inoculating 2×10^6 cells i.v. to syngeneic mice. The mice were killed when they looked moribund and examined macroscopically for evidence of metastatic growth in different organs.

b. Invasive capacity in confluent 10T1/2 fibroblast cultures was estimated by counting underlying, flattened, invasive lymphoma cells. Monolayer invasion is expressed as the logarithm of the number of flattened lymphoma cells per cm^2 (invasion index).

c. Invasive capacity in hepatocyte cultures was estimated by counting the number of infiltrating cells per hepatocyte nuclei (i.e. the infiltration index). The infiltration index is considered as a measure for the extent of invasion.

ACKNOWLEDGEMENTS

This work was supported by a research grant of the A.S.L.K. cancer fund. P. De Baetselier is a N.F.W.O. fellow.

REFERENCES

1. G. Barski, and J. Belehradek, Inheritance of malignancy in somatic cell hybrids, Somatic cell genet., 5:897 (1979).
2. G. Barski, S. Sorieul, and F. Cornefert, "Hybrid" type cells in combined cultures of two different mammalian cell strains, J. Nat. Cancer Inst., 26:1297 (1961).
3. E. J. Stanbridge, Suppression of malignancy in human cells, Nature, 260:17 (1976).
4. E. P. Evans, M. D. Burtenshaw, B. B. Brown, R. Henrion, and H. Harris, The analysis of malignancy by cell fusion. IX. Re-examination and clarification of the cytogenetic problem, J. Cell. Sci., 26:113 (1982).
5. H. Harris, Cell fusion and the analysis of malignancy, Proc. R. Soc. Lond. B, 179:1 (1971).
6. E. J. Stanbridge, R. R. Flandermeyer, D. W. Daniels, and W. A. Nelson-Rees, Specific chromosome loss associated with the expression of tumorigenicity in human cell hybrids, Somatic cell genet., 7:699 (1981).
7. G. Klein, and E. Klein, Evolution of tumors and the impact of molecular oncology, Nature, 315:190 (1985).
8. F. Wiener, E. M. Fenyo, G. Klein, and H. Harris, Fusion of tumour

cells with host cells, Nature (New Biology), 238:155 (1972).
9. R. Ber, F. Wiener, and E. M. Fenyo, Proof of in vivo fusion of murine tumor cells with host cells by universal fusers, J. Nat. Cancer Inst., 60:931 (1978).
10. N. B. Atkin, Premature chromosome condensation in carcinoma of the bladder: presumptive evidence for fusion of normal and malignant cells, Cytogenet. Cell genet., 23:217 (1979).
11. G. Poste, and I. J. Fidler, The pathogenesis of cancer metastasis, Nature, 283:139 (1980).
12. D. M. Goldenberg, R. A. Pavia, and M. C. Tsao, In vivo hybridization of human tumour and normal hamster cells, Nature, 250:649(1974).
13. P. De Baetselier, E. Gorelik, Z. Esschar, Y. Ron, S. Katsav, M. Feldman, and S. Segal, Hybridization between plasmacytoma cells and B lymphocytes confers metastatic properties on a non-metastatic tumor, J. Nat. Cancer Inst., 67:1079 (1981).
14. P. De Baetselier, E. Roos, L. Brys, L. Remels, and M. Feldman, Generation of invasive and metastatic variants of a non-metastatic T-cell lymphoma by in vivo fusion with normal host cells, Int.J. Cancer, 34:731 (1984).
15. R. S. Kerbel, E. A. Lagarde, J. W. Dennis, and T. P. Donaghue, Spontaneous fusion between normal host and tumor cells: Possible contribution to tumor progression and metastases studied with a lectin resistant mutant tumor, Mol. Cell. Biol., 3:523 (1983).
16. L. Larizza, V. Schirrmacher, L. Graf, E. Pflüger, M. Peres-Martinel, and M. Stöhr, Suggestive evidence that the highly metastatic variant Esb of the T-cell lymphoma Eb is derived from spontaneous fusion with a host macrophage, Int. J. Cancer, 34:695 (1984).
17. P. De Baetselier, E. Roos, L. Brys, L. Remels, M. Gobert, D. Dekegel, S. Segal, and M. Feldman, Nonmetastatic tumor cells acquire metastatic properties following somatic hybridization with normal cells, Cancer Metastasis reviews, 3:5 (1984).
18. E. Roos, and I. V. Van de Pavert, Antigen-activated T lymphocytes infiltrate hepatocyte cultures in a manner comparable to liver-colonizing lymphosarcoma cells, Clin. Exp. Metast., 1:173 (1983).
19. Y. Naparstek, I. R. Cohen, Z. Fuks, and I. Vlodavsky, Activated T lymphocytes produce a matric-degrading heparan sulphate endoglycosidase, Nature, 310:241 (1984).
20. E. Roos, G. La Riviere, J. G. Collard, M. J. Stukart, and P. De Baetselier, T-cell hybridomas: invasiveness in vitro and metastasis formation by bloodborne cells, Cancer Res., in press.
21. I. A. Ramshaw, S. Carlsen, H. C. Wang, and P. Badenoch-Jones, The use of cell fusion to analyse factors involved in tumour cell metastases, Int. J. Cancer, 32:471 (1983).
22. E. Sidebottom, and S. R. Clark, Cell fusion segregates progressive growth from metastasis, Br. J. Cancer, 47:399 (1983).
23. V. Schirrmacher, P. Altevogt, and K. Bosslet, Spontaneous phenotypic shifts from low to high metastatic capacity, in: "Biochemical and Biological Markers of Neoplastic Transformation", P. Chandra, ed., Plenum Press, New York, London, pp. 121 (1983).
24. D. M. Williams, C. D. Scott, and T. M. Beck, Premature chromosome condensation in human leukemia, Blood, 47:687 (1976).
25. G. Hovacs, and A. Georgh, Spontaneous cell fusion in human malignancies: possible mechanism leading to heterogeneity, Lancet, 2:350 (1985).
26. L. Larizza, and V. Schirrmacher, Somatic cell fusion as a source of genetic rearrangement leading to metastatic variants, Cancer Metastasis reviews, 3:193 (1984).
27. Y. Ron, P. De Baetselier, and S. Segal, Involvement of the spleen in murine B cell differentiation, Eur. J. Immunol., 11:94 (1981).
28. Y. Ron, P. De Baetselier, and S. Segal, Involvement of the spleen in the control of the immunogenic and phagocytic function of thio-

glycolate-induced macrophages, Eur. J. Immunol., 11:608 (1981).
29. S. Slavin, S. Morecki, and L. Weiss, The role of the spleen in tumor growth: kinetics of the murine B cell leukemia (BCL1), J. Immunol., 124:586 (1980).
30. P. L. Witte, and R. Ber, Improved efficiency of hybridoma ascites production by intrasplenic inoculation in mice, J. Nat. Cancer Inst., 70:575 (1983).
31. V. Schirrmacher, Shifts in tumor cell phenotypes induced by signals from the microenvironment. Relevance for the immunobiology of cancer metastasis, Immunobiology, 157:85 (1980).

ONCOGENES AND TYROSINE KINASE ACTIVITIES AS A FUNCTION OF THE METASTATIC PHENOTYPE

Lea Eisenbach and Michael Feldman

Department of Cell Biology
The Weizmann Institute of Science
PO Box 26, Rehovot, 76100, Israel

INTRODUCTION

Cancer cell dissemination resulting in metastasis formation is a multistage process. It involves detachment of cells from the primary tumor (1), penetration of tumor cells through the host's intercellular matrices and capillary basement membranes (2), dissemination via the blood circulation, extravasation into the target organ (3), and progressive growth in the target organs (4). At each of these steps the interaction between the disseminating cells and the host's immune system may determine the probability of survival of the disseminating tumor cells. The generation of new antigenic variants (5) that are resistant to natural killer cells (6) or to cytotoxic T cells (7,8) was demonstrated in many tumor systems. We have shown that changes in the expression of genes of the major histocompatibility complex (MHC) control the immunogenicity and thereby the metastatic phenotype of individual clones of two murine tumors, the Lewis lung carcinoma and the T10 sarcoma (9-12). Are the immunogenic properties the only one to play a role in the metastatic phenotype or do other mechanisms play a role in the metastatic process of these two tumors? To explore additional components which control the metastatic phenotype of 3LL and T10 clones we tested the generation of metastases in the absence of an active immune system.

DIFFERENCES IN METASTATIC PROPERTIES OF TUMOR CLONES EXIST IN IMMUNE DEFICIENT HOSTS

We inoculated high metastatic and low metastatic cells of the 3LL and T10 tumors in immunocompetent and in 550-R irradiated recipients. Table 1 demonstrates that low metastatic cells can form metastases in the absence of an active immune system but the level of metastatic growth, expressed as lung weights, is lower in the low metastatic clones IC9, A9 as, compared to high metastatic clones IB9, IE7, D122, 3LL even in immune deficient mice. To identify the non immune mechanisms which may be involved in the metastatic process, we analysed the invasive properties of low metastatic and high metastatic clones of 3LL and T10 sarcoma, as manifested by two proteolytic enzymes, plasminogen activator and collagenase type IV which were implicated in tumor dissemination (13,14).

Table 1. Spontaneous Metastatic Potential of T10 and 3LL Cloned Cells

Tumor	Clone	Normal recipients	Irradiated recipients
T10 sarcoma	IC9	173±63	255±52
	D6	292±103	n.d.
	IB9	427±133	687±163
	IE7	464±208	691±164
	T40	369±963	481±181
3LL Carcinoma	A9	239±35	356±165
	D122	657±445	832±207
	3LL	404±180	603±200

Weight of lungs (mg)± S.D. Single Cell suspensions of 10^5 cells of each clone were inoculated intra footpad into untreated and sublethally irradiated (550 rad co) syngeneic recipients (C57BI/6) for 3LL carcinoma clones and (C3HxC57BI)/6)F1 for T10 sarcoma clones. When the tumor diameter reached a size of 8-10 mm, the tumor bearing leg was amputated. In each experimental group, mice were killed 21 days post-amputation and their lungs removed and assessed for metastatic load by weights and by number of nodules counted after fixation in Bouin's solution (data not presented). n.d. - not determined.

COLLAGENASE TYPE IV ACTIVITY IN SUBCELLULAR FRACTIONS OF CLONED TUMOR CELL POPULATIONS

Collagenase type IV is a protease that degrades specifically basement membrane of blood vessels. Production of this enzyme by human and murine invasive tumors was shown to be a property of metastatic, as distinct from benign tumors (15-17). We tested whether there is a correlation between the metastatic potency of individual clones from a metastatic tumor and the level of synthesis and secretion of collagenase IV. Using ^{14}C proline labeled collagen IV purified from EHS sarcoma we found that both low and high metastatic cells secreted low levels of enzyme into the media of cell cultures (Table 2 entry 1,2). When the same cells were injected subcutaneously into mice, and the solid tumors were cultured as an organ culture, much higher levels of enzyme were secreted indicating that interactions taking place in vivo of tumor cells with other cellular components, might induce secretion of collagenase type IV (Table 2, entry 9,10). Although all clones of both the 3LL and T10 tumors manifested similar enzyme activities, the regulation of the active enzyme production is by no means simple. Testing subcellular fractions of cells we found that active enzmye is localized in the cytoplasmic proteins (Table 2, entry 3,4). Plasma membrane vesicles show no activity (Table 2, entry 5,6). In fact, reconstitution of cytoplasmic fractions with membrane vesicles resulted in a marked inhibition of enzyme activity (Table 2, entry 7,8). Thus, the secreted enzyme is not the free form of collagenase, but rather the inhibited form, at least under tissue culture conditions. Moreover, in detergent lysates the inhibitory factor seemed to have been released, and it completely abolished the degradation activity (14). The ability of non metastatic cells within the parental metastatic tumors to produce and secret proteolytic enzymes is probably an indication that induction of specific collagenases is one of the early tumor progression steps, taking place before the segregation to metastatic and non-metastatic sub-sets.

Table 2. Basement membrane collagen degradation activity in secreted and subcellular fractions of tumor cells

Entry No.	Clone[a]	Subcellular fraction	Comparable No. of cells or tissue weight	Net degradation %
1	D6(tc)	Conditioned media, enriched by 25-50 % $(NH_4)_2SO_4$	$10^5 - 8 \times 10^6$	0-15
2	IE7(tc)	Conditioned media, enriched by 25-50 % $(NH_4)_2SO_4$	$10^5 - 8 \times 10^6$	0-15
3	D6(tc)	Cytoplasm	8×10^6 8×10^5 8×10^4	63.7 12.7 2.7
4	IE7(tc)	Cytoplasm	8×10^6 8×10^5 8×10^4	57.4 7.2 2.7
5	D6(tc)	Plasma membrane vesicles	$10^5 - 2 \times 10^6$ (10-200 µg protein)	0
6	IE7(tc)	Plasma membrane vesicles	$10^5 - 2 \times 10^6$ (10-200 µg protein)	0
7	D6(tc)	Cytoplasm and membranes	8×10^6 + 200 µg protein	5.4
8	IE7(tc)	Cytoplasm and membranes	8×10^6 + 200 µg protein	6.1
9	D6(in vivo)	Conditioned media from organ culture, enriched by 25-50 % $(NH_4)_2SO_4$	0.2 g tissue 0.02 g tissue	43.6 4.1
10	IE7(in vivo)	Conditioned media from organ culture, enriched by 25-50 % $(NH_4)_2SO_4$	0.2 g tissue 0.02 g tissue	39.3 3.3
11	D6(in vivo)	Crude homogenate	0.0001 - 0.2 g	0-12
12	IE7(in vivo)	Crude homogenate	0.0001 - 0.2 g	0-12

[a] Clones A9, D122, F2, parental 3LL, clones IC9, and IB9, and parental T8 and T40 were analyzed in the same way; all showed similar results.
tc = tissue culture propagated.
Collagen type IV substrate was prepared from EHS sarcoma (14). Solid EHS tumor tissue was minced, washed, and labeled by 5 µCi/ml ^{14}C proline in proline free DMEM supplemented by 20 % dialysed FCS, 75 µg/ml ascorbate, and 50 µg/ml 3-aminopropionitrile fumarate for 4-5 hours,

at 37 °C and 5 % CO_2. The collagen IV was purified as described by Liotta et al. (16) to a specific activity of 10^5 cpm/mg protein. Conditioned media, cytoplasmic fractions and plasma membrane vesicles were prepared from 3LL carcinoma and T10 sarcoma clones as described (14).
Collagenase activities in conditioned media and subcellular fractions were determined by release of radioactive peptides from ^{14}C collagen IV: enzmye samples in 50 mM Tris HCl, 0.2 M NaCl, and 5 mM $CaCl_2$, pH 7.6, were activated by 0.01 % (0.25 vol of sample) Trypsin for 5 min. at 37 °C. Trypsin was inhibited by 0.05 % soybean trypsin inhibitor (0.25 vol) PMSF 0.5 mM and N-ethylmaleimide 2 mM were added to inhibit activities of serine proteases and sulfhydryl proteases. 50 μg of ^{14}C collagen IV were added and degradation reaction proceeded at 34 °C for 16 hr. The reaction was stopped by addition of 20 μg BSA plus 2 % TCA and 0.1 % tannic acid. Mixtures were inucbated for 90 min at 4 °C, centrifuged at 3000xg for 15 min. to precipitate undigested collagen IV. Supernatants were mixed with aqualuma and counted. Degradation by bacterial collagenase was counted as 100 %. For negative controls samples containing buffer with trypsin and soybean trypsin inhibitor were processed and counted.

PLASMINOGEN ACTIVATOR IN CLONES OF LEWIS LUNG CARCINOMA AND T10 SARCOMA

Plasminogen activator (PA) is a highly specific serine proteinase, the best studied activity of which is the proteolytic conversion of the inactive plasma zymogen plasminogen into plasmin. Plasmin has been shown to degrade laminin, a major protein of basement membranes (18) and fibronectin, and therefore has been suggested to be an important enzyme in tumor invasion through the BM (17). In fact, PA activity has been correlated with various aspects of the malignant state, including anchorage independent growth, tumorigenicity in nude mice, following viral transformation, metastatic potentiation of melanoma cells and tumor promoter treatment of virus-transformed cell cultures (19-23). Yet in other systems a dissociation between the malignant phenotype and PA was observed (24,25). We studied the PA activities in clones cell populations of 3LL and T10 tumors. Our results showed a good correlation between PA content and secretion and the metastatic phenotype, when cell homogenates or conditioned media from tissue culture grown cells were tested. On the other hand, homogenates of in vivo grown tumors, from both low and high metastatic clones cells, all showed high levels of PA activity (13). The high PA activity in solid tumors from low metastatic clones could be attributed to a high infiltration of macrophages into the local tumors produced by the low metastatic clones (13). Local tumors from high metastatic clones hardly showed infiltrated macrophages. We have purified those macrophages and demonstrated that they exhibited high levels of PA. Thus, if PA is important in the first step of detachment and penetration of cells into the circulation, then both metastatic and non metastatic cells have high PA activity in their microenvironment. This suggests that invasion of tumor cells at the primary stages of dissemination is not a rate limiting factor in the metastatic process of these two tumor systems.

Having found that invasive activities are manifested by both low and high metastatic cells of metastatic tumors, we have turned to analyse enzymatic aspects of later steps in the metastatic process. The growth of disseminating tumor cells in the target organs is the final limiting step of the metastatic cascade. Such growth could depend on organ specific growth factors and tumor cell surface receptors for such factors. Since many growth factor receptors have a protein kinase activity (26), we tested phosphorylation processes in the plasma membranes of low and high metastatic clones of our malignant tumors.

PROTEIN KINASE ACTIVITIES IN PLASMA MEMBRANE OF TUMOR CLONES

Purified plasma membrane vesicles were prepared from three 3LL clones and five T10 clones by hypotonic shock and sucrose gradient centrifugation. As control cells we used NIH-3T3 cells, Primary embryonal fibroblasts, and spleen cells. Optimal conditions for incorporation of ^{32}P γ-ATP were established. In Figure 1. C,D we demonstrate the time curves for ^{32}P incorporation into membrane vesicles of the different clones. Membranes of D122, 3LL, IB9, IE7, and T40 cells, all high metastatic tumor clones of the 3LL carcinoma and the T10 sarcoma show a very high incorporation of labeled phosphate, compared to membranes of the low metastatic clones A9 (3LL), IC9, and D6 (T10). Proliferating, non transformed cells, like NIH-3T3 or primary fibroblasts show a low activity, while normal adult spleen cells (C57BI origin) show almost no activity. When the ^{32}P labeled membrane vesicles were dissociated in SDS and ß mercaptoethanol and electrophoresed on acrylamide gels (Fig. 1, A,B) a number of endogenous phosphorylated proteins were observed. In the three carcinoma clones A9, D122 and 3LL 14 common proteins were observed. In addition to these 4 proteins of molecular weight 30-40k are phosphorylated only in high metastatic D122 and 3LL cells but not in low metastatic A9 cells. Out of the 14 common proteins, 4 major phosphoproteins in A9 cells, are equivalent to phosphoproteins of D122 and the parental 3LL cells, while the other 10 proteins show a heavier label in high metastatic cells than in low metastatic cells (Fig. 1A). In the T10-sarcoma (Fig. 1B), membrane proteins of the non metastatic IC9 clone were phosphorylated at a very low level on a ca. 55 kd protein and on a ca. 85 kd protein (long exposure, not shown). The moderately metastatic clone, D6, showed 10 phosphorylated proteins whereas the metastatic IB9, IE7, T40 showed 14-15 phosphorylated proteins. The ten phosphoproteins observed in D6 cells were also phosphorylated in the three high metastatic cell membranes. In addition a 50 kd Protein and a group of 30-40 kd proteins were observed only in the high metastatic cells (Fig. 1B).

Cellular protein kinases can be divided into functional subclasses, according to biochemical properties like ion dependence, cyclic nucleotide dependence and according to substrate and amino acid specificities (27). Biochemical characterization of the kinase(s) in tumor cell membranes could therefore serve as a clue to the origin of the differential phosphorylation in high and low metastatic cells.

BIOCHEMICAL PROPERTIES OF THE PHOSPHORYLATION REACTION

Autophosphorylation of membranes of the different cell clones was tested at different temperatures, buffers, and divalent ions. Time dependence reactions at 0 °C, 15 °C, 25 °C, and 37 °C (data not shown) showed that at higher temperatures maximal incorporation was achieved at 2-20 min and declined at 20-60 min, while at 0 °C no decline was usually observed before 45-60 minutes. This was probably due to higher activity of phosphatases at higher temperature. Neutral buffers, Tris, Hepes,m Tes, Pipes did not affect total incorporation or substrate specificities, pH optimum of the reaction was at 7.4-7.5. A marked reduction in incorporation was observed under pH 6.5 or above 8.5 (data not shown). Testing ^{32}P incorporation and phosphoprotein profiles at different concentrations of Mg^{++} and Mn^{++} (Fig. 2), showed a strict dependence on Mn^{++} ions (Fig. 2A,C). In the presence of Mn^{++} (20 mM) no dependence on Mg^{++} ions was observed (Fig. 2D). Also the phosphoprotein profile on SDS-PAGE is not affected by changes in Mg^{++} concentrations (Fig. 2B). Both Ca^{++} ions or Zn^{++} (inhibitors of membrane associated phosphatases, 28), did not further elevate the phosphorylation

Phosphorylation reaction mixtures, 200 μl, contained membrane vesicles (20 μg protein), BSA and 0.2 mg/ml, 20 mM $MnCl_2$, 6 mM $MgCl_2$, 5 mM NaF in 10 mM Tris, 140 mM NaCl, pH 7.5. Reactions were initiated by 10 μCi $\gamma\ ^{32}P$ ATP (3000 Ci/mmole) for 0-60 min, in an ice bath. At given time

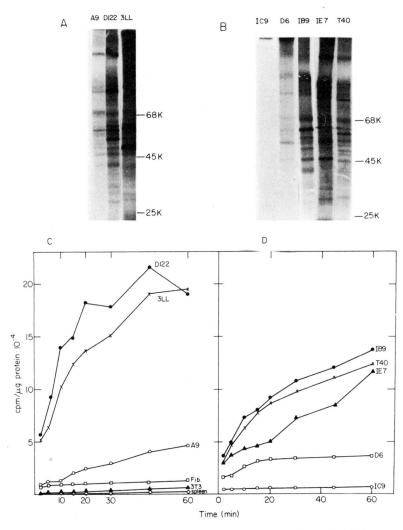

Fig. 1. Membrane phosphorylation of low metastatic and high metastatic clones of 3LL and T10.

intervals 20 μl aliquots were withdrawn, spotted on 2 cm squares of Whatman 3 mm paper and dropped immediately into cold 10 % TCA containing 0.01 M sodium pyrophosphate. Filters were washed 4 times in TCA, twice in ethanol and once in ethanol:ether, dried and monitored in a β counter. For SDS-PAGE 20 μl of reaction mixtures were removed after 30 min reaction, mixed with an equal volume of 2fold sample buffer and electrophoresed on 10 % acrylamide gels.

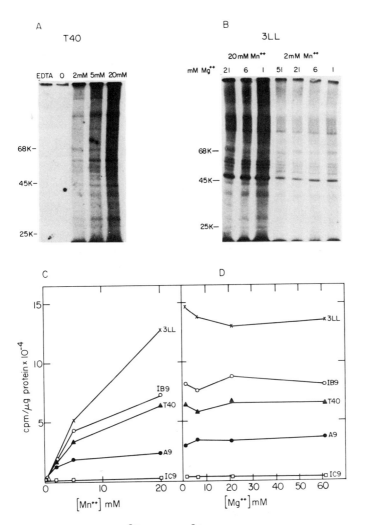

Fig. 2. Effect of Mn^{2+} and Mg^{2+} concentrations on membrane phosphorylation.

Standard phosphorylation assays were carried out as described in Fig. 1 using various divalent ion concentrations. For each Mn^{2+} and Mg^{2+} concentration a full time dependence curve was drawn. The values in figures C and D represent values for 20 min reactions. The samples for SDS PAGE (A,B) were withdrawn after 30 min reactions.

activity. Cyclic AMP and cyclic GMP at concentrations ranging from 2×10^{-7} M to 2×10^{-5} M were added to reaction mixtures and full time kinetics (0-60 min) were performed. Fig. 3B,C shows ^{32}P incorporation after 15 min reactions. C-AMP had no effect on the reactions at 2×10^{-7} M and 2×10^{-6} M and only a slight inhibitory effect was observed at 2×10^{-5} M. C-GMP had no effect on kinase reactions. Phosphoprotein profiles in presence and absence of cyclic nucleotides were compared (Fig. 3A). No changes in autologous substrate specificities were observed.

AMINO ACID SPECIFICITY OF THE MEMBRANAL KINASE

Many membrane associated kinases which are Mn^{++} dependent and cyclic nucleotide independent belong to a class of tyrosine protein kinases (E.C. 2.7.1.37)(29). Phosphorylation on tyrosine residues rather than on serine or threonine residues is associated with kinases of receptors to growth factors or to plasma membrane proteins encoded by many oncogenes. We compared the relative amount of phosphotyrosine, phosphoserine and phosphothreonin in membranes of low metastatic to that of high metastatic tumor clones by acid digestion and two dimensional thin layer electrophoresis. Equal numbers of ^{32}P counts of each tumor clone labeled Membranes (A9 "CPM" was twice that of the others) were taken for the assay. Figure 4 shows that primary fibroblasts were labeld mainly on serine (Fig. 4a). The low metastatic sarcoma clone IC9 had a low level of tyrosine phosphorylation (Fig. 4b) compared to the high metastatic sarcoma clone IB9 (Fig. 4c). In the 3LL clones the relative amount of phosphotyrosine in the low metastatic A9 clone is less than 40 % that of the high metastatic cells D122 and 3LL (Fig. 4e,f). Taking in account that the absolute amounts of membrane proteins used in these assays were much higher for the low metastatic clones (40 μg of IC9 compared to 2.7 μg of IB9; 18 μg of A9 compared to 2 μg of D122 and 3LL) we concluded that in the high metastatic clones of the T10 sarcoma the phosphotyrosine per cell is about 70fold more abundant than in the low metastatic clones. Similarly in high metastatic 3LL carcinoma cells phosphotyrosine is about 10fold more abundant than in the low metastatic A9 cells. What is then the identity of this protein tyrosine kinase? Does it specify a receptor to a known growth factor? Is it a product of a known oncogene?

We have tested the binding of ^{125}I EGF to low and high metastatic cells. The result was that both types of cells had a similar density of EGF receptors. Testing for insulin receptors we studied both binding of insulin and the effect of insulin on tyrosine kinase activity in partially purified (on lectin columns) lysates of carcinoma and sarcoma clones. All cells had a moderate to high activity of insulin receptor kinase yet no major differences were found among the different clones. We subsequently turned to screen our tumor clones for organization and expression of different onc-genes.

ONCOGENES RELATED TO PROTEIN TYROSINE KINASE IN TUMOR CELL CLONES

Many of known proto-oncogenes are coding for proteins that are enzymatically protein tyrosine kinases or show sequence homology with tyrosine kinases (30). Among these are abl, erbB, fes/fps, fgr, fms, ros, src and yes. Two other oncogenes, mos and mil/raf code for protein serine/threonine kinases. We prepared DNA and poly A selected m-RNA from seven carcinoma clones, six sarcoma clones and from normal mouse tissues (spleen and liver). DNAs were restricted by 5-6 restriction enzymes, separated on agarose gels and Southern blots were prepared. m-RNAs were electrophoresed in 1 % formaldehyde-agarose gels and Northern

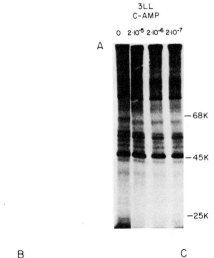
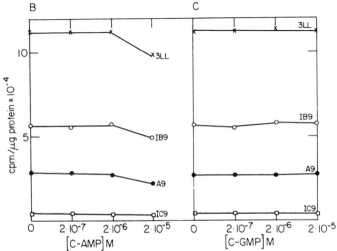

Fig. 3. Effect of cyclic AMP and cyclic GMP on membrane phosphorylation

Standard reaction mixtures containing 20 mM Mn^{2+}, 1 mM Mg^{2+} and 5 mM NaF were analysed at different times in the presence or absence of C-AMP and C-GMP (2×10^{-5} - 2×10^{-7} M). (A) 20 μl (2 μg protein) of reaction mixtures were taken after 30 min reaction and analysed on SDS-PAGE. Autoradiograms were developed after 48 hr.
(B,C) TCA precipitates counted at 15 min reaction at 0 °C.

200 µg membrane vesicels were labeld by γ ^{32}P ATP as described. Labeled membranes were precipitated in 45 % TCA overnight, washed twice in TCA and once in ethanol:ether and dried. After hydrolysis in 0,5 ml 6 N HCl for 3 hr at 110 °C a 10 µl aliquot was counted, the samples dried by

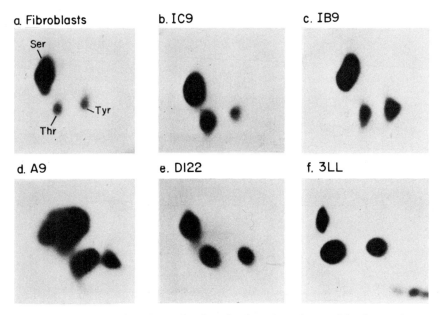

Fig. 4. Two dimensional analysis of phosphoamino acids in membrane vesicles.

lyophylisation and resuspended to yield 10,000 cpm/µl in a mixture of cold phosphoamino acids. Electrophoresis at pH 1.9 in the first dimension and pH 3.5 in the second dimension were performed. (A) Normal embryonal fibroblasts, 10,000 cpm, 12 µg membrane protein, (B) IC9, low metastatic clone of T10 sarcoma, 10,000 cpm, 23 µg protein, (C) IB9, high metastatic clone of T10 sarcoma, 10,000 cpm, 1.2 µg protein, (d) A9, low metastatic clone of 3LL carcinoma, 20,000 cpm, 6.9 µg protein, (e) D122, high metastatic clone of 3LL carcinoma, 10,000 cpm, 0.8 µg protein, (f) 3LL, parental, high metastatic clone of 3LL carcinoma, 10,000 cpm, 1 µg protein.

blots were prepared. Hybridization to nick translated probes of the various oncogenes revealed the following: a) On the DNA level, a rearrangement of the fgr gene is observed in all 3LL clones tested, however, only low expression of fgr was observed on the RNA level, no rearrangement was observed in T10 sarcoma clones. b) The abL and raf genes are expressed in both 3LL and T10 clones, especially high levels of raf transcripts are present in both metastatic and nonmetastatic clones of both tumors. No rearrangement or amplification of abL and raf were observed in Southern blots. c) Hybridization of Northern blots to a 3' region of the v-fms oncogene, which included thy tyrosine kinase domain revealed a 13.7 kb transcript which exists in low levels in the low metastatic clones (A9, A9FA1, IC9) and is highly expressed in the metastatic clones D122, 3LLFB4, 3LLFL5, 3LLFL6 and IE7 (Fig. 5). The normal c-fms 3.7 kb transcript in mice did not appear in any of the clones. The origin of this transcript is not clear as yet. We are studying the possible relations between this fms related transcript, thy tyrosine protein kinase we described and the CSF1 receptors (which were shown to be related to the fms gene) (31) on our tumor cells.

OTHER ONCOGENES IN 3LL AND T10 CLONES

We have screend the tumor clones described before for additional oncogenes, among them myc, myb, fos, P-53, Ha-Ras, Ki-Ras, N-Ras, sis, rel, and Ets. We found a) That in the 3LL carcinoma the c-myc was 60fold amplified on both DNA and RNA level. This finding is of particular interest since many of the human small cell lung carcinomas show an amplification of one of the myc gene family c-myc, N-myc or L-myc (32). We cloned this gene from a genomic 3LL library and preliminary mapping showed changes in the 5' region as compared to normal mouse c-myc. b) The fos oncogene, whose function was implicated in both mitogenesis and differentiation of myelocytic leukemia (33-35) and of the PC12 pheochromocytoma (36) is highly expressed in the low metastatic clones of the 3LL but not in the high metastatic ones. The possible relation between fos expression and changes in MHC genes in a number of systems is under investigation (Barzily et al., unpublished results). c) In T10 sarcoma clones, we found a high expression of the P-53 proto-oncogene. Chromosomal analysis of cells of metastatic clones and of an early passage of the parental tumor revealed that the T10 sarcoma is triploid for chromosome 11 that carries the p53 gene. This is probably related to tumor promotion rather than to progression of the neoplastic cells to a metastatic phenotype. Similarly we found by DNA transfection into NIH-3T3 of total DNA from low and high metastatic T10 sarcoma clones that a ki-Ras gene is transferred repeatedly and causes foci formation in 3T3 cells. Again, this activated ki-Ras appears in low metastatic, high metastatic and parental cells and is probably related to early transformation events.

CONCLUSIONS

We have shown that high metastatic carcinoma and sarcoma clones can be characterized by a high activity of a protein tyrosine kinase and by phosphorylation of endogenous membrane proteins (30 - 40 kd) that are not phosphorylated in low metastatic cells. We investigated the possibility that this enzyme is related to EGF or insulin receptors as well as the possibility that this enzyme is a product of a known tyrosine kinase related oncogene. Our results indicate that a fms related m-RNA is differentially expressed in low and high metastatic cells. The nature of the m-RNA and its relation to the membranal tyrosine kinase are studied. Screening of oncogene in clones of 3LL carcinoma and T10

Northern blots, containing either 50 µg of total RNA (A9, D122, A1, B4, L5, L6) or 5 µg of Poly A$^+$ purified mRNA (liver, 3LL, IC9, IE7, D6, T40) were hybridized to a nick translated V-fms probe (ATCC 41017) insert, that represents the 3' end of the structural gene and contains the tyrosine kinase region of the gene. As molecular weight markers 28S and 18S r-RNA were used.

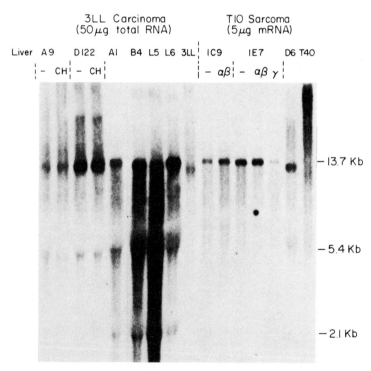

Fig. 5. Expression of fms related mRNA in high and low metastatic clones of T10 sarcoma and 3LL carcinoma

sarcoma revealed amplification of c-myc in 3LL clones and p53 in T10 clones, as well as an activated ki Ras in T10 clones. These oncogenes seem to be related to tumorigenesis rather than to progression of single clones to a metastatic phenotype. An inverse correlation was also found between c-fos expression and the metastatic competence in 3LL clones. The relation between fos expression and the differentiation state of different tumor systems is under investigation.

REFERENCES

1. I.J. Fidler, D.M. Gersten, and I.R. Hart, Adv. Cancer Res., 28: 149-250 (1978).
2. P.A. Jones, and Y.A. Declerck, Cancer Metastasis Rev., 1: 289-317 (1982).
3. L.A. Liotta, S. Garbiza, and K. Tryggvasson, in: "Tumor invasion and metastasis" (L.A. Liotta, and I.R. Hart, eds.), The Hague, Boston and Londong, pp. 319-333 (1982).
4. S.Paget, Lancet, 1:571-573 (1979).
5. I.S. Coon, R.S. Winstein, and I.L. Summers, Amer. J. Clin. Path., 77: 692-699 (1982).
6. E.Gorelik, R.H. Wiltrout, K. Okumura, S. Habu, and R.D. Herberman, Int. J. Cancer, 30: 107-112 (1982).
7. H.Festenstein, and W. Schmidt, Immunol. Rev., 60: 85-127 (1981).
8. K.Hui, F. Grosveld, and H. Festenstein, Nature, 311:750-752 (1984).
9. L.Eisenbach, S. Segal, and M. Feldman, Int. J. Cancer, 32:113-120 (1983).
10. L.Eisenbach, N. Hollander, L. Greenfeld, H. Yakor, S. Segal, and M. Feldman, Int. J. Cancer, 34:567-573 (1984).
11. P.De Baetselier, S. Katzav, E. Gorelik, M. Feldman, and S. Segal, Nature, 288:179-181 (1980).
12. R.Wallich, N. Bulbuc, G.J. Hammerling, S. Katzav, S. Segal, and M. Feldman, Nature, 315:301-305 (1985).
13. L.Eisenbach, S. Segal, and M. Feldman, J. Natl. Cancer Inst., 74: 77-85 (1985).
14. L.Eisenbach, S. Segal, and M. Feldman, J. Natl. Cancer Inst., 74: 87-93 (1985).
15. L.A.Liotta, K. Tryggvasson, S. Garbiza, I.R. Hart, C.M. Foltz, and S. Shafie, Nature, 284:67-68 (1980).
16. L.A.Liotta, K. Tryggvasson, S. Garbiza, P.G. Robey, and S.Abe, Biochemistry, 20:100-104 (1981).
17. L.A. Liotta, U.P. Thorgeirsson, and S. Garbiza, Metastasis Rev., 1:277-288 (1982).
18. L.A. Liotta, R.H. Goldfarb, R. Brundage, G.P. Siegal, V. Terranova, and S. Garbiza, Cancer Res., 41:4629-4636 (1981).
19. E.Reich, in: "Biological Markers of Neoplasia. Basic and Applied Aspects" (R.W. Ruddon, ed.). Elsevier/North Holland, Amsterdam, pp.491-500 (1978).
20. R.H. Goldfarb, and J.P. Quigley, Cancer Res., 38:4601-4609 (1978).
21. J.P. Quigley, B.M. Martin, R.H. Goldfarb, C.J. Scheiner, and W.D. Muller, in: "Proceedings of the Conference on Cell Proliferation" (G.H. Sato, and R. Ross, eds.), Cold Spring Harbor, N.Y., Vol.6, pp.219-238 (1979).
22. B.S. Wang, G.A. McLaughlin, J.P. Richie, and J.A. Mannick, Cancer Res., 40: 288-292 (1980).
23. L.Ossownski, and E. Reich, Cancer Res., 40:2300-2309 (1980).
24. J.C. Barrett, S. Sheela, K. Ohki, and T. Kakunaga, Cancer Res., 40: 1438-1442 (1980).
25. D.M. Mott, P.H. Fabisch, B.P. Sani, and S. Sorof, Biochem. Biophys. Res. Commun., 61:621-627 (1974).

26. T. Hunter, and J.A. Cooper, in: "Evolution of hormone/receptor systems". UCLA Symposia on Molecular and Cellular Biology, New Series, Vol. 6 (R.A. Bradshaw, and G.N. Gill, eds.) pp.369-382 (1983).
27. D.A. Flockhart, and J.D. Corbin, CRC Critical Reviews in Biochemistry, 133-186 (1982).
28. N.D. Richert, D.L. Blithe, and I.H. Pastan, J. Biol. Chem., 257: 7143-7150 (1982).
29. B.M. Sefon, and T. Hunter, in: "Advances in Cyclic Nucleotide and Protein Phosphorylation Research" (P. Greengard, and G.A. Robison, eds.), Raven Press, N.Y., Vol. 18, 195-226 (1984).
30. R.A. Weinberg, Science, 230:770-776 (1985).
31. C.J. Sherr, C.W. Rettenmier, R. Sacca, M.F. Roussel, A.T. Look, and E.R. Stanley, Cell, 41:665-676 (1985).
32. M.M. Nau, B.J. Brooks, J. Battey, E. Sausville, A.F. Gazdar, I.R. Kirsch, O.W. McBride, V. Berness, G.F. Hollis, and J.D. Minna, Nature, 318:69-73 (1985).
33. M.E. Greenberg, and Ziff, Nature, 311: 433-438 (1984).
34. R. Mitchell, L. Zokas, R.D. Schreiber, and I.M. Verma, Cell, 40: 209-217 (1985).
35. E. Sariban, T. Mitchell, and D. Kufe, Nature, 316:64-66 (1985).
36. W. Kruijer, D. Schubert, and I.M. Verma, Proc. Natl. Acad. Sci., 82: 7330-7334 (1985).

IMMUNE REACTIVITIES DURING THE PRECANCER AND EARLY CANCER PERIODS:
NOVEL APPROACHES FOR IMMUNOMODULATION

Isaac P. Witz[*], Liora Agassy-Cahalon[*], Benjamin Fish[**], Yaron Lidor[**], Yardena Ovadia[*], Haim Pinkas[**], Maya Ran, Michael Schickler[***], Nechama Smorodinsky[*], Benjamin Sredni[***] and Ilana Yron

[*] The Department of Microbiology and the Moise and Frida Eskenasy Institute for Cancer Research, The Goerge S. Wise Faculty of Life Sciences, Tel Aviv Univ., Tel Aviv
[**] The Department of Obstetrics and Gynaecology The Beilinson Medical Center, Sakler Faculty of Medicine, Tel Aviv University, Tel Aviv
[***] The Department of Life Sciences, Bar Ilan University Ramat Gan

INTRODUCTION

It is widely accepted that transformation of normal cells is associated with the activation or alteration of cellular oncogenes (1). Transformed cells have also, in many cases, the ability to propagate via an autocrine growth pathway, i.e. to synthesize and utilize their own growth factors (2). Activated onc genes and autocrine growth, being intrinsic characteristics of the transformed clone, may not suffice for such a clone to progress into a fully fledged cancer. The in-vivo progression of the transformed clone would depend, at least so some extent, also on its ability to survive in the milieu provided by the host. Such environmental factors may include angiogenesis (3); interaction with neighbouring tissues and cells (4); supply of growth promoting factors such as hormones (5) and immunological pressure (6). Conceivably, only those transformed cells which are endowed with the capacity to adapt themselves to host-derived environmental factors would be able to progress towards malignancy.

From an immunological point of view one could conceive the idea that only those transformed cells that would be resistant to host-immunity effector functions could proliferate and progress further. The evolvement of a resistant state to host immunity could be due to activation of "evasion" genes as postulated by Snyderman and Cianciolo (7).

An alternative pathway for a transformed clone to proliferate and progress is if certain immune functions of the host would be suppressed or otherwise disregulated. Suppressed or disfunctioning immune reactivities following exposure to biologic, chemical and physical oncogenic insults have been indeed reported (7-12).

It is also not inconceivable that transformed cells may be able to influence and possibly alter host-derived immunity. Such alterations may, under certain circumstances, facilitate the survival and further progression of these clones. Immunomodulating substances are indeed secreted by tumors (13,14).

In the present paper we wish to summarize the present state of knowledge of 3 ongoing studies related to the immunology of the precancer period. Within the frame of this research area we address the following problems:

1) Can we detect disorders in the immune functions of cancer-bearing individuals and of populations at high risk to develop a specific type of cancer? If such disorders are indeed detected, can we determine the role played by the carcinogen in inducing these disorders? Here we discuss our results of suppressed IL-2 production in patients with localized endometrial carcinoma and the role of estrogen, a putative cocarcinogen for this type of cancer, in this suppression.

2) Can we alter those immune functions which seem to be involved (positively or negatively) in primary carcinogenesis? In this report we discuss the effects of altering the levels of tumor-reactive naturally occurring antibodies on urethan-induced carcinogenesis. These antibodies seem to be involved in primary tumor-induction by this carcinogen (15).

3) Can we provide experimental proof for the assumption that transformed cells are able to alter the "normal" state of the immune system? We report preliminary evidence that non-lymphoid cells are able to induce the proliferation of certain T cells.

EXPERIMENTAL

IL-2 Production by Peripheral Blood Leukocytes of Endometrial Carcinoma Patients

Endometrial Carcinoma (E.C.) is a cancer amenable for studies focusing on the precancer period because of 2 major reasons:

1) E.C. is usually diagnosed at early stages of the disease. The vast majority of the cases are localized stage 1 tumors curable by surgery (16). Furthermore there is a well defined high risk group for this disease: Postmenopausal women, who usually present bleeding, and in whom a hyperplastic endometrium was diagnosed (16,17). This high risk group can be followed longitudinally and prospectively.

2) Estrogens are postulated to play a cocarcinogenic (if not a carcinogenic) role in E.C. by causing endometrial proliferation (18). As argued in the introduction (see above) carcinogens may act on the immune system of exposed individuals, in addition to their carcinogenic activity on the target cells. In view of this possibility we studied estrogen-mediated effects on the immune system of E.C. patients and of individuals belonging to the high risk group.

Measuring IL-2 production by PBL of E.C. patients we found that compared to age matched healthy women, E.C. patients have a decreased ability to produce IL-2 (Table 1). As expected, PBL from E.C. patients were also deficient in their ability to be stimulated by PHA (Table 1). We found a good correlation in individual E.C. patients between the ability of their PBL to proliferate and to produce IL-2 (data not shown).

Table 1. The Proliferative Response to PHA of PBL from Patients with Localized Endometrial Cancer and Controls

Parameter Measured	Controls(27)[a]	E.C.(20)[a]
IL-2 Production[b]	13.9 ± 9.0	7.3 ± 5.2
Proliferation[c]	47.8 ± 24.3	25.7 ± 21.2

[a] The values in parentheses indicate number of age matched healthy controls and of E.C. patients assayed.

[b] The values represent arbitrary units of IL-2 calculated in reference to a standard IL-2 preparation.

[c] The values represent stimulation index (CPM ^{3}H-TdR incorporated by PHA-treated PBL divided by CPM ^{3}H-TdR incorporated by PBL incubated in culture medium).

Attempting to elucidate the mechanism responsible for the decreased ability of PBL from E.C. patients to be transformed by PHA and to produce IL-2 we addressed the question whether or not suppressor cells are involved in this suppression. In view of studies showing that monocytes are able to suppress IL-2 production (19) we tested possible monocyte involvement in the suppression of IL-2 production in E.C. patients. This was performed by depleting such cells from patient's PBL by adherence to plastic surfaces and by panning on surfaces coated with antibodies directed against Ig. The results summarized in Table 2 show that T cell enriched populations produced more IL-2 than unfractionated PBL. The enhancement of IL-2 production by PBL from E.C. patients was, on the average, higher than IL-2 production by PBL from controls. We then performed experiments in which monocytes from E.C. patients were mixed with T cells from healthy controls as well as the reciprocal mixing experiments. The results of these mixing experiments indicated that although monocytes have the ability to suppress IL-2 production, especially in E.C. patients, the monocyte-mediated suppression is not the major factor responsible for the decreased IL-2 production in these patients.

We investigated the possibility that autorosetting T-cells known to down-regulate several expressions of immunity (20) including response to PHA (Sredni et al., submitted for publication) are involved in suppression of the capacity of PBL from E.C. patients to produce IL-2. Although we do not have a definite answer to this question we have found that PBL of E.C. patients contain a higher proportion of con A activated autorosetting T cells (17 % - average value of 8 patients) than PBL of healthy age matched controls (7 % - average value of 9 controls). We next assayed the effect of estrogen (17β estradiol) or other sex hormones (progesterone and testosterone) on the frequency of autorosetting cells when added in-vitro to con A activated PBL from controls or E.C. patients. The results show that estrogen but not progesterone or testosterone increased the incidence of autorosettes in PBL of controls and to a lesser degree in PBL of E.C. patients (data not shown).

These results indicate that the possibility raised in the introduction, namely that carcinogens (estrogen in the particular case

Table 2. The Effect of Removing Adherent and Surface Ig Positive Cells on IL-2 Production by PHA-stimulated PBL from E.C. Patients and Controls

Source of PBL	Units of IL-2 Produced by PBL[a]		% Enrichment
	Unfractionated	T-Cell Enriched	
Controls(6)	11.3 ± 5.2	13.9 ± 2.1	23 ± 2
E.C. Patients(7)	7.5 ± 3.8	11.0 ± 5.4	47 ± 4

[a] See Footnote b Table 1.

of E.C.) may affect several types of cellular systems (immunocytes in the present case) in addition to their ability to transform the appropriate target cells (endometrial cells in the case of E.C.).

It remains to be seen whether or not there is a cause-effect relationship between E.C. formation (involving most probably estrogen as a cocarcinogen) and estrogen-mediated activation of suppressor auto-rosetting T cells.

The Ability of Naturally-Occurring Lymphoma Reactive Antibodies to Modulate Chemical Carcinogenesis

In a previous study we found that the titers of naturally-occurring IgM antibodies which interact in-vitro with L5178-Y cells are different in urethan-treated mice which developed lung adenomas after a precancer period of 4-5 months and in urethan-treated mice which did not develop tumors within this period (15). These results prompted us to prepare L5178-Y reactive monoclonal antibodies from unimmunized BALB/C mice. A preliminary characterization of some of the naturally-occurring L5178-Y reacting monoclonals was reported (21). In view of the results showing different titers of L5178-Y reactive antibodies in urethan-treated tumor-bearers and in urethan-treated mice which did not develop tumors (15), we addressed the question whether there is a causal relationship between these 2 phenomena. This problem was approached by passively administering naturally-occuring L5178-Y reactive monoclonals (21) to urethan-treated mice during the precancer period. Three monoclonals were selected: one which interacts with several cell types of syngeneic, allogeneic or xenogeneic cells (Table 3, group I), another which has a more restricted interaction pattern (Table 3, group II), and the third which interacts primarily with lymphatic tumors (Table 3, group III). We also administered IgM secreted from plasmacytoma MOPC 104E. This protein reacts with dextran (22).

Table 4 shows that the passive administration of monoclonal 1.80 (Table 3, group III) decreased adenoma incidence as compared to urethan-treated controls but the difference was not statistically significant. However, tumor load (taking into account the number of tumor foci and their size) was significantly decreased as compared to controls. The MOPC 104E-derived IgM increased tumor incidence (non-significantly) and tumor load (significantly).

Table 3. The Reactivity Pattern[a] of L5178-Y Binding Naturally-Occurring Monoclonal Antibodies

Monoclonal Group	Binding to			
	Lymphoma[c] Cells	Non-lymphoid[c] Tumor Cells	Normal Lymphoid Cells	Non-lymphoid Normal Cells
I (1.91, 2.2)[d]	yes	yes	yes	yes
II (1.67)[d]	some	some	yes	yes
III (1.80)[d]	some	no	no	no

[a] The reactivity pattern was measured by an indirect radioimmunobinding assay utilizing target cells treated first with L5178-Y binding monoclonal IgM produced by hybridomas from non-immunized BALB/c mice followed by ^{125}I-labelled rabbit antibodies directed against mouse IgM.

[b] The naturally-occurring monoclonal antibodies used could be grouped into 3 groups according to their reactivity pattern.

[c] The tumor target cells were of syngeneic, allogeneic or xenogeneic origin.

[d] The numbers in parentheses are the designations of individual monoclonals.

Table 4. The In-vivo Effect of Naturally-Occurring Monoclonal IgM Antibodies on Urethan Carcinogenesis

Antibody Injected	Tumor Incidence[a]	Average Tumor Load[b]
None	29/52 (56 %)	2.3 ± 3.8
104E(control)	26/42 (62 %)	4.1 ± 4.8
2.2	15/35 (43 %)	2.1 ± 4.2
1.67	16/42 (38 %)	1.9 ± 3.4
1.80	20/46 (43 %)	1.1 ± 1.4

[a] Tumor incidence (number of urethan-treated mice with lung tumors over total number of urethan-treated mice in the group). Tumor incidence and load were determined 4-5 months after urethan treatment (15).

[b] A manuscript describing the method to determine tumor load was submitted for publication.

These results show that urethan-mediated chemical carcinogenesis can be decreased by immunological intervention during the precancer period. This was achieved by the passive administration of naturally-occurring antibodies.

These results obtained thus far are obviously not complete as yet and protocols should be designed to optimize the antibody-mediated anti carcinogenesis treatment. Such protocols can only be devised after elucidating the modus-operandi of the anti carcinogenesis effect of the passively administered monoclonals.

The Induction of Fc-receptor Expressing T Cells Proliferation by a Fibroblast-derived Factor

One of the immunological events occurring in tumor-bearers is the systemic increase in Fc-receptor (FcR) expression on their immunocytes. This effect could have been brought about by an increase in the number of FcR positive immunocytes or an increase in the amount of FcR per cell (23). Furthermore, increased FcR expression of host-derived immunocytes correlates positively with increased malignancy and poor prognosis of breast cancer and colon carcinoma patients (Ran et al., submitted for publication).

In previous papers we hypothesized that tumor progression may be facilitated by the ability of malignant cells to bring about an increase in the levels of cell bound or circulating receptors for the Fc domain of γ chains (FcγR) (24,25). First, such receptors are expressed on suppressor T cells (26) and increased levels of FcγR may indicate an increased level of immunosuppression in tumor-bearers. Secondly, cell free FcR receptors or immunoglobulin-binding factors (IBF) are potent immunoregulators (27). Increased levels of such molecules may modulate the physiological balance of immune response creating circumstances that could facilitate or at least not interfere with tumor growth. Thus, transformed cells capable of inducing increased FcR expression in the host would enjoy a selective advantage. Transformed cells able to cause an increased release of IBF from immunocytes (either by enhancing the propagation of FcγR positive cells or by increasing the levels of expression of such receptors on the host's immunocytes, or both) would also enjoy such an advantage.

The goal of the present study was to test this assumption. We used as a representative for host immunocytes the T2D4 T cell hybridoma (28). These hybridoma cells express FcR and release IBF constitutively. In addition T2D4 cells can be induced to express isotype-specific FcR and release the corresponding IBF by the appropriate Ig isotype (27,28).

We exposed T2D4 cells to H-ras transformed NIH-3T3 cells (29) or to soluble factors thereof, and measured FcR expression on the T2D4 cells by using a radioimmunobinding assay (30) and monoclonal antibodies directed against the murine Fcγ_2b/γ_1 receptor (31). Table 5 shows that coculturing T2D4 cells with H-ras transformed NIH-3T3 cells (ras-3T3), or exposing them to culture supernatants or water extracts of such cells brings about a mitogenic response in the target cells. FcR expression per cell did not alter considerably (data not shown). FcR-negative BW5147 thymoma cells which are the parental cells of the T2D4 hybridoma were not induced to proliferate by ras-3T3 extracts. The ability to induce proliferation in T2D4 cells was not restricted to transformed ras-3T3 cells. Untransformed NIH-3T3 cells or even embryonic fibroblasts induced proliferation of T2D4 cells although to a lesser extent than ras-3T3 cells.

Table 5. The Mitogenic Response of FcR Positive T2D4 cells[a] to Mitogens Derived from H-ras Transformed NIH-3T3 Cells

Source of Mitogen	Stimulation Ratio[b]	Cell Number Ratio[c]
Coculture[d]	1.38; 1.33; 2.86	
Culture Supernatant[e]		
50 %	4.05	
25 %	2.11	
12.5 %	1.45	
Extract[f]	1.63	3.4

[a] T2D4 targets were at the stationary phase of the cell cycle.

[b] Stimulation ratio is the ratio between CPM ^{3}H-TdR incorporated into T2D4 cells in the presence of the mitogenic stimulus and CPM ^{3}H-TdR incorporated into cells grown under control conditions.

[c] Cell number ratio is the ratio between the number of T2D4 cells grown in the presence of extract and the number of cells grown in culture medium.

[d] T2D4 cells were cocultured with non-confluent H-ras transformed 3T3 cells for 24 hrs and then harvested. The controls were T2D4 cells cultured without the transformed 3T3 cells. The results of three experiments are presented.

[e] Culture supernatants were removed from confluent cultures of H-ras 3T3 cells and added at the indicated concentration to T2D4 cells. Culture supernatants of T2D4 cells were used as controls.

[f] Water extracts of H-ras 3T3 cells were dialized against culture medium and added to T2D4 cells. Extracts of a certain number of H-ras 3T3 cells were added to an equal number of T2D4 cells.

An interesting feature of the ras-3T3 derived mitogenic factor is that it seems to operate via the FcR of the target T2D4 cells. Table 6 shows that blocking the FcR on these cells by monoclonal anti FcR antibodies inhibits the mitogenic response of T2D4 cells. Conversely, preexposure of T2D4 cells to ras-3T3-derived extract decreased significantly the binding of the anti FcR antibodies to the cells (Table 7).

These preliminary results indicate that non lymphoid cells are able to interact with and induce the proliferation of FcR-positive lymphocytes and that this interaction apparently takes place via the FcR. It is not unlikely that an increased proliferation of the FcR-positive lymphocytes would occur if more of the inducing factor would be released. An increased release of the mitogenic factor could occur as a result of a transformation-induced increased proliferation of the factor-releasing cells.

Table 6. Monoclonal Antibodies Against Murine FcR Inhibit the Mitogenic Response of T2D4 Cells to ras-3T3 Extracts

Inhibitor[a]	^{3}H-TdR CPM Control[b]	Extract	No. Cells x 10^{-6} Control[b]	Extract
none	7014 ± 18	11097 ± 352	0.94	2.22
medium[c]	8967 ± 339	13240 ± 352	1.62	3.42
FcR[d]	7432 ± 273	8738 ± 1223	1.60	1.80

[a] The T2D4 cells were preincubated with the inhibitor, washed and then incubated with ras-3T3 extracts.

[b] Control cells were incubated with culture medium.

[c] The medium was that used to propagate the 2.4G2 hybridoma cells which secrete the monoclonal anti FcR antibody (31).

[d] Monoclonal antibodies directed against murine $Fc\gamma_2 b/\gamma_1$ receptor (αFcR) secreted by the 2.4G2 hybridoma.

Table 7. Extracts of ras-3T3 Cells Inhibit the Binding of Monoclonal Antibodies Against Murine FcR to T2D4 Cells

Assay	Experiment No.	Binding (CPM) Control	Extract	% Inhibition
RIB[a]	1	2889	1645	43
	2	2330	1479	37
Flow Cytometry[b]	1	376	19	95
	2	2208	665	70

[a] The radioimmunobinding assay (RIB) was performed as described in (30). The values indicated CPM of ^{125}I-αFcR bound to 10^5 T2D4 cells preincubated either with culture medium (control) or with ras-3T3 extracts.

[b] T2D4 cells preincubated either with culture medium or with ras 3T3 extracts were stained with FITC conjugated αFcR antibodies. Average fluorescence was determined using an Ortho flow cytometer.

A CONCLUDING REMARK

The results of the three ongoing studies presented in this manuscript indicate that certain immune functions are altered during the precancer and early cancer period. It seems that certain immunocytes are regulated by factors without the immune system. Moreover, there is a possibility to modulate primary carcinogenesis by immune intervention during the precancer period.

ACKNOWLEDGEMENTS

Isaac P. Witz is the incumbent of the David Furman Chair of Immunobiology of Cancer.

This work was supported by a grant awarded by Concern Foundation in conjunction with the Cohen-Applebaum-Feldman Families Cancer Research Fund, Los Angeles, and by the Fainbarg Family Fund, Orange County, CA.

REFERENCES

1. J. M. Bishop, Retroviruses and cancer genes, Adv. Cancer Res.,37:1-32 (1982).
2. M. B. Sporn, and A. B. Roberts, Autocrine growth factors and cancer, Nature, 313:745-747 (1985).
3. J. Folkman, Tumor Angiogenesis, Adv. Cancer Res., 19:331-356 (1974).
4. I. J. Fidler, D. M. Gersten, and I. R. Hart, The biology of cancer invasion and metastasis, Adv. Cancer Res., 28:149-250 (1978).
5. M. Kodama, and T. Kodama, Relation between steroid metabolism of the host and genesis of cancers of the breast, uterine, cervix and endometrium, Adv. Cancer Res., 38:77-119 (1983).
6. P. C. Doherty, B. B. Knowles, and P. J. Wettstein, Immunological surveillance of tumors in the context of major histocompatibility complex restriction of T cell function, Adv. Cancer Res., 42:1-65 (1984).
7. R. Snyderman, and G. J. Cianciolo, Immunosuppressive activity of the retroviral envelope protein P15E and its possible relationship to neoplasia, Immunol. Today, 5:240-244 (1984).
8. O. Stutman, Immunodepression and Malignancy, Adv. Cancer Res.,22:261 - 433 (1975).
9. R. Ehrlich, M. Efrati, A. Bar-Eyal, M. Wolberg, G. Schiby, M. Ran, and I. P. Witz, Natural cellular reactivities mediated by splenocytes from mice bearing three types of primary tumors, Int. J. Cancer, 26:315-323 (1980).
10. E. Gorelik, and R. B. Herberman, Inhibition of mouse NK cells by urethan, J. Nat. Cancer Inst., 66:543-548 (1981).
11. R. Ehrlich, M. Efrati, E. Malatzky, L. Shochat, A. Bar-Eyal, and I. P. Witz, Natural host defence during oncogenesis: NK activity and dimethylbenzanthracene carcinogenesis, Int. J. Cancer, 31: 67-73 (1983).
12. D. R. Parkinson, R. P. Brightman, and S. D. Waksal, Altered natural killer cell biology in C57BL/6 mice after leukomogenic split-dose irradiation, J. Immunol., 126:2129-2135 (1981).
13. I. P. Witz, E. Harness, and M. A. Pikovski, Suppressed plaque and rosette formation by spleen cells of mice injected with tumor extracts, in: "Microenvironmental Aspects of Immunity", B. D. Jankovic, and K. Isakovic, eds., Plenum Publ. Co., New york, 461-468 (1972).
14. I. Kamo, and H. Friedman, Immunosuppression and the role of suppressive factors in cancer, Adv. Cancer Res., 25:271-321 (1977).
15. I. P. Witz, M. Yaakubowicz, I. Gelernter, Y. Hochberg, R. Anavi, and M. Ran, Studies on the level of natural antibodies reactive with various tumor cells during urethane carcinogenesis in BALB/c mice, Immunobiol., 166:131-145 (1984).
16. M. L. Berman, S. C. Ballon, L. D. Lagasse, and W. G. Watring, Pro-

gnosis and treatment of endometrial cancer, Amer. J. Obst. Gynaecol., 136:679-688 (1980).
17. A. L. Sherman, and S. Brown, The precursors of endometrial carcinoma, Amer. J. Obst. Gynaecol., 135:947-956 (1975).
18. B. E. Henderson, R. K. Ross, M. C. Pike, and J. T. Casagrande, Endogeneous hormones as a major factor in human cancer, Cancer Res., 42:3232-3239 (1982).
19. S. Chouaib, and D. Fradelizi, The mechanism of inhibition of human IL-2 production, J. Immunol., 129:2463-2468 (1982).
20. S. Kumagai, I. Scher, and I. Green, Autologous Rosette-forming T cells regulate responses of T cells, J. Clin. Invest., 68:356-364 (1981).
21. I. P. Witz, M. Efrati, R. Ehrlich, B. Gonen, L. Kachalom, O. Sagi, E. Sahar, L. Shochat, N. I. Smorodinsky, S. Yaakov, M. Yaakubowicz, and I. Yron, Natural defense and chemical carcinogenesis, in: "Haematology and Blood Transfusion", Vol. 29, Modern Trends in Human Leukemia VI, Neth, Gallo, Greaves, Janka, eds., Springer-Verlag, Berlin, Heidelberg, New York, 492-497 (1985).
22. K. R. McIntire, R. M. Asofsky, M. Potter, and E. L. Kuff, Macroglobulin producing plasma cell tumor in mice. Identification of a new light chain, Science, 150:361-363 (1965).
23. J. Bray, and T. A. McPherson, Fc receptor-bearing blood mononuclear cells in breast cancer patients: a possible marker for tumor burden and prognosis, Clin. Exp. Immunol., 44:629-637 (1981).
24. M. Ran, and I. P. Witz, FcR derived from without the immune system - A potential escape mechanism for cells propagating in hostile immunological environment, Contr. Gynec. Obstet., 14:9-14 (1985).
25. I. P. Witz, and M. Ran, Could Fc receptors facilitate the escape of immunogenic premalignant cells from host defence? A hypothesis, Ann. Inst. Pasteur, 136C:423-428 (1985).
26. W. H. Fridman, C. Rabourdin-Combe, C. Neauport-Sautes, and K. H. Gisler, Characterization and function of T cell Fc receptor, Immunol. Rev., 56:51-58 (1981).
27. M. Daeron, and W. H. Fridman, Towards an isotypic network, Ann. Inst. Pasteur, 136C:383-387 (1985).
28. C. Neauport-Sautes, C. Rabourdin-Combe, and W. H. Fridman, T cell hybrids bear Fc receptors and secrete suppressor immunoglobulin-binding factor, Nature, 277:656-668 (1979).
29. Y. Yuasa, S. K. Srivastava, C. Y. Dunn, J. S. Rhim, E. Premkumar-Reddy, and S. A. Aaronson, Acquisition of transforming properties by alternative point mutations within c-bas/has human proto-oncogene, Nature, 303:775-779 (1983).
30. M. Ran, L. Dux, R. Anavi, N. I. Smorodinsky, and I. P. Witz, A radioimmunoassay with monoclonal antibodies for the detection of antigenic cell-free Fc receptor, J. Immunol. Meth., 68:275-284 (1984).
31. J. C. Unkeless, Characterization of a monoclonal antibody directed against mouse macrophage and lymphocyte Fc receptors, J. Exp. Med., 150:580-596 (1979).

IMMUNOMODULATION OF TUMOR METASTASES

L. Eisenbach, S. Katzav and M. Feldman

Department of Cell Biology
The Weizmann Institute of Science
Rehovot, Israel

The local tumor cell population seems to be diverse with regard to the metastatic competence of its individual cells (1). Cells capable of generating metastases, have to detach themselves from the local cell population, to penetrate basement membranes of blood capillaries, to migrate via the blood circulation, while undergoing cell aggregation, to extravasate at the target organs, and to induce angiogenesis for the metastatic growth.

This complex multistep process may fail, if at any of these successive stages, the disseminating cells are recognized by the host's immune system. Such recognition, if it results in the activation of cytotoxic cells, may eliminate the invading tumor cells. Recognition by effector T cells of tumor cell surface molecules was shown to be restricted by class I antigens of the major histocompatibility complex (MHC), expressed on the tumor cell surface (2). It was therefore conceivable that quantitative or qualitative alterations in the expression of H-2 genes on the tumor cell membrane may play a determining role in tumor immunogenicity, and consequently in controlling metastatic spread. We initially studied, the 3LL Lewis lung carcinoma - a spontaneous metastatic mouse tumor, originating in a C57BL ($H-2^b$) mouse. Unlike all normal cells, or most tumors, this carcinoma could grow across H-2 barriers (3); in fact it grew progressively in mice of any strain tested. Yet, metastases appeared only in syngeneic recipients (4). To test what are the minimal genetic identities between the tumor's strain of origin and the allogeneic recipients, which are required for the development of metastases, we tested various H-2 recombinants as recipient mice. We discovered that if the recipients expressed the $H-2D^b$ determinant and the non-$H-2^b$ C57BL background, this was sufficient for the generation of spontaneous metastases by transplants of the 3LL tumor. The $H-2K^b$ was completely irrelevant (5). We subsequently revealed that identity at the $H-2K^b$ was unnecessary for the generation of metastases because the 3LL cells hardly expressed the $H-2K^b$ glycoprotein on their membrane (6). It consequently seemed that the association of tumor cell surface epitopes with the expressed $H-2D^b$ molecules was either non-immunogenic or it elicits the generation of suppressor cells which enables the 3LL cells to grow in an incompatible recipient. This could be causally related to the capacity of this tumor to metastasize in syngeneic hosts. The same tumor and cell-surface epitopes, if associated with $H-2K^b$ molecules, could be expected to confer immunogenicity on the 3LL cells and this would prevent growth in allogeneic mice, and prevent the generation of metastases in syngeneic animals.

Imbalance in $H-2K^b/H-2D^b$ ratios of 3LL clones correlates with their metastatic potential

To test these notions, we cloned in soft agar the 3LL cells and tested individual clones for MHC expression and metastatic properties. Thirty clones were analysed in details, using monoclonal antibodies 28-13-3 and 20-8-4 which identify $H-2K^b$ molecules and antibody 28-14-8 which identifies $H-2D^b$ molecules (7). Three types of assays were used, a direct binding radioimmunoassay, immunoprecipitation of ^{125}I-labeled cell extracts and a fluorescent antibody binding assay performed by a cell sorter (FACS II). The metastatic phenotype of the individual clones was tested in C57BL/6 mice. Table 1 shows that 15 of the tested clones were of a low metastatic phenotype. These could be divided into two subgroups, one (3 clones) having a low level of K^b and D^b expression and another (12 clones) exhibiting high densities of both K^b and D^b molecules. The moderately metastatic group (5 clones) always showed a much higher D^b than K^b expression and the high metastatic group (10 clones) showed low levels of K^b and high levels of D^b molecules. Thus, the lower the $H-2K^b$ versus $H-2D^b$ ratio, the higher was the metastatic potential of the individual clone (Fig. 1).

The relative expression of $H-2K^b/H-2D^b$ genes is correlated with the immunogenic competence of the tumor cells

To analyse whether the low H-2K/H-2D ratio controls the expression of the metastatic phenotype because it decreases the immunogenic potency of the tumor cells, we studied two clones of the 3LL tumor. A low metastatic clone A9 that expresses both $H-2K^b$ and $H-2D^b$ glycoproteins and a high metastatic clone D122 that expresses mainly $H-2D^b$ molecules. We first tested whether the low metastatic, K^b-positive clone would differ from the high metastatic, K^b-negative cells in their capacity to grow across allogeneic barriers. We inoculated cells of clones A9 and D122 into groups of C57BL/6J ($H-2^b$), (C57BL/6JxC3Heb)F1 ($H-2^b \times H-2^k$), C3Heb ($H-2^k$) and BALB/c ($H-2^d$) mice. The cells of clone A9 grew progressively in all syngeneic C57BK/J or F1 mice, but were completely rejected in BALB/c or C3H/eb mice at 10^6; 3×10^5; 10^5 or 3×10^4 cell inocula. On the other hand, the D122 clone grew progressively in mice of all four strains. Clone A9 is nonmetastatic in syngeneic or F1 mice whereas clone D122 metastasizes only in syngeneic mice. It thus appeared that the metastatic potential is correlated with the immunogenic effect in the allogeneic recipients.

To test whether the immune rejection of cells of the A9 clone can be attributed to immunogenicity of class I antigens, 10^5 A9 or D122 cells were inoculated to groups of H-2 recombinant mice on a C57BL/10 background. We used B10.HTG (K^dD^b), B10.D2 (K^dD^d) and B10.A(4R) (K^kD^b) mice. Clone A9 (K^bD^b) grew in 9/10 B10.HTG with a slower growth rate than in C57BL/6J. Only partial and slow growth was observed in B10.D2 mice (5/11) and in only one of nine mice of the B10.A(4R) strain did the A9 grow. In contrast, D122 cells grew in C57BL/6J and in the three recombinant strains at a similar rate. Testing for metastases, we found that D122 metastasized in C57BL/6J (K^bD^b), B10.HTG (K^dD^b) and B10.A(4R) (K^kD^b) but not in B10.D2 (K^dD^d) mice. Thus the higher immunogenic effect of clone A9 compared to clone D122, is a function of the $H-2K^b$ determinant.

Table 1. Range of H-2 expression and metastases of low, moderate and high metastatic groups of clones

	No. of clones	Positive cells (gain 8) (%)			Range of average lung wt (mg) (b)
		K^b range	D^b range	K^b-D^b range (a)	
Low metastatic I	3	5-13	5-15	0.75-1.00	217-243
Low metastatic II	12	30-61	42-87	0.54-1.00	192-236
Moderate metastatic	5	14-42	45-89	0.31-0.53	259-367
High metastatic	10	6-22	47-96	0.12-0.26	429-947

Tumor cells (2 to 3 x 10^6) were incubated at 4 °C for 30 minutes in purified anti-K^b or anti-D^b monoclonal antibody, washed twice in phosphate-buffered saline (1.0 % bovine serum albumin and 0.2 % sodium azide), and reincubated in fluorescein isothiocyanate-labeled rabbit anti-mouse Ig. Fluorescence-activated cell sorter (FACS) II analysis (Becton Dickinson, Sunnyvale, Calif.) was performed with the photomultiplier tube set at 550 V. Each clone was analyzed three times. The range is of the average of each clone in the group.
a) The K^b-D^b ratio was determined for each clone individually.
b) Ten C57BL/6J male mice were inoculated intra-footpad with 10^5 cells of each clone. When the primary tumor reached 8 mm in diameter, the tumor-bearing legs were amputated. The metastatic load was determined 21 days postamputation, and the average lung weight was determined for each clone individually. The Range of these averages is given.

To examine whether the metastatic phenotype is controlled via the immunogenic properties of the 3LL clones, we analysed the immune responses evoked by the A9 and D122 clones in syngeneic mice. C57BL/6J mice were immunized by three weekly intraperitoneal injections of 10^7 irradiated and mitomycin C treated A9, D122 or 3LL cells. Ten days later immunized mice and controls were challenged by A9 or D122 cells. We found that immunization by A9 cells significantly slowed the growth rate of a second A9 tumor but did not affect the growth rate of the metastatic D122 tumor. Immunization by clone D122 or by 3LL did not retard the growth of a second A9 or D122. Similar results were obtained when mice were immunized intradermally by living A9 or D122 cells (8).

Following these in vivo observations, we tested the cytotoxic T lymphocyte (CTL) responses evoked by cells of A9 and D122 clones in syngeneic hosts (Table 2). A9 induced high levels of cytotoxic activity, manifested against A9 cells, and to a lower extent against D122 target cells. D122 cells induced a lymphocyte population that manifested only a low cytotoxic activity against D122 or A9 target cells. Thus, the in vitro interaction of A9 immune lymphocytes with A9 cells led to the destruction of the tumor cells, whereas lymphocytes interacting with D122 cells are significantly less efficient in destroying the tumor cells. It thus appears that the nonmetastatic phenotype of cells of the A9 clone may be determined by their immungenicitiy in syngeneic mice and their susceptibility to the lymphocyte cytotoxicity they elicit (8).

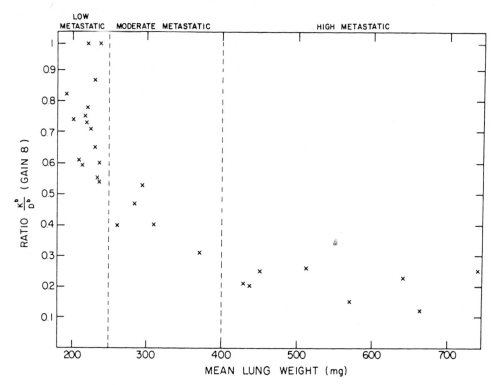

Figure 1. MHC expression and spontaneous metastases were tested as described in Table 1.

Table 2. Correlation of cytotoxicity and susceptibility to cell mediated cytolysis with $H-2K^b$ expression on 3LL clones

	Target cells % specific lysis							
	A9				D122			
Effector:target Immunization	50:1	25:1	12:1	6:1	50:1	25:1	12:1	6:1
A9 10^5	68	48	36	27	38	16	8	2
A9 5×10^4	66	45	32	27	32	15	10	2
D122 10^5	27	15	9	7	12	2	0	0
D122 5×10^4	25	9	6	4	12	2	0	0

C57BL/6J mice were immunized intradermally with 5×10^4 or 10^5 A9 or D122 cells. On day 12 spleens were removed and lymphocytes were restimulated in vitro on irradiated and mitomycin treated A9 or D122 cells. Indium-111 labeled A9 and D122 cells reacted with these lymphocytes in a 16 hr assay.

Modulation of H-2K/H-2D expression modulates the metastatic phenotype of 3LL clones

To test whether the relative expression of class I antigens of the MHC is causally related to its metastatic phenotype, we attempted to induce alterations in the $H-2K^b/H-2D^b$ ratio and test whether such alterations would concomitantly result in changes in the metastatic potency of the tumor cells. We focused on the low metastatic A9 clone and the high metastatic D122 clone. We pretreated, in vitro, cells of these two clones with either interferon α + β, interferon γ or retinoic acid. Treatment by retinoic acid increased significantly the cell surface expression of the $H-2D^b$ gene, but did not affect $H-2K^b$ expression. This treatment converted cells of the low metastatic A9 clone to a high metastatic phenotype and increased the metastatic mass produced by D122 cells (7,9). Interferons α + β and γ stimulate the expression of both genes coding for class I antigens, but exerted different effects on $H-2K^b$ and $H-2D^b$ genes. Interferons α + β showed a major effect on $H-2D^b$ genes and induce to a lesser extent the expression of $H-2K^b$ gene. In contrast, interferon γ increased dramatically the expression of $H-2K^b$ on both A9 and D122 cells. Table 3 shows the effects of both interferon on cell surface expression of H-2 subregions, and on the metastatic potential of these cloned cell populations (10).

Table 3. The effect of interferons (α+β; γ) on MHC cell surface expression and metastasis

Treatment by interferon	Conc. (U/ml)	% positive cells (gain 8)			Number of mice with metastases out of total injected	
		K^b 28-13-3	K^b 20-8-4	D^b 28-14-8	Spontaneous (IFP)	Experimental (IV)

Clone A9						
-	0	51.8	52.3	66.2	2/10	2/10
α+β	100	58.5	58.8	93.8	6/9	7/10
α+β	500	61.4	69.9	96.9	7/9	4/10
γ	25	80.7	83.3	97.3	2/9	3/10
γ	50	90.4	92.8	96.4	n.d.	1/9
γ	100	92.7	95.1	96.9	1/10	2/10

Clone D122						
-	0	7.1	7.7	47.1	8/8	9/9
α+β	100	29.8	30.2	70.2	8/10	10/10
α+β	500	34.1	35.3	92.9	10/10	7/10
γ	25	82.1	88.2	97.7	8/9	4/10
γ	50	90.5	90.1	98.8	n.d.	1/10
γ	100	94.3	94.6	98.8	5/9	1/9

Cells were treated for 6 days in various concentrations of interferons. MHC expression and spontaneous metastases; after intra-food pad injections were tested as described in Table 1. Experimental metastases were tested by injection of 5×10^5 cells I.V. Animals were sacrified 20 days post-injection.

Interferon α + β that mainly increased $H-2D^b$ expression converted cells of clone A9 from a low metastatic clone to a high metastatic phenotype. This effect was more pronounced in the generation of spontaneous metastases as compared to experimental metastases, especially when a high dose of α + β interferon (500 U/ml) was used. Interferon γ had little effect on the properties of A9 clone. On the other hand the high metastatic clone D122 was not affected by treatment with interferon α + β, but treatment with interferon γ that increased significantly the K/D ratio, decreased the metastatic potency of D122 cells. The major effect of the suppression of metastases by γ interferon was observed in experimental metastases, while the generation of spontaneous metastases was affected more by high dose treatment (100 U/ml). In summary, interferon α + β elevated the metastatic potential of the low metastatic A9 clone, and this effect was increased in long term testing of spontaneous metastases, while interferon γ decreased the metastatic potential of the high metastatic D122 clone and its effect was registered in short term testing of experimental metastases. These differential effects could be the result of a) a different time kinetics of induction and disappearance of H-2 molecules on tumor cells, b) different cellular mechanism exerted by both interferons on H-2 genes. To clarify the basis of the effects of interferon on H-2 gene expression, we tested a) the stability of H-2 expression in vitro, after withdrawal of interferons from culture medium: Cells that were cultured for 6 days in 100 U/ml of γ interferon were tested up to 6 days after transfer to unsupplemented medium. A sharp decline was observed in H-2K and H-2D expression already after 3 days, but the level of expression was still higher than that of nontreated cells even after 6 days. Interferon α + β, although mainly inducing the H-2D gene expression, showed an increased level of H-2 expression in vitro even 6 days after withdrawal (C. Gelber et al., unpublished results). b) Effects of interferon treatments on DNA methylation of $H-2K^b$ genes in A9 and D122 cells. Genomic DNA of A9 and D122 cells before and after treatment by interferons were digested by MspI and HpaII restriction enzymes, separated on agarose gels and bloted to nitrocellulose. Hybridization to a H-2K specific probe revealed that (i) methylation patterns were different between A9 and D122 cells, (ii) that γ interferon did not change the methylation patterns of DNA in the treated cells, (iii) interferon α + β induced a changed MspI profile in both A9 and D122 cells. Since restriction enzyme MspI cleaves both methylated and nonmethylated sequences, it seems that these interferons induce a genomic change (base substitution?) in the cells (D. Fleksin et al., unpublished results). However the effects of interferons is both transient and could act on multiple components of the cells. Since interferons also affect genes other than those of the MHC, we studied the effects of H-2K expression <u>per se</u> on the metastatic phenotype, by transfecting tumor cells with the H-2K gene.

Abrogation of the metastatic phenotype following transfection with H-2K genes

We studied a second metastatic tumor, the T10 sarcoma which was induced by methylcholanthrene in a mouse of a (C3HxC57BL)F1, $(H-2^k \times H-2^b)$. Deducing from its original genotype, cells of the T10 sarcoma should have expressed four distinct class I antigens: K^b, D^b, K^k, and D^k. Yet we found that both metastatic and nonmetastatic clones lacked the cell surface expression of the K^k and K^b molecules. Nonmetastatic clones expressed only the $H-2D^b$ molecules, whereas metastatic clones expressed both $H-2D^b$ and $H-2D^k$ molecules (11). Experiments were carried out to test whether the expression of both the D^k and D^b molecules are required for the metastatic competence of tumor cells. Cells of metastatic clones were transferred in homozygous mice of the parental strains. The result was that the metastatic clone (IE7) which to begin with, expressed both the H-2D^b and the H-2D^k - when serially transplanted in C3H mice (H-2^k), lost the cell surface expression of the D^b molecule and retained only the D^k molecules. This resulted in an increase of the metastatic competence. It thus appeared that products of the H-2D^k gene determined the metastatic potency of the T10 sarcoma cells (12). We subsequently examined in colloboration with Dr. G. Hammerling's group in Heidelberg whether transfection of clones of the T10 sarcoma with cloned K-genes would alter the malignant properties of the tumor cell (13). A metastatic clone (IE7) and a nonmetastatic clone (IC9) were selected from the two sets of the T10 sarcoma clones. Cells were transfected with H-2K^b plasmid DNA (13) and plasmid pAG60 containing the neomycin resistant gene (neo^r), then selected in medium containing geneticin. Resistant clones were expanded and tested for expression of the transfected H-2 genes using specific monoclonal H-2 antibodies. Positive cells were recloned and the H-2 phenotype was determined using four different assays: 1) Cytofluorography, 2) Cellular radioimmunoassay, 3) Lysis by alloreactive CTL and 4) immunoprecipitation by monoclonal H-2 antibodies. All yielded identical results. Table 4 shows a typical cellular radioimmunoassay using monoclonal H-2 antibodies and ^{125}I-labeled protein A. Parental IC9 cells or IC9 transfected with neo^r alone, expressed only the H-2D^b molecule, whereas IC9 cells transfected with H-2K^b or H-2K^k genes synthesized also K^b or K^k antigens. Similarly, the transfected IE7 cells expressed the parental D^b and D^k antigens and in addition K^b, K^k or both K^b and K^k (Table 4). Serological analysis with more than 20 monoclonal antibodies against distinct H-2 determinants and two-dimensional gel electrophoresis showed that the K^b molecules on the transfected cells were identical to those of typical H-2^b cells. Testing the generation of spontaneous lung metastases by IE7 cells following intra-footpad inoculation of 10^4 - 10^6 cells into syngeneic (C3HxC57BL/6)F1 mice, we observed that the expression of either K^b, K^k or both resulted in the loss of their metastatic competence (Table 4). Such H-2K transfected IE7 clones produced lung metastases when grafted in animals which had been exposed to 550 rad total body irradiation, a dose which suppresses the mouse immune reactivity. This suggested that the abrogation of the metastatic phenotype following expression of H-2K antigens, is a function of the acquisition of immunogenic properties conferred on these cells by the H-2K^b and H-2K^k molecules. To explore the immunogenicity acquired by IE7 and IC9 cells following H-2K transfection we applied two tests: The first demonstrated that syngeneic mice transplanted with K positive IE7 cells, manifested after tumor excision, followed by re-challenge with transfected IE7 cells retarded tumor growth (13). The second test demonstrated that immunization with irradiated cells of the K^b positive IE7 and IC9 clones elicited cytotoxic T cells which lysed only K^b transfected IE7 and IC9 target cells, but not the parental or K^k transfected IE7 or IC9 cells (Table 5). Neither did such CTL lyse cells of an unrelated K^b positive EL4 tumor cells. Thus, the CTL elicited by the transfected cells are specific for T10 tumor associated antigen shared by both the metastatic IE7 and nonmetastatic IC9 cells, restricted bei the K^b molecule. An immune response against such antigens seemed to prevent the generation of metastases in syngeneic animals by H-2K transfected IE7 cells.

DISCUSSION

The multistep nature of the metastatic process, makes the host-tumor interactions controlled by gene products of the MHC, only one of the molecular properties which determine the probability of metastatic

Table 4. H-2 phenotype and metastatic potential of H-2K transfected IE7 cells (13)

Clone	Transfected genes	Anti-D^b	Anti-K^b	Anti-D^k	Anti-K^k	Weight of lungs (mg)
IE7	-	27,849	580	6,438	697	397 $\pm$ 70
IE7 neo^r-1	neo^r	27,069	270	6,072	294	420 $\pm$ 15
IE7 K^b-1	neo^r, K^b	10,447	9,350	3,433	851	168 $\pm$ 8
IE7 K^b-2	neo^r, K^b	27,780	14,877	6,505	419	182 $\pm$ 50
IE7 K^b-3	neo^r, K^b	17,350	12,631	4,817	913	180 $\pm$ 5
IE7 K^k-1	neo^r, K^k	16,461	949	4,964	5,785	186 $\pm$ 8
IE7 K^b,K^k-1	neo^r, K^b,K^k	16,160	13,664	5,539	6,992	170 $\pm$ 3

Radio immunoassay: 10^6 cells were incubated in microtiter plates with anti H-2 monoclonal antibody (anti D^b B22-249; anti K^b K10-56; anti K^k 16-3-22; anti D^k 15-5-5S). After washing, ^{125}I-labeled protein A was added and the radioactivity bound determined. Spontaneous metastases were tested as described in Table 1.

Table 5. Generation of CTL specific for transfected tumors in (C3HxC57BL/6)F1 mice (13)

	target cells (% specific lysis)							
Immunization	IE7	IE7 K^b-1	IE7 K^k-1	IC9	IC9 K^b-1	IC9 K^b-3	EL4	L929
IE7	0	0	0	0	0	0	0	0
IE7 K^b-1	0	35	0	0	0	40	0	0
IE7 K^k-1	0	n.d.	59	0	0	0	0	5
IC9	0	0	0	0	n.d.	0	0	0
IC9 K^b-3	0	28	0	0	0	45	0	0

(C3H x C57BL) F1 mice were immunized three times at weekly intervals with 10^6 tumor cells irradiated by 3000 R. Seven to 10 days later spleen cells were restimulated in vitro with the same tumor cells irradiated at 10,000 R. Cytolytic activity was determined at different effector to target ratios against ^{51}Cr labeled target cells including EL4($H-2^b$) and L929 ($H-2^k$) as controls. Data shows percent specific lysis obtained with an effector/target ratio of 50:1. n.d.: not determined.

formation. Yet, with regard to the unique properties of cells which are "programmed" to metastasize, within the diverse local tumor, the cell-surface expression of MHC genes may represent one of the specific molecular properties which characterize the metastatic phenotype. Thus, metastatic tumors may differ from non-metastatic tumors in synthesizing and secreting higher levels of collagenase IV, yet we observed that within a given tumor, metastatic and non-metastatic clones do not differ with regard to the level of collagenase IV (14). The same seems to apply to other enzymes, which may participate in penetration of basement membranes (endoglycosidases, Vlodavsky, personal communication). We observed that plasminogen activator is secreted by metastatic clones at higher levels than by non-metastatic clones, yet this did not seem to us to confer on the metastatic phenotype, a higher probability of achieving the initial stages of tumor dissemination (15).

It appears therefore that metastatic and non-metastatic phenotypes within the local tumor may share many properties functionally relevant to the complex metastatic process, and only a few may characterize just the metastatic cells. The properties conferred by the different expression of MHC genes may be one of them.

Imbalanced expression of class I genes, conferring on the neoplastic cells different immunologicl properties, was observed in other neoplastic systems. Thus, Festenstein's group demonstrated that cells of AKR lymphoma, line K36.16 manifest at their cell surface, a very low level of H-2K/H-2D, compared to normal lymphocytes. Cells of such lines are significantly less immunogenic than H-2K positive leukemic cells (16). They are incapable of eliciting, in syngeneic animals, CTL, neither are they susceptible to CTL produced in response to H-2K positive lymphoma cells. Such CTL seem to recognize the Gross viral antigen only in the context of the H-2K product. Transfection of such cells with the H-2K gene resulted in a phenotype which was non-tumorigenic: it was rejected immunologically by syngeneic recipients (17).

In other systems, although the immunological basis of the observation was not studied, a correlation between a low H-2K/H-2D ratio, and the neoplastic properties of the tumor cells, was observed. Rigbi et al. demonstrated that BALB/c 3T3 cells, which normally express very low H-2 gene products, following transformation with SV 40 - enhanced significantly the expression of class I antigens, and these were confined to the H-2D region (18,19). Thus the transformation seems to have been associated with a reduction in H-2K/H-2D ratio, due to induction of H-2D gene expression. It should be pointed out that in addition to the H-2D, Hood's group demonstrated that the closely located Tla class I gene was activated in a metastatic SV40 transformed 3T3 originated tumor SVT2 (20).

An intriguing question is whether there is a correlation between the expression of MHC genes in the metastatic phenotype, and the expression of other genes, the products of which might contribute to the malignant properties. Of obvious interest here are the oncogenes. Screening for the expression of oncogenes we observed a 60x amplification of the c-myc gene, yet this characterized both the metastatic and non-metastatic clones of the 3LL tumor. On the other hand, the c-fos gene was expressed only in the non-metastatic cells of the 3LL tumor (10). To what extent transfection of the metastatic D122 cells with the c-fos, would render the latter non-metastatic, and if so would such transfection co-activate the expression of the $H-2K^b$ gene, is under investigation in our laboratory. Deducing from our results of interferon γ effects on our metastatic cells, increasing the relative expression of H-2K and decreasing the metastatic potency, co-regulation of the c-fos with the MHC expression seems an inviting possibility.

ACKNOWLEDGEMENTS

Supported by US Public Health Service grant No. CA-28139 from the National Cancer Institute, Department of Health and Human Services.

References

1. J. J. Fidler, and M. L. Kripke, Science, 197: 893-895 (1977)
2. R. M. Zinkernagel, and P. C. Doherty, J. Exp. Med., 141: 1427-1436 (1975)
3. N. Isakov, M. Feldman, and S. Segal, Invasion metastasis, 2: 12-32 (1982)
4. N. Isakov, M. Feldman, and S. Segal, J. Natl. Cancer Inst., 66: 919-926 (1981)
5. N. Isakov, M. Feldman, and S. Segal, Transp. Proc., 13: 778-782 (1981)
6. N. Isakov, S. Katzav, M. Feldman, and S. Segal, J. Natl. Cancer Inst. 71: 139-145 (1983)
7. L. Eisenbach, S. Segal, and M. Feldman, Int. J. Cancer, 32: 113-120 (1983)
8. L. Eisenbach, N. Hollander, L. Greenfeld, H. Yakor, S. Segal, and M. Feldman, Int. J. Cancer, 34: 567-573 (1984)
9. L. Eisenbach, N. Hollander, S. Segal, and M. Feldman, Transp. Proc., 17: 729-734 (1985)
10. L. Eisenbach, and M. Feldman, In: "Modern trends in human Leukemia" R. Neth, Ed., Vol. 29, 499-507 (1985)
11. P. De Baetselier, S. Katzav, M. Gorelik, M. Feldman, and S. Segal, Nature, 288: 179-181 (1980)
12. S. Katzav, S. Segal, and M. Feldman, Int. J. Cancer, 33: 407-415 (1984)
13. R. Wallich, N. Bulbuc, G.J. Hammerling, S. Katzav, S. Segal, and M. Feldman, Nature, 315: 301-305 (1985)
14. L. Eisenbach, S. Segal, and M. Feldman, J. Natl. Cancer Inst., 74: 87-93 (1985)
15. L. Eisenbach, S. Segal, and M. Feldman, J. Natl. Cancer Inst., 74: 77-85 (1985)
16. W. Schmidt, and H. Festenstein, Immunogenetics, 16: 257-264 (1982)
17. K. Hui, F. Grosveld, and H. Festenstein, Nature, 311: 750-752 (1984)
18. P. M. Brickell, D. S. Latchman, D. Murphy, K. Willison, and P.W.J. Rigby, Nature, 306, 756-760 (1983)
19. P. M. Brickell, D.S. Latchman, D. Murphy, K. Willison, and P.W.J. Rigby, Nature, 316: 162-163 (1985)
20. R. Robinson, S. Hunt, and L. Hood, in preparation

MONOCLONAL ANTIBODY TARGETING OF ANTI-CANCER AGENTS

R. W. Baldwin

Cancer Research Campaign Laboratories
University of Nottingham
Nottingham NG7 2RD, U.K.

The concept of targeting cytotoxic agents to tumour sites following linkage to antibodies has become more acceptable following the development of monoclonal antibodies which react with tumour associated antigens (1). These include anti-CEA monoclonal antibodies in colorectal cancer as well as antibodies designated 791T/36, 19-9 and 17-1A, which react with colon carcinoma-associated membrane antigens. Monoclonal antibodies have also been produced which react with other types of human cancer including malignant melanoma, bone and soft tissue sarcomas, and carcinomas of breast, ovary and lung (1).

Theoretically it is desirable to target highly cytotoxic agents such as plant toxins or their A chain moieties which can kill cells following internalization of only a few molecules (2). In this case, the antibody vector should be highly specific for the tumour target cell. An alternative approach with cytotoxic agents is to target drugs which are already in clinical use and where toxic side effects are considered acceptable although limiting with respect to therapy. In this approach the improved therapeutic response may results from specific drug targeting and in addition toxic side effects will be minimized (3-4). Finally tumour localizing monoclonal antibodies may be used to target biological response modifiers the objective being to enhance host mediated reactions within tumour deposits (5).

These approaches to antibody targeted therapy can be illustrated by investigations with monoclonal antibody 791T/36 which was originally produced against a human osteogenic sarcoma but which also reacts with colorectal, ovarian and breast carcinoma (6).

Localization of Monoclonal Antibody 791T/36 in Colorectal Cancer

The localization of monoclonal antibody 791T/36 in colorectal cancers has been established in imaging trials where ^{133}I- and ^{111}In-labelled preparations have shown consistent localization in primary and metastatic tumours (7-9). This is illustrated in Table 1 which summarizes the first 33 patients with primary colon carinomas imaged with ^{131}I-labelled antibody. Considering tumours within the pelvis (rectum and rectosigmoid) 5 of 13 patients (42 %) imaged. This compares with localization of ^{131}I-791T/36 antibody in patients who had lesions outside the pelvis where 8 of 11 (67 %) were positive. The difficulty in imaging tumours in the pelvis is due to the presence of ^{131}I excreted through the urinary tract. Table 1 also summarizes the target (tumour): non-target ratios of radioactivity determined by analysing regions of interest (ROI) from the ^{131}I-images (8). This ranged from 2.1:1 to 5.0:1 (mean 4.4:1).

Table 1. Imaging of Primary Colon Cancers with ^{131}I-labelled 791T/36 Monoclonal Antibody (8)

Tumour Site	Number of Patients	Number Imaged	Number Positive		Mean T : NT* after subtraction
Caecum and Ascending Colon	4	4	4		3.5 : 1
Transverse and Descending Colon	2	2	1	8/11	2.5 : 1
Sigmoid Colon	7	5	3		3.9 : 1
Rectosigmoid	5	3	1	5/13	3.0 : 1
Rectum	15	10	4		5.0 : 1
	33	24	13		

* Determined by analysis of regions of interest of gamma camera images (7).

Eighteen patients with recurrent or metastatic colorectal cancer (26 sites) were also imaged following injection of ^{131}I-791T/36 (Table 2). In all there were 26 sites of metastases and of these 21 (85 %) were positively identified by imaging. These included 13 patients with liver metastases and positive imaging was achieved in 11 patients, the smallest deposits visualized being multiple 1 cm-diameter lesions in one patient (8).

In a number of patients imaged pre-operatively, localization of ^{131}I-labelled 791T/36 was demonstrated by radiochemical analysis of resected tumour and normal colonic mucosae. This is illustrated in Figure 1 which shows the tumour and normal tissue distribution of ^{131}I-labelled 791T/36 and ^{123}I-labelled normal mouse IgG2b injected simultaneously into a colon carcinoma patient. Specific uptake of antibody into tumour compared with non-tumour tissue is clearly obtained, but this was not the case with ^{123}I-normal mouse IgG2b (7).

Binding of Monoclonal Antibody 791T/36 to Tumour Cells Derived from Primary and Metastatic Colon Carcinomas

For effective delivery of agents linked to monoclonal antibodies to tumour the conjugates should bind specifically to tumour cells. This has been demonstrated by flow cytometric analysis of 791T/36 antibody binding to tumour cells obtained by collagenase disaggregation of tumour

Table 2. Imaging of Metastatic and Recurrent Colorectal Cancer with ^{131}I-labelled 791T/36 Monoclonal Antibody (8)

Tumour Site	Number Imaged	Number Positive	Meant T : NT After subtraction
Liver	13	11	$3.5 \pm 1.2 : 1$
Pelvis/perineum	7	5	$4.0 \pm 2.0 : 1$
Intra abdominal	3	3	$7.4 \pm 0.8 : 1$
Lung	2	1	Not quantifiable
Brain	1	1	$4.0 : 1$
	26	21	

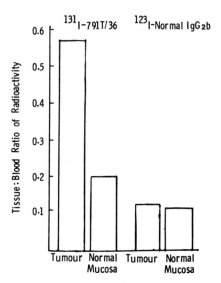

Fig. 1. Localisation in colon carcinoma and normal mucosa. ^{131}I-791T/36 and ^{123}I normal mouse IgG2b injected simultaneously into a patient.

tissue derived from primary colorectal cancers as well as lymph node and hepatic metastases (10). Target cells were reacted with 791T/36 antibody and bound mouse IgG detected by reaction of cells with fluorescein isothiocyanate (FITC)-conjugated rabbit anti-mouse IgG. The mean linear fluorescence (MLF) of cells in the size range of malignant populations was then determined by flow cytometry. A typical analysis of 791T/36 antibody binding to cells derived from a primary colon carcinoma and lymph node and liver metastases taken from the same patient is illustrated in Figure 2. This shows that 791T/36 binds to target cells derived from both the primary and metastatic tumour deposits. Flow cytometry analyses have been carried out with target cells derived from 50 primary colorectal tumours giving a mean linear fluorescence/cell of 134. Likewise, target cells derived from 17 lymph node metastases (MLF 139) and 17 hepatic/peritoneal metastases (MLF 139) confirmed that metastatic tumours continued to express the antigen defined by 791T/36 antibody (111).

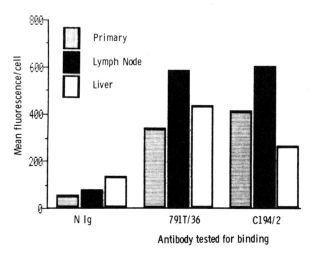

Fig. 2. Binding of monoclonal antibodies to primary and metastatic colon carcinoma cells as determined by flow cytometry.

Monoclonal Antibody-Drug Conjugates

Conjugation of drugs to monoclonal antibodies aims to introduce the maximum number of drug residues under conditions which ensure optimal retention of both drug and antibody reactivities. In this respect, only a limited number of drug residues can be conjugated by direct linkage to antibody without producing loss of antibody reactivity. This will be dependent upon the antibody and the physical and chemical properties of the conjugated compound. In general substitution ratios of greater than 10:1 with respect to IgG antibodies results in an unacceptable loss of antibody reactivity (3,12).

In order to increase the amount of drug coupled to antibody 'drug-carrier' systems are being designed (3,12). Here the drug moiety is firstly linked to carrier molecules such as dextran and human serum albumin and then the complex is linked to antibody. Conjugates produced in this fashion can have 'a drug-loading' at least 10 times greater than that produced by direct drug-antibody linking. This can be illustrated by the synthesis of methotrexate (MTX) conjugates with 791T/36 where products containing more than 4 moles MTX/mol antibody had reduced antibody reactivity (13). Conjugates synthesized using human serum albumin (HSA) as a carrier for MTX (MTX-HSA-791T/36) contain of the order of 30-40 MTX residues/antibody and still retain antibody reactivity (13,14).

Antibody conjugation of drugs may also lead to modification of drug cytotoxicity. In many cases e.g. anthracycline-conjugates (12) cytotoxicity is considerably reduced, but in some situations conjugates are as active, or even more active than free drug. This is illustrated in Figure 3 which compares the inhibition of colony formation of sarcoma 791T cells following exposure to free methotrexate or MTX-HSA-791T/36 antibody conjugates (14). This shows that free MTX was less effective than the antibody conjugate the doses producing 50 % inhibition of colony growth being 3 ng/ml for MTX on a 0.5 ng (as MTX) for the conjugate. This is relevant with respect to MTX-antibody conjugate therapy, since it implies that a major pathway of entry of MTX into cells is effected through antibody mediated reactions. In this respect it has also been found that MTX-791T/36 conjugates were much more cytotoxic than free MTX for cultured colon carcinoma cells derived from surgical specimens (15). With three colon carcinoma target cells, they were all relatively resistant to free MTX, doses of 1000 to 8000 ng/ml culture medium being required to produce 50 % inhibition of tumour cell survival (IC_{50}). This is considerably greater than that required with MTX-791T/36, IC_{50} values ranging from 63 to 400 ng/ml medium.

Therapeutic Activity of Drug-Antibody Conjugates

Therapy trials using human tumours growing as xenografts in immunodeficient mice have been designed to determine efficacy of drug-antibody conjugates in suppressing tumour growth compared with responses achieved with free drug (1,3). Whilst many of these trials have not been fully evaluated, evidence is growing to show the therapeutic effectiveness of drug-antibody conjugates. This is illustrated in Fig. 4 which summarizes tests over a range of doses with methotrexate-HSA-791T/36 monoclonal antibody conjugates against xenografts developing following injection of osteogenic sarcoma 791T cells. When expressed as a ratio of tumour weights in treated compared with control mice (T/C ratio) two weeks after the final treatment, free MTX exerted a significant effect at 20 mg/kg body weight and overall these data indicated a T/C ratio of 0.5 at 24 mg/kg body weight. MTX-HSA-791T/36 antibody conjugates elicited a more pronounced therapeutic response than free drug with a T/C ratio of 0.5 being obtained at a dose of methotrexate of 14 mg/kg body weight. In one test virutally complete tumour suppression was achieved at 18 mg/kg body weight of MTX in antibody conjugated form whereas the maximum dose of free MTX tested (60 mg/kg) only reduced tumour growth in treated mice (T/C 0.30).

Similar trials are in progress with a range of drug-monoclonal antibody conjugates and whilst few are as advanced as the tests with MTX-791T/36 conjugate, they indicate that antibody conjugates have therapeutic potential (3,12). Vindesine (VDS) linked to several monoclonal antibodies including anti-melanoma antibody (96.5), anti-CEA (11.285.14) and 791T/36 suppressed growth of xenografts of melanoma, colon carcinoma and osteogenic sarcoma cells (15). In this study the most marked effect was with VDS-conjugated to the anti-CEA antibody (11.285.14) where almost complete suppression of growth of colon carcinoma xenografts was produced.

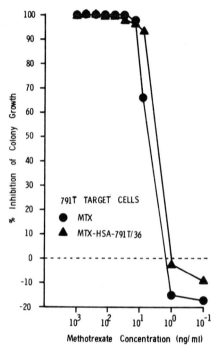

Fig. 3. Cytotoxicity of MTX-HSA-791T/36 conjugate against 791T oestogenic sarcoma cells as measured by colony inhibition assay.

In addition to improving in vivo therapeutic responses drug conjugation to monoclonal antibodies frequently results in a significant reduction of drug toxicity. For example with daunomycin (12,16), the LD50 for mice given twice weekly injections is 14 mg/kg body weight, while drug conjugated to monoclonal antibody 791T/36 showed not toxicity at doses up to 30 mg/kg weight (in terms of daunomycin). Similarly acute toxicity tests with vindesine (VDS) indicated an LD50 of 6.7 mg/kg whereas no marked toxicity was observed with VDS-conjugated to monoclonal antibody (11.285.14) at doses up to 90 mg/kg in terms of VDS content (16).

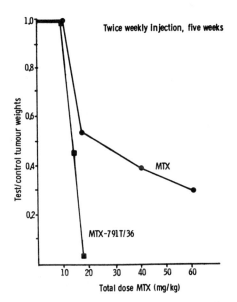

Fig. 4. Influence of 791T/36-MTX conjugate on growth of 791T osteogenic sarcoma xenografts. Groups of 8-10 athymic mice were injected subcutaneously with 791T cells, and twice weekly therapy with MTX or 791T/36-MTX conjugate initiated three days later. Mice were killed after seven weeks and a test/control tumour weight determined as:

$$\frac{\frac{\text{weight of all tumours in treated mice}}{\text{total no. of treated mice}}}{\frac{\text{weight of all tumours in control mice}}{\text{total no. of control mice}}}$$

CONCLUSIONS

Considerable progress has been achieved in producing murine monoclonal antibodes which react with human tumours (1). A number of these have been found to localize in vivo in primary tumours and more important metastatic lesions so that they can be considered for targeting antitumour agents (1,2,12). This has led to the design of procedures for conjugating antibodies to cytotoxic agents either by direct linkage or through drug-carrier systems. Already a range of cytotoxic drugs have been used to produce antibody conjugates including agents known to have some clinical effect. These include methotrexate, adriamycin, vinca alkaloids, cis-platinum and interferons (1,3,4,12). Although efficacy trials are still in progress with many antibody conjugates, several including those containing methotrexate, vindesine have proven reactivity when tested against human tumour xenografts. These xenograft studies provide the basis for developing clinical trials and studies with methotrexate-conjugates are in progess in colorectal cancer and non-small cell lung cancer.

ACKNOWLEDGEMENTS

These studies were supported by grants from the Cancer Research Campaign, U.K.

REFERENCES

1. R. W. Baldwin, and V. S. Byers, "Monoclonal Antibodies for Tumour Detection and Drug Targeting", Academic Press, 1985.
2. J. W. Uhr, Immunotoxins: Harnessing Nature's Poisons. J. Immunol., 133:1 (1984).
3. R. W. Baldwin, Design and development of drug-monoclonal antibody 791T/36 for cancer therapy, in: "Monoclonal Antibody Therapy of Human Cancer", K. Foon, and A. C. Morgan, eds., Martinus Nijhoff, Boston, p. 23 - 56 (1985).
4. G. F. Rowland, and R. G. Simmonds, Effects of monoclonal antibody-drug conjugates on human tumour cell cultures and xenografts, in: "Monoclonal Antibodies for Tumour Detection and Drug Targeting", R. W. Baldwin, and V. S. Byers, eds., Academic Press, (1985).
5. J. M. Pelham, J. Dixon-Gray, G. R. Flannery, M. V. Pimm, and R.W. Baldwin, Interferon conjugation to human osteogenic sarcoma monoclonal antibody 791T/36. Cancer Immunol. Immunother., 15:210 (1983).
6. M. J. Embleton, B. Gunn, V. S. Byers, and R.W. Baldwin, Antitumour reactions of monoclonal antibody against a human osteogenic sarcoma cell line, Br. J. Cancer, 43:582 (1981).
7. N. C. Armitage, A. C. Perkins, M. V. Pimm, P. A. Farrands, R. W. Baldwin, and J. D. Hardcastle, The localisation of an anti-tumour monoclonal antibody (791T/36) in gastrointestinal tumours. Br. J. Surg., 71:407 (1984).
8. N. C. Armitage, A. C. Perkins, J. D. Hardcastle, M. V. Pimm, and R. W. Baldwin, Monoclonal antibody imaging in malignant and benign gastrointestinal diseases, in: "Monoclonal Antibodies for Tumour Detection and Drug Targeting", R. W. Baldwin, and V. S. Byers, eds., Academic Press (1985).
9. N. C. Armitage, A. C. Perkins, M. V. Pimm, R. W. Baldwin, and J.D. Hardcastle, Imaging of primary and metastatic colorectal cancer using an ^{111}In-labelled antitumour monoclonal antibody (791T/36). Nucl. Med. Comm., In press (1985).

10. L. G. Durrant, R. A. Robins, N. C. Armitage, A. Brown, R.W. Baldwin, J. D. Hardcastle, Association of antigen expression and DNA ploidy in colorectal tumours, Cancer Res., In press (1985).
11. K. C. Ballantyne, L. G. Durrant, N. C. Armitage, R. A. Robins, R.W. Baldwin, and J. D. Hardcastle, Monoclonal antibody binding to primary and metastatic colorectal cancer, Gut, 26:Abstr. 1154 (1985).
12. R. W. Baldwin, M. J. Embleton, J. Gallego, M. Garnett, M.W. Pimm, and M.R. Price, Monoclonal antibody drug conjugates for cancer therapy, in: "Monoclonal Antibodies for Diagnosis and Therapy of Cancer", J. Roth, ed., Futura Publishing Co. (1986).
13. M. C. Garnett, M. J. Embleton, E. Jacobs, and R. W. Baldwin, Preparation and properties of a drug-carrier-antibody conjugate showing selective antibody-directed cytotoxicity in vitro, Int. J. Cancer, 31:661 (1983).
14. M.C. Garnett, and R. W. Baldwin, An improved synthesis of a methotrexate-albumin-791T/36 monoclonal antibody conjugate cytotoxic to osteogenic sarcoma cell lines, Cancer Res., In Press (1986).
15. G. F. Rowland, C. A. Axton, R. W. Baldwin, J. P. Brown, J. R. F. Corvalan, M. J. Embleton, V. A. Gore, I. Hellstrom, K. E. Hellstrom, E. Jacbos, C. H. Marsden, M. V. Pimm, R. G. Simmonds, and W. Smith, Anti-tumour properties of vindesin-monoclonal antibody conjugates, Cancer Immunol. Immunother., 19:1 (1985).
16. J. Gallego, M. R. Price, and R. W. Baldwin, Preparation of four daunomycin-monoclonal antibody 791T/36 conjugates with anti-tumour activty, Int. J. Cancer, 33:737 (1984).

TREATMENT OF HUMAN B CELL LYMPHOMA WITH MONOCLONAL ANTI-IDIOTYPE ANTIBODIES

E. M. Rankin[*] and A. Hekman

Netherlands Cancer Institute
Antoni von Leeuwenhock Huis, Plesmanlaan 121
1066 Cx Amerstam

INTRODUCTION

Human malignant B cell tumours are thought to arise from the proliferation of a single clone of cells (1-4). The immunoglobulin (Ig) that is expressed and in some cases secreted by the tumour cells is limited to the expression of one single V_H and V_L region, and to a single light chain. The unique variable region of the Ig, the idiotype, may be considered a model tumour marker, since it is found only on the malignant cells. Antibodies directed against this target, anti-idiotype antibodies, would have the potential to destroy malignant tissue while leaving the residual normal lymphoid tissues intact.

Monoclonal anti-idiotype antibodies have been used by Levy and colleagues for the treatment of eleven patients with non-Hodgkin's lymphoma; a durable complete remission was obtained in one patient in whom treatment was free from side effects; partial remissions were induced in five patients (5,6). In several patients problems with anti-mouse antibodies, antigen modulation or tumour cell heterogeneity were encountered. Polyclonal anti-idiotype antibodies raised in sheep have been given to four patients with B cell leukaemia (7,8) and to three patients with lymphoma (9). There were transient reductions in the level of circulating lymphocytes, but no lasting anti-tumour effect was seen. After each infusion, there was evidence of antigen modulation with alterations in the density and distribution of the surface antigen, and resistance to complement-mediated lysis.

We have developed a fast and reliable method for the manufacture of mouse monoclonal anti-idiotype antibodies (10). Studies of the effects in vivo of these antibodies are being performed (11) and the results of treatment in two of the four patients treated to date are presented here. The aims of the study are threefold: 1) to determine the immunological effects of treatment and the method by which a response, if any, is produced; 2) to determine the toxic effects of the administration of mouse immunoglobulin; 3) to establish the best schedule for treatment. Such trials are of crucial importance, since even if

---

[*] E.M.R. is a research fellow of the European Organization for the Research on Treatment of Cancer. Present address: Medical Oncology Clinic, Guy's Hospital, London.

the monoclonal antibodies are ineffective as antitumor agents when used alone, they may still have a role as carriers of toxins, radioisotopes or chemotherapeutic agents.

MATERIAL AND METHODS

Manufacture of Monoclonal Anti-Idiotype Antibodies

The method by which the monoclonal anti-idiotype antibodies were raised has been described previously (10). Immunization with spleen cells of patient TOP yielded the antibody T2; after immunization with patient KOS' lymph node cells, two anti-idiotype antibodies, K1 and K2, were obtained. The specificity of the antibodies was proven by the lack of reaction with a variety of normal cells, 14 cell lines representing different stages of lymphoid and myeloid cell differentiation and a wide range of different non-Hodgkin's lymphomas. Each antibody precipitated immunoglobulin heavy and light chains from surface-labelled homologous tumour cell lysates, and their target antigen was capped by anti-light chain antibodies.

From the two antibodies against KOS idiotype, K1 was chosen for treatment, since it had a higher affinity for the tumour cells than K2, as judged by immunofluorescence. Also, unlike K2, K1 did not modulate the antigen as tested by visualizing bound antibody with fluorescein-labelled anti-mouse Ig after various times of incubation. Generally, monoclonal antibodies against idiotype or light chain were found to be less capable of induction of capping and modulation of surface Ig than polyclonal antisera, possibly because of a lower degree of crosslinking of the antigens. T2 and K1 were IgG2a immunoglobulins.

Antibody Purification

Monoclonal antibodies were purified from ascitic fluid of pristane-primed BALB/c mice injected with 2 to 5 x 10^6 hybridoma cells. Ascitic fluid was collected aseptically. T2 antibodies were isolated by ion exchange chromatography on DEAE Sephadex A50 (Pharmacia, Uppsala, Sweden) in 50 mmol/L Tris/HCl buffer, pH 7.5, with 0.1 mol/L NaCl. Under these conditions, the antibodies did not bind to the column. K1 antibodies were prepared by precipitation with ammonium sulphate at 40 % saturation. Both antibody preparations were dialyzed against 0.14 mol/L NaCl, centrifuged for 45 minutes at 100,000 g to remove aggregates, passed through a 0.22-μm filter and stored at 4°C. The antibodies were 80 % to 90 % pure by sodium dodecyl sulphate-polyacrylamide gel electrophoresis (SDS-PAGE). Cultures for aerobic and anaerobic microorganisms were negative. Absence of endotoxins was tested by the Limulus amoebocyte lysate assay (Malinkroft, Bethesda, Md) and by pyrogenicity tests in vivo in rabbits. Before administration, the antibodies were again passed through a 0.22-μm filter, then diluted to the desired concentration with 0.14 mol/L NaCl with human serum albumin (Central Laboratory of the Netherlands Red Cross Blood Transfusion Service, Amsterdam) as a protein carrier.

Patients

Patient TOP. In July 1980, a 71-year old woman was found to have stage IV poorly differentiated diffuse non-Hodgkin's lymphoma with involvement of the lymph nodes, spleen, bone marrow and a peripheral lymphocytosis (7 x 10^9/L). A partial remission of disease was obtained after several different chemotherapy regimens had been given; systemic treatment was stopped in September 1981. Two courses of radiotherapy were given to the spleen, and one course was given to enlarged nodes in the neck. The spleen was removed in December 1982.

When anti-idiotype therapy was begun in October 1983, there was massive abdominal disease with ascites and generalized lymphadenopathy. The patient complained of fatigue: her Karnofsky performance status was 70. The haemoglobin was 7.7 mmol/L, the platelet count was 230 x 10^9/L, and the white cell count was 19 x 10^9/L with 66 % lymphocytes. All surface Ig-positive cells in blood, bone marrow, and lymph node also stained with the anti-idiotype antibody T2. The serum biochemistry, liver function, and renal function were normal except for a creatinine clearance of 40 mL/min. There was no paraprotein band in the serum and no excess of monoclonal light chains in the urine.

Patient KOS. A 62-year old man was seen in October 1979 and a stage IV non-Hodgkin's lymphoma was diagnosed. Cervical node biopsy showed a nodular well-differentiated pattern, and the bone marrow showed a diffuse poorly differentiated lymphoma. A partial remission of disease was obtained with a regimen that combined chemotherapy with radiotherapy to the site of presenting bulk disease.

In March 1983, a node biopsy, performed because of increasing lymphadenopathy, showed a change to a higher-grade malignancy with a diffuse poorly differentiated pattern. A variety of investigational chemotherapy agents were tried without success.

When treatment with anti-idiotype antibody was begun in December 1983, the patient had rapidly advancing disease with diffuse lymphadenopathy and with massive involvement of the para-aortic and mesenteric groups, obstruction of the inferior vena cava, and considerable ascites. The patient's Karnofsky performance status was 40. The blood count at the time treatment was begun was: Hb 5.9 mmol/L, platelet count 255 x 10^9/L, white cell count 6.5 x 10^9/L, of which 37 % were lymphocytes; of these 16 % had -IgM surface immunoglobulin, and 12 % stained with the anti-idiotype antibody K1. K1 reacted with 31 % of bone marrow cells. Of the cells isolated from the ascites fluid before treatment, 63 % were idiotype positive tumour cells. Serum biochemistry, liver function, and renal function tests were normal; the creatinine clearance was 70 mL/min, and the serum protein immunoelectrophoresis was normal. There was no serum paraprotein band and no excess of monoclonal light chains in the urine.

Study Parameters

Blood samples were taken at regular intervals during and after therapy and the following measurements were made: full differential blood count, serum fibrinogen and fibrin degradation products, prothrombin and partial prothrombin times, complement (C1q globulin, C3 globulin, C4 globulin, C3d, CH50 titer, C1q binding test), serum electrolytes, renal function and liver function tests, and serum protein immunoelectrophoresis. Free idiotypic Ig, unbound mouse anti-idiotype and human anti-mouse Ig antibody were measured as described below. Radiographs of the thorax and ultrasound scans of other sites of evaluable disease were taken at intervals. Computerized axial tomography (CAT scan) was performed before treatment began and when indicated thereafter. Creatinine clearance was measured regularly, and the urine was examined daily for the presence of blood, proteins, and casts.

The study protocol was approved by the Ethical Committee of the Netherlands Cancer Institute. Both patients gave written informed consent to the trial.

Antibody Administration

Thirty minutes before treatment was begun, the patient was given 500 mg aspirin and 25 mg benadryl orally; allopurinol was given to the patients throughout the treatment period in case of massive cell lysis. Before each antibody infusion, a skin test for immediate hypersensitivity to mouse protein was performed with a 0.1-mL solution containing 0.002 mg of mouse antibody. The patients were carefully monitored for any evidence of anaphylaxis or complications of cell destruction such as hypotension, fever, disseminated intravascular coagulation, ect.

Assays for Idiotypic Ig and Anti-Idiotype Antibody

The level of circulating free idiotype in the serum or ascites fluid was determined by a solid phase sandwich enzyme immunoassay with biotinyl conjugates of the anti-idiotype antibody. Idiotypic IgM was isolated from KOS serum by affinity chromatography on Sepharose-linked K1, using elution with 0.1 mol/L glycine/HCl, pH 2.5. This was used as reference preparation to quantify the assay of the idiotype in the serum.

The presence of circulating mouse anti-idiotype antibody in serum or ascites was detected by enzyme immunoassay on lymphoma cells fixed in Terasaki microtiter plates. Titration of the antibody preparations used for treatment showed that the limit of detection was 0.2 g/mL. Both methods have been described previously (10).

Immunofluorescence

Mononuclear cells were isolated from peripheral blood, bone marrow, ascites fluid, and cell suspensions of lymph node by centrifugation over Ficoll/Hypaque (Nyegaard, Oslo). The following monoclonal antibodies and polyclonal antisera were used for direct and indirect immunofluorescence, which was performed as described previously (10), fluorescein-conjugated rabbit anti-human IgD, IgG, IgM (Dako, Copenhagen), fluorescein-conjugated goat $F(ab)_2$ anti-mouse IgG (Tago, Burlingame, Calif.), fluorescein-conjugated goat anti-mouse Ig (Nordic Diagnostics, Tilburg, The Netherlands), fluorescein-conjugated rabbit anti-human complement (C3c + C3d) (Central Laboratory of The Netherlands Red Cross Blood Transfusion Service), OKT3, OKT4, OKT8 (Orthodiagnostics N.V., Beerse, Belgium) and fluorescein-labelled anti-T3 antibody (Dr. H. Spits, Netherlands Cancer Institute, Amsterdam). The degree of cell saturation in vivo by the anti-idiotype after treatment was determined by comparing the staining obtained by fluoresceinated anti-mouse antibody with and without a prior exposure in vitro to the anti-idiotype antibody.

Punch Biopsy of Lymph Nodes

Fine needle aspirations of involved lymph nodes were made using a 23-gauge needle attached to a 20-mL disposable syringe held in a pistol grip (Cameco, Sweden). A portion of the material was smeared directly, air dried and stained routinely with Giemsa; the rest was suspended in Dulbecco's modified Eagle's medium/fetal calf serum (DMEM/FCS) with heparin (1,000 U heparin per millilitre of medium).

Cell Kinetic Studies

Suspensions of lymphocytes from blood, bone marrow, lymph node, and ascites fluid were prepared, and the percentage of S-phase cells was determined by DNA staining with ethidium bromide and flow cytometric analysis (12) by Dr. L. Smets, Department of Experimental Cytology.

Monocyte Functional Assays

Monocytes were isolated from whole patient blood using the elutriation technique (13), and were tested for their capacity to mediate antibody-dependent cellular cytotoxicity with anti-D antibody-coated human red blood cells as the target (13). Spontaneous cellular cytotoxicity using K562 cells as a target was measured in a ^{51}Cr release assay (14). Activity of monocytes was measured by chemiluminescence. Five million monocytes were suspended in 250 µL RPMI supplemented with 10 % FCS and incubated in plastic scintillation vials for 30 minutes at 37°C. After this period, 200 µL leucigenin (92 mg/L) and 50 µL serum-opsonized Zymosan (10 mg/mL), both dissolved in RPMI, were added to the vials. Chemiluminescence was measured every ten minutes in a liquid scintillation counter (Packard Tricarb, set in the coincidence "off" mode) for a total period of 70 minutes. Peak count rates were determined after subtraction of the chemiluminescence measured in control vials (without Zymosan).

Indium Labelling of Lymphocytes

Lymphocytes were isolated from defibrinated peripheral blood by centrifugation over Ficoll-metrizoate (Lymphoprep, Nyegaard, Oslo), washed twice, and resuspended in 2 mL phosphate buffered physiological saline, pH 7.4. Viability of the cells, measured by trypan blue exclusion, was 95 %. Labelling was performed with 1.48 MBq (40 µCi) indium-111 oxine (Byk Mallinckrodt, The Netherlands) per 10^8 lymphocytes. Labelling efficiency measured after 15 minutes' incubation at room temperature was 90 %. The total injected activity varied from 5.2 to 8.3 MBq (142 - 224 µCi).

Serial Blood Sampling

Samples of blood of 3 mL were taken from a peripheral vein at regular intervals after reinjection of the indium-labelled lymphocytes. On the first occasion whole blood was counted. On the second the sample was centrifuged at 1400 g and the plasma and cell pellet were separately counted. All the samples were stored at 4°C and counted at the end of the sampling period in a gammacounter.

Gammacamera Imaging

Scintigrams were made with a double headed gammacamera with a large field of view (Siemens Rota 275ZLC) fitted with parallel hole and medium energy collimators and connected to an on line computer system (MDS A2). We used a dual window setting over the energy peaks of 171 and 245 kiloelectron volts. Digital scintigrams were recorded in a 256 x 256 matrix. After intravenous injection of autologous lymphocytes labelled with 5.2 MBq (141 µCi) indium-111, static anterior view scintigrams were made at intervals of three minutes, 30 minutes, one hour 45 minutes, 18 hours, and 120 hours. At the onset of infusion of antibody 8.3 MBq (224 µCi) autologous lymphocytes labelled with indium-111 were injected, and dynamic acquisition of thorax, liver and upper abdomen was performed during the first 30 minutes. Thereafter static anterior and posterior view scintigrams, collecting counts over 15 minutes, were recorded simultaneously at intervals of 30 minutes and two, 28, 24, 44, and 49 hours.

Antibodies Against Mouse Immunoglobulin

Passive hemagglutination was used to test the patient's serum for the presence of antibodies against moust immunoglobulins. Human blood group 0 erythrocytes were fixed with 0.06 % glutaraldehyde and coated with mouse immunoglobulin (of the same isotype as that used for treatment) by exposure to 1.2 mmol/L $CrCl_3$. Twenty microliter-aliquots of the red cell suspension were incubated with 20 µL of test serum dilutions for 90 minutes at room temperature in V-bottomed microtiter plates (Greiner, Nurtingen, FRG). The plates were then centrifuged for two minutes at 225 g, rested at an angle of 60° for ten minutes and then read. In each test, normal human serum, rabbit anti-mouse immunoglobulin (Nordic), normal rabbit serum, and uncoated erythrocytes were used as controls. Rabbit sera were absorbed with uncoated human erythrocytes before the test. A positive hemagglutination reaction was indicated by the persistent formation of a clump of erythrocytes, and a negative reaction by a diffuse band of erythrocytes in the bottom of the well.

RESULTS

The Effect of Different Schedules of Administration on the Circulating Lymphocytes and Demonstration of Homing of the Antibody of All Tumour Sites

Immediately before treatment, immunofluorescence studies showed that 80 % of the lymphocytes were B cells with IgM IgD IgG surface immunoglobulin. All B cells stained with the anti-idiotype antibody T2, although as with anti-K, there was a considerable variation between cells in the strength of the fluorescent reaction. Circulating free antigen was barely detectable in the serum when measured either by the sandwich enzyme immunoassay or by the ability of the serum to inhibit the binding of T2 to TOP cells in immunofluorescence experiments. The effect of different schemes of antibody administration was determined by monitoring the level of circulating lymphocytes in the blood (Fig. 1).

The effects of three different schedules of administration were explored. In the first, the antibody was given as an infusion over six hours, the dose being doubled daily. Transient falls in the level of circulating lymphocytes were seen but this was not dose-dependent since 20 mg produced the same proportional fall as did 160 mg. In the second schedule (on day 20) the dose was rapidly escalated from 10 mg/h to 150 mg/h. The lymphocytes fell continuously through the period of the infusion but rose again to pretreatment level when therapy was stopped. In an attempt to prevent this rise at the end of the infusion, a third regimen explored the effects of a rapid bolus dose of 150 or 300 mg followed by a low-dose continous infusion of 20 mg/h (Fig. 1, treatments a, b, and c). Again the fall in the number of lymphocytes, although substantial (on one occasion the count fell from 34×10^9/L to 8.6×10^9/L) was only temporary.

Ten milligrams of the antibody was sufficient to coat the blood lymphocytes in vivo as detected by reaction with fluorescein-labelled goat anti-mouse Ig; the tumour cells in the lymph nodes and bone marrow were saturated with T2 during the third regimen. Cryostat sections of a lymph node removed at the end of the infusion on day 63 were examined for immunofluorescence and immunoperoxidase staining with goat anti-mouse Ig antibodies. Only the malignant tumour cells had been coated in vivo with the anti-idiotype antibody. After 1,300 mg of antibody, the cells

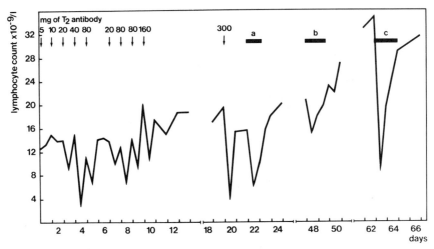

Fig. 1. Effect of administration of T2 antibody on the level of circulating lymphocytes.

* The dose of 300 mg on day 20 consisted of 10, 20, 40, 80, and 150 mg over consecutive hours; (A) 150 mg T2 over two hours, followed by 20 mg/h for 27 hours (total 690 mg); (B) 150 mg T2 over two hours, followed by 20 mg/h for 40 hours (total 950 mg); (C) 300 mg T2 over two hours, followed by 22 mg/h for 46 hours (total 1,300 mg).

in the ascites were coated with T2 and they remained so even three days after the end of the infusion. Free T2 antibody was detectable in the serum after the second regimen, declining over 24 hours from a maximum of 12 µg/mL at the end of the infusion. After the third regimen, free T2 was detectable for three days in the serum, but never attained measurable levels in the ascites.

With the aid of fluorescence activated cell sorter (FACS) analysis, fluctuations in the amount and degree of antibody coating of the tumour cells were seen at different times during and after treatment, the reaction frequently being stronger 24 hours after the end of the infusion than immediately at the end of the infusion. The proportional number of T cells in the circulation rose and fell in an inverse relationship with the B cells. There was no change in the fluorescence intensity of the cells stained with OKT3 during or after treatment.

The third schedule was applied three times with total doses of 700, 950, and 1,300 mg. As there was no clinically detectable response, and the peripheral lymphocyte count was rising inexorably, further treatment with anti-idiotype antibody was stopped. Unexpectedly, the lymph nodes afterwards decreased in size, and two weeks later were 10 % smaller than they were before treatment was begun. However, the patient's condition had deteriorated, and she was started on prednisolone. Because steroids themselves may exert an antitumour effect, all the data presented her refer to events before the introduction of steroids. The patient died three months after the end of treatment; permission for postmortem was not given.

Lack of Toxicity of Anti-Idiotype Antibody Therapy

The patient was closely monitored throughout the treatment period. Apart from one febrile episode of short duration, no symptoms developed that were attributable to the treatment, in particular there were no chills, bronchospasm, or hypotension. There was no haematological toxicity and no disturbance of liver or renal function; creatinine clearance remained at the pretreatment level of 40 mL/min. There was no evidence of activation of the complement pathway or of disseminated intravascular coagulation.

Absence of Host Response to Mouse Protein

A total of 3,800 mg of T2, a mouse immunoglobulin of the IgG2a subclass, were given to patient TOP. No antibodies to the mouse protein were detectable even one month after the last treatment.

Anti-Idiotype Antibody Did Not Modulate the Antigen In Vivo

Before treatment, the peripheral blood lymphocytes showed considerable variation of the strength of the fluorescent reaction with anti-K and anti-IgM antisera, and with the T2 antibody. No antigenic modulation was observed in vitro (10). At the end of the infusions, only cells showing a moderate staining were detectable, but no antigen-negative tumour cells were observed. Twenty-four hours later brightly staining cells were again detectable. To test whether these changes in antigen density were caused by changes in cell populations or by partial modulation, the weaker staining cells found at the end of therapy, when there was excess circulating anti-idiotype, were washed and put into short-term culture in medium. Their reaction with anti-K or with T2 was tested at various intervals up to 24 hours, but no increase in the strength of fluorescent reaction was seen.

Observations on the Changes of Tumour Cell Populations

It was important to establish whether the circulation was repopulated as a result of cell division or by migration of cells from an extravascular reservoir into the blood. Blood lymphocytes showed no change in DNA synthesis by tritium-labelled thymidine incorporation or autoradiography at the time of maximal rebound (data not shown). The percentage of S-phase cells in the blood before and after treatment was unaltered at 2 %. In the lymph node, the percentage varied, 7,5 % of the cells were in S-phase before treatment, 2 % at the end of treatment, and 14 % of the cells were in S-phase, most of them in early S-phase, four hours after the end of the infusion (Fig. 2). Because at this time a substantial part of the repopulation of the blood had already occured, it was unlikely that this was caused by the increased proliferation of the lymph node cells.

It is possible that the transient fall in peripheral lymphocytes seen during treatment was the result of alterations by the antibody of the circulation pathways of the tumour cells. This was investigated using peripheral lymphocytes labelled with Indium-111 oxine (15).

Blood disappearance curves. Figure 3 shows the rate of disappearance of the lymphocytes labelled with Indium-111 oxine reinjected at a time when the patient was not receiving treatment. Initially activity fell rapidly until 12 hours, when a secondary rise occurred followed by a more gradual decrease. Activity was still detectable at 118 hours. The disappearance curve for the period of treatment (Fig. 4) was very different. Rapid fluxes of cells into and out of the circulation occurred

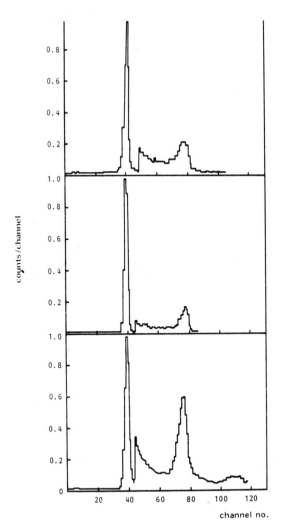

Fig. 2. Flow cytometer histogram showing the number of S-phase cells in the lymph node at different times during the treatment period. (A) Before treatment: 7.5 % of the cells are in S-phase; (B) after the infusion of 1,300 mg T2 antibody: 2 % of the cells are in S-phase; (C) four hours after the end of an infusion of 950 mg of T2 antibody: 14 % of the cells are in S-phase. After the G1 peak, the scale of the ordinate is increased tenfold.

during the first six hours after which the activity stayed low and constant with a minor rise corresponding to the end of the infusion of the antibody. In contrast, counts in the plasma varied little indicating that the cells were not being destroyed intravascularly. These fluxes in the circulation did not parallel the changes in the total number of lymphocytes in the blood (Fig. 5). These fell from 20.7×10^9/L to a nadir at 3.5 hours of 14×10^9/L, gradually rising over the next few hours to reach 20.6×10^9/L at 12 hours and 28.1×10^9/L at 140 hours.

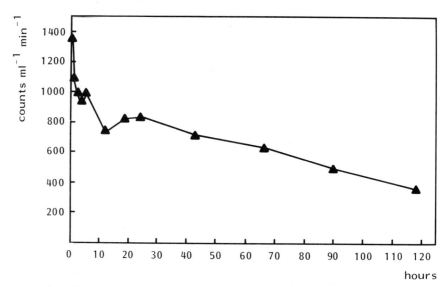

Fig. 3. Blood disappearance curve for lymphocytes labelled with Indium-111 reinjected at a time when the patient was not receiving treatment.

△ whole blood activity.

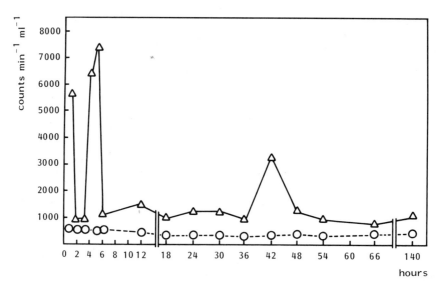

Fig. 4. Blood disappearance curve for lymphocytes labelled with Indium-111 during infusion of T2 anti-idiotype antibody.

△ activity in cell pellet
○ activity in plasma.

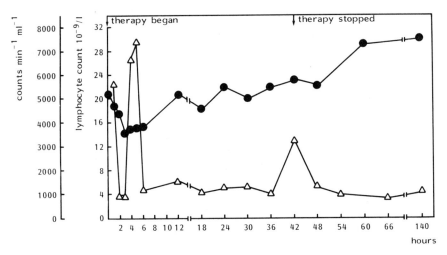

Fig. 5. Relation between the count of circulating lymphocytes labelled with Indium-111 (Δ), counted with a gamma counter and total number of lymphocytes (o) counted with a Coulter counter, in blood during treatment with T2 anti-idiotype antibody.

Gammacamera images. Figure 6 shows some of the gammacamera images obtained when the patient was not being treated. At 30 minutes after reinjection of lymphocytes the uptake in the liver was already appreciable, that in the lungs much less. By 18 hours the uptake by bone marrow in the pelvic girdle could be seen clearly and the retroperitoneal mass of nodes below the liver was delineated. Figure 7 shows dynamic scans taken during the first 30 minutes of infusion of antibody and Figure 8 shows serial scans taken at intervals thereafter. Infusion of the antibody through a cannula in the right superior vena cava began at the same time as the labelled lymphocytes were reinjected into the right arm. Considerable lung activity occured at one minute. This decreased thereafter as the liver activity increased. By 18 minutes three hot spots were visible in the thorax; these corresponded to: the site of insertion of the long linge; a right pleural effusion; and a mass of tumour measuring 5 x 6 cm in the left breast. At two hours the tumour in the mediastinum was visible and by 18 hours uptake by bone marrow was apparent and activity in the retroperitoneal nodes could be seen.

A remarkable effect of the antibody was seen within the lymph node. Serial aspiration biopsies (kindly performed by Dr. P. van Heerde), showed a clear increase in the percentage of cells showing lysis during the period of anti-idiotype therapy. The results are shown in Table 1 and Figure 9.

Treatment With Anti-Idiotype Antibody Improved Monocyte Function

At intervals before, during, and after the treatment period, a monocyte fraction of the peripheral blood was prepared. The functional properties of these monocytes were measured in two ways, by antibody-

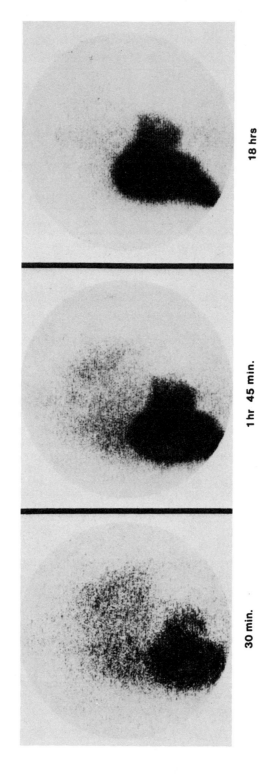

30 min.   **1 hr 45 min.**   **18 hrs**

Fig. 6. Serial gammacamera scans taken when not receiving treatment, after reinjection of indium-111 labelled autologous lymphocytes.

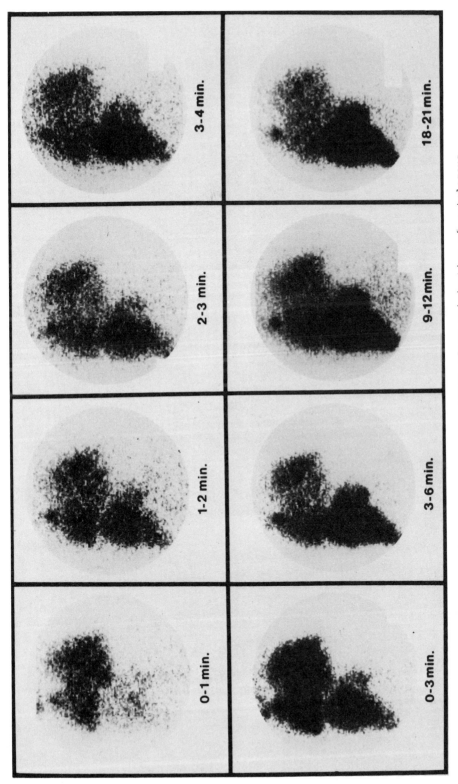

Fig. 7. Dynamic gammacamera scans taken after reinjection of autologous lymphocytes labelled with indium-111 during 30 minutes of infusion with antibody. Injection and start of infusion were simultaneous.

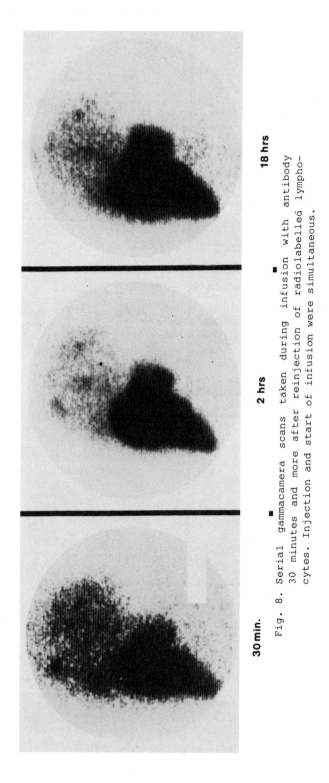

30 min. ■ **2 hrs** ■ **18 hrs**

Fig. 8. Serial gammacamera scans taken during infusion with antibody 30 minutes and more after reinjection of radiolabelled lymphocytes. Injection and start of infusion were simultaneous.

Table 1. The Relation Between Anti-Idiotype Therapy, Cell Lysis, and Recruitment Into S-Phase in Tumour Cells From Lymph Nodes.

Day of Treatment	Amount of Antibody Infused (mg)	No. of Lytic Cells* (%)	No. of Cells in S-Phase* (%)
0	None	1	7.5
20	300 mg	1	Not tested
23	720 mg	1	Not tested
40	None	5 - 10	2
50+	950 mg	20 - 25	14

* These two indices were measured on samples taken at the same time from the same lymph node.

+ Biopsy was done four hours after the end of the infusion, in the other cases, biopsy was done immediately after therapy was finished.

dependent cellular cytotoxicity (ADCC) using lysis of rhesus-positive red cells in the presence of an anti-rhesus antiserum as the test system; and by chemiluminescence induced by phagocytosis. As can be seen from Figure 10, there was a dramatic improvement in the efficiency with which the monocytes killed the cells in the ADCC. After therapy, the monocyte function measured in this system was normal, whereas before treatment it was negligible. The phagocytic function of the monocytes also improved during treatment, but had not reached normal levels by the time therapy was stopped (data not shown).

PATIENT KOS

When therapy was begun, there was massive abdominal disease causing considerable mechanical problems with ascites and pitting oedema to the level of the umbilicus. The patient's poor condition precluded extensive investigation for research purposes. We were concerned that he should have the opportunity to benefit from any therapeutic effect resulting from antibody administration. Our previous experience with patient TOP had been encouraging, since anti-idiotype therapy in that case had been free from side effects.

Before treatment, the patient's serum contained 300 µg idiotypic immunoglobulin per millilitre. This was used to monitor the effects of treatment, since the number of circulating malignant cells was too low (12 % of 2.4×10^9/L lymphocytes) for this purpose. Our plan was to establish as rapidly as possible the amount of antibody required to remove all the free idiotype, to produce saturation of the lymph node cells and an excess of mouse antibody in the circulation.

No untoward effects were seen with a four-hour trial infusion of 5 mg K1 antibody. The first dose of 50 mg was given over one hour. All subsequent treatments were given at a rate of between 100 to 200 mg antibody per hour. It can be seen from Figure 11 that removal of free idiotype from the circulation correlated with the detection of free K1 antibody in the blood. K1 was detectable in the circulation at a level of 20 µg/mL at the end of a total dose of 1,200 mg. This free K1 had nearly disappeard by the following morning when a very small amount of free

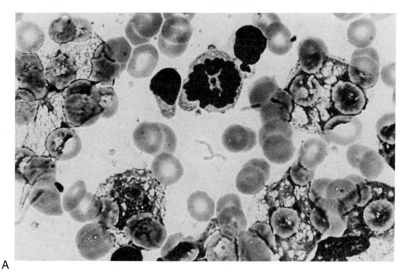

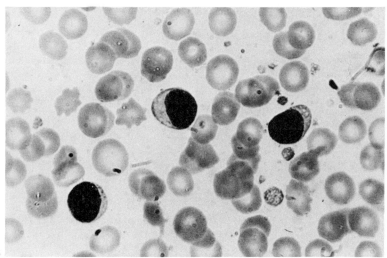

Fig. 9. Cytological smears from punch biopsies of the lymph node of patient TOP. (A) Before treatment began. Three tumour cells can be seen with large nuclei, dense nucleoli, and small amounts of cytoplasm. (B) At the end of the treatment period. Bizarre cells in various stages of lysis can be seen ( ). One cell is undergoing mitosis.

idiotype was again identified. At the end of the 1,200-mg dose, cells from a lymph node were saturated with K1 antibody in vivo. However, tumour cells from the ascites were not coated with K1, and no free K1 was detectable in the ascites.

During a break from treatment over the Christmas period, the level of free idiotype returned almost to the pretreatment level. Three more treatments of 800 mg, 600 mg and 1,000 mg of K1 were given. Transient falls in the level of free idiotype were seen. Free idiotype returned to much higher levels over 24 hours than in the first treatment period. Free K1 antibody was found in the circulation after therapy, reaching a maximum of 5 µg/mL, but it had always disappeared within 24 hours of the end of treatment. The lymph node cells were coated but not saturated with K1 on day 14, but after the 1,100 mg dose on day 15,

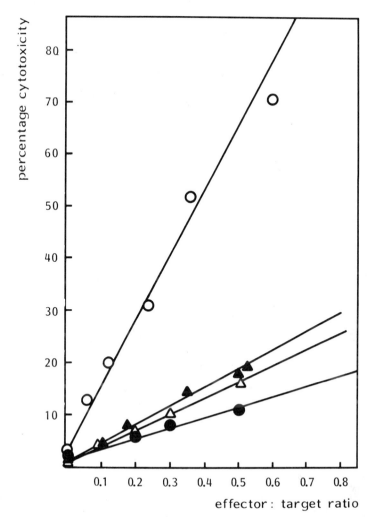

Fig. 10. Cytotoxic activity of monocytes isolated from the blood of patient TOP at different times during treatment; o, before treatment; ▲, day 12; △, day 26; o, day 52.

no anti-idiotype could be detected on the lymph node cells. This time the ascites were weakly stained.

During this period of treatment, there was no evidence of antigen modulation by the antibody. Before and after treatment, all B cells reacted with K1; thus, there was no immunoselection. At no time during or following treatment was there any detectable antibody response to the large amount of mouse protein administered. Serial biopsies showed that there was no significant increase in the number of necrotic cells in the lymph nodes after antibody treatment.

No change in the intra-abdominal tumour was seen on tomography, but the femoral and iliac nodes were about 10 % smaller after 5.8 g of

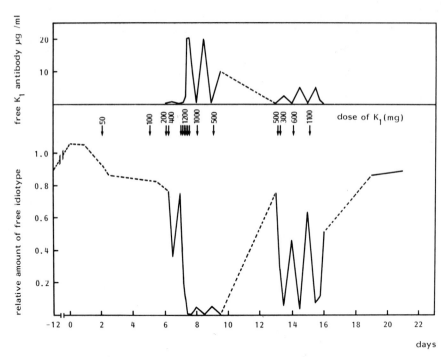

Fig. 11. Relation between the levels of circulating KOS idiotype and unbound K1 antibody in the serum, and the dose of K1 administered. Before treatment, the serum contained 300 µg free idiotype per millilitre.

K1 antibody. Treatment with anti-idiotype antibody was stopped, since increasingly high doses of antibody were required for the same effect. Six weeks later, the patient was started on steroids. He died four months after the anti-idiotype treatment.

DISCUSSION

In this study, we report the results of a trial of treatment with monoclonal anti-idiotype antibodies in two patients with advanced lymphoma in an aggressive phase. We have defined the effect of various schedules of administration on the malignant cells, demonstrating homing of the antibody to the different tumour sites. Transient falls in the level of circulating malignant cells and free antigen were produced, and unbound monoclonal antibody was identified in the serum of both patients. In both cases, a minimal tumour response was seen.

Other investigators have observed that the effect of anti-idiotype antibodies has been limited by antigenic modulation, heterogeneity of the tumour or the development of antibodies against mouse protein (6,8). These problems were not encountered in our patients.

Apart from fever of short duration on one occasion, no toxicity was seen in either patient. This contrasts with the experience of others with the use of monoclonal antibodies in vivo, which has been associated sometimes with temporary alterations in renal function (16-20) and fever, chills, hypotension or respiratory distress (6,19-21).

We have been unable to explain why the treatments given to patient KOS on day 14, 15 and 16 were less effective than those given earlier. The level of free idiotype had not substantially altered, there was no increase in the level of circulating tumour cells and there was no anti-mouse antibody in the serum. Antibody titre and affinity were not different.

The phenomenon of transient falls in the circulating tumour cells resulting from antibody treatment is well documented (6-9,17,18,20-22). In some patients, the level of circulating lymphocytes had risen above the pretreatment baseline by the end of treatment (17,20,21). It is crucially important to know wheter the antibody is merely altering the circulation pathways of the cells by temporarily sequestering them in an extravascular compartment, from which they emerge near or after the end of treatment, or whether the cells are removed altogether from the blood. If radiolabelled cells are injected at the time antibody treatment begins, their fate can be followed; if cells returning when the cell count rebounds during treatment do not carry the radiolabel, they are new cells and not those that disappeared.

The patterns of lymphocyte circulation in the blood of patient TOP were radically altered during the infusions of anti-idiotype antibody. The lack of correlation between the blood disappearance curves during treatment (Fig. 4) and the changes in the blood lymphocyte counts - a transient fall followed by a slow inexorable rise (Fig. 5) - suggests that the antibody permanently removed circulating tumour cells and that the circulation was repopulated by lymphocytes from an extravascular compartment. The rapid repopulation of the blood by tumour lymphocytes, a phenomenon also seen after leukopheresis (7) is consistent with the presence of an extravascular pool of accessible tumour cells in the tissues that can exchange with the blood (23-25). Evidence from gamma-camera images suggests that the cells that disappeared during treatment were cleared by the reticuloendothelial system in the liver. Thus the recirculation pathways were altered because the lymphocytes were destroyed.

The low numer of S-phase cells in the blood and the low tritium-labelled thymidine incorporation suggests that mature, non-proliferating cells migrated into the blood (26). The percentage of S-phase cells was higher in the lymph node several hours after the end of treatment. But since the number of malignant cells in the blood had already increased when the lymph node cells were still in early S-phase, it seems unlikely that the repopulation was caused by this enlarged dividing fraction. Rather, the increase in the number of S-phase cells in the lymph node after treatment can be explained by the fact that reduction of malignant cell density increases the proportion of cells in the growth fraction (27).

The mechanisms by which anti-idiotype antibodies exert their anti-tumour effects are not established. It is known that macrophages mediate the killing of tumour cells in vitro, in the presence of murine IgG2a monoclonal antibody to the tumour (28). Both antibodies T2 and K1 were of the IgG2a subclass. The indium-labelling experiments suggested that the reticuloendothelial system was responsible for the removal of circulating tumour cells, a finding in accordance with that of others (16,18, 21). This would explain the persistence of antibody-coated cells in the ascitic fluid. It may be that the massive tumour cell destruction seen during therapy resulted in a shortage of macrophages, thereby producing an 'effector cell shortage' limiting the benefits of the antibody infusion (29). However, time was allowed between the different treatments to permit recovery of the reticuloendothelial system. Unexpectedly, tests of macrophage and monocyte function showed improvement in patient TOP during the period of treatment, with monocyte function returning to normal (Fig. 10).

In the case reported by Miller et al. (5) a lymph node biopsy before treatment showed infiltration by reactive T cells, more than two thirds of which were 'helper' cells. A biopsy of the regressing lymphoma lesions after therapy showed that there were large numbers of macrophages and activated T cells present (30). In our patients, no change was seen in the low numbers of T cells and macrophages in the nodes. Nevertheless in Figure 9 the lysis of tumour cells in a lymph node induced by treatment is clearly seen and is suggestive of a direct lethal effect of the antibody on the malignant cell population. It is unlikely that this effect was mediated by complement, since in vitro the antibodies were not cytotoxic with human complement. Studies in vivo showed there was no alteration in the complement status of either patient, and human complement (C3c and C3d) was not detectable on the surface of antibody-coated tumour cells during treatment. Levy and colleagues have recently observed that the kinetics of response in one of their patients treated with an anti-idiotype antibody also suggested a direct effect of the antibody on the tumour cells (6).

The development of anti-idiotype antibodies has allowed examination of some of the fundamentals of tumour biology. Using these highly specific tools it has been shown that some follicular lymphomas have more than one clone of cells, each expressing a different idiotype (31). In addition somatic mutations in the V region can lead to idiotype heterogeneity within a given B cell clone (32,33). In some cases it may be necessary to use a battery of antibodies each directed against a different idiotype if tumour eradication is to be achieved. An essential question requiring further investigation is raised by the studies of Kubagawa et al. (34). They have shown that clonal involvement in several different B cell malignancies could be traced back to the pre-B cell stage, before the expression of membrane Ig. If this is the case, malignant precursor cells will escape from the effect of anti-idiotype antibodies. It is of obvious importance for future therapeutic applications of anti-idiotype antibodies to establish the nature of the tumour stem cell in B cell lymphoma.

ACKNOWLEDGEMENTS

We thank Dr. R. Somers and Dr. W. W. ten Bokkel Huinink for allowing us to study their patients and the staff of the Divisions of Immunology and Haematology and the Clinical Research Unit of the Netherlands Cancer Institute for their help.

REFERENCES

1. R. Levy, R. Warnke, R. F. Dorfman, and J. Haimovich, The monoclonality of human B cell lymphomas, J. Exp. Med., 145:1014 (1977).
2. P. J. Fialkow, E. Klein, G. Klein, and S. Singh, Immunoglobulin and glucose-6-phosphate dehydrogenase as markers of cellular origin in Burkitt lymphoma, J. Exp. Med., 138:89 (1973).
3. S. E. Salmon, and M. Seligmann, B. cell neoplasia in man, Lancet, 2: 1230 (1974).
4. K. R. Schroer, D. E. Briles, J. A. Van Boxel, and J. M. Davie, Idio-

typic uniformity of cell surface immunoglobulin in chronic lymphocytic leukaemia. Evidence for monoclonal proliferation, J. Exp. Med., 140:1416 (1974).
5. R. A. Miller, D. G. Maloney, R. Warnke, and R. Levy, Treatment of B cell lymphoma with monoclonal anti-idiotype antibody, N. Engl. J. Med., 306:517 (1982).
6. T. C. Meeker, J. Lowder, D. G. Maloney, R. A. Miller, K. Thielemans, R. Warnke, and R. Levy, A clinical trial of anti-idiotype therapy for B cell malignancy, Blood, 65:1344 (1985).
7. T. J. Hamblin, A. K. Abdul-ahad, J. Gordon, F. K. Stevenson, and G.T. Stevenson, Preliminary experience in treating lymphocytic leukaemia with antibody to immunoglobulin idiotypes on the cell surfaces, Br. J. Cancer, 42:495 (1980).
8. J. Gordon, A.K. Abdul-Ahad, T. J. Hamblin, F. K. Stevenson, and G.T. Stevenson, Mechanisms of tumour cell escape encountered in treating lymphocytic leukaemia with anti-idiotype antibody, Br. J. Cancer, 49:547 (1984).
9. F. R. Macbeth, F. K. Stevenson, G. T. Stevenson, and J.M.A. Whitehouse, Anti-idiotype antibody therapy of patients with non-Hodgkin's lymphoma. Proceedings of the Second European Conference on Clinical Oncology and Cancer Nursing, Amsterdam, p. 75 (abstr.) (1983).
10. E. M. Rankin, and A. Hekman, Mouse monoclonal antibodies against the idiotype of human B-cell non-Hodgkin's lymphomas: Production, characterization and use to monitor the progress of disease, Eur. J. Immunol., 14:1119 (1984).
11. E. M. Rankin, A. Hekman, R. Somers, and W. W. ten Bokkel Huinink, Treatment of two patients with B cell lymphoma with monoclonal anti-idiotype antibodies, Blood, 65:1373 (1985).
12. L. A. Smets, J. Taminiau, K. Hahlen, F. De Waal, and H. Behrendt, Cell kinetic responses in childhood acute nonlymphocytic leukaemia during high-dose therapy with cytosine arabinoside, Blood, 61:79 (1983).
13. C. G. Figdor, W. S. Bont, I. Touw, J. de Roos, E. S. Roosnek, and J. E. De Vries, Isolation of functionally different human monocytes by counterflow centrifugation elutriation, Blood, 60:46 (1982).
14. J. E. De Vries, J. Mendelsohn, and W. S. Bont, The role of target cells, monocytes and Fc-receptor bearing lymphocytes in human spontaneous cell mediated cytotoxicity and antibody-dependent cellular cytotoxicity, J. Immunol., 125:396 (1980).
15. E. M. Rankin, A. Hekman, C. A. Hoefnagel, and M. R. Hardeman, Dynamic studies of lymphocytes labelled with indium-111 during and after treatment with monoclonal anti-idiotype antibody in advanced B cell lymphoma, Br. Med. J., 289:1097 (1984).
16. R. A. Miller, D. G. Maloney, J. McKillop, and R. Levy, In vivo effects of murine hybridoma monoclonal antibody in a patient with T cell leukaemia, Blood, 58:78 (1981).
17. R. A. Miller, and R. Levy, Response of cutaneous T cell lymphoma to therapy with hybridoma monoclonal antibody, Lancet, 2:226 (1981).
18. L. M. Nadler, P. Stashenko, R. Hardy, W. D. Kaplan, L. N. Button, D. W. Kufe, K. H,. Antman, and S. F. Schlossman, Serotherapy of a patient with a monoclonal antibody directed against a human lymphoma-associated antigen, Cancer Res., 40:3147 (1980).
19. A. B. Cosimi, R. C. Burton, R. B. Colvin, G. Goldstein, F. J. Delmonico, M. P. LaQuaglia, N. Tolkoff-Rubin, R. H. Rubin, J. T. Herrin, and P. S. Russell, Treatment of acute renal allograft rejection with OKT3 monoclonal antibody, Transplantation, 32: 535 (1981).
20. R. A. Miller, A. R. Oseroff, P. T. Stratte, and R. Levy, Monoclonal

antibody therapeutic trials in seven patients with T cell lymphoma, Blood, 62:988 (1983).
21. R. O. Dillman, D. L. Shawler, R. E. Sobol, H. A. Collins, J. C. Beauregard, S. B. Wormsley, and I. Royston, Murine monoclonal antibody therapy in two patients with chronic lymphocytic leukaemia, Blood, 59:1036 (1982).
22. E. D. Ball, G. M. Bernier, G. G. Cornwell, O. R. McIntyre, J. F. O'Donnell, and M. W. Fanger, Monoclonal antibodies to myeloid differentiation antigens: In vivo studies of three patients with acute myelogenous leukemia, Blood, 62:1203 (1983).
23. J. Manaster, J. Fruhling, and P. Stryckmans, Kinetics of lymphocytes in chronic lymphocytic leukemia. 1. Equilibrium between blood and a readily accessible pool, Blood, 41:425 (1973).
24. H. Theml, F. Trepel, P. Schick, W. Kaboth, and H. Begemann, Kinetics of lymphocytes in chronic lymphocytic leukemia: Studies using continous ^{3}H-thymidine infusion in two patients, Blood, 42:623 (1973).
25. J. L. Scott, R. McMillan, J. V. Marino, and J. G. Davidson, Leukocyte labelling with ^{51}Chromium. IV. The kinetics of chronic lymphocytic leukemia lymphocytes, Blood, 41:155 (1973).
26. T. S. Zimmerman, H. A. Godwin, and S. Perry, Studies of leukocyte kinetics in chronic lymphocytic leukemia, Blood, 31:277(1968).
27. A. M. Mauer, S. B. Murphy, A. B. Hayes, and G. V. Dahl, Scheduling and recruitment in malignant cell populations, in: "Growth Kinetics and Biochemical Regulation of Normal and Malignant Cells", B. Drewinko, R. M. Humphrey, eds., Williams and Wilkins, Baltimore (1977).
28. Z. Steplewski, M. D. Lubeck, and H. Koprowski, Human macrophages armed with murine immunoglobulin IgG2a antibodies to tumours destroy human cancer cells, Science, 221:865 (1983).
29. H. S. Shin, J. S. Economou, G. R. Pasternack, R. J. Johnson, and M. L. Hayden, Antibody mediated suppression of grafted lymphoma: IV. Influence of time of tumour residency in vivo and tumour size upon the effectiveness of suppression by syngeneic antibody, J. Exp. Med., 144:1274 (1976).
30. R. Levy, and R. A. Miller, Tumor therapy with monoclonal antibodies. Fed. Proc., 42:2650 (1983).
31. J. Sklar, M. L. Cleary, K. Thielemans, J. Gralow, R. Warnke, and R. Levy, Biclonal B cell lymphoma, N. Engl. J. Med., 311:20 (1984).
32. M. Raffeld, L. Neckers, D. L. Longo, and J. Cossman, Spontaneous alteration of idiotype in monoclonal B cell lymphoma, N. Engl. J. Med., 312:1653 (1985).
33. T. Meeker, J. Lowder, M. L. Cleary, S. Stewart, R. Warnke, J. Sklar, and R. Levy, Emergence of idiotype variants during treatment of B cell lymphoma with anti-idiotype antibodies, N. Engl. J. Med., 31:1658 (1985).
34. H. Kubagawa, M. Mayumi, W. D. Gathings, J. F. Kearney, and M. D. Cooper, Extent of clonal involvement in B cell malignancies, in: "Leukaemia Research: Advances in Cell Biology and Treatment", S. B. Murphy and J. R. Gilbert, eds., Elsevier, Amsterdam p. 65 (1983).,

HUMAN MONOCLONAL ANTIBODIES DIRECTED AGAINST

HuMTV ANTIGENS FROM HUMAN BREAST CANCER CELLS

I. Keydar[*], I. Tsarfaty[*], N. Smorodinsky[*], E. Sahar[**],
Y. Shoenfeld[***] and S. Chaitchik[****]

Department of Microbiology and Biotechnology[*], Tel Aviv University; Dept. of Medicine D, Beilinson Medical Center, Petach Tikva[***] and Tel Aviv Medical Center Department of Oncology[****]

INTRODUCTION

Tumor specific markers for human malignancies are invaluable for diagnostic, prognostic as well as therapeutic purposes. The study reported here focuses on the possible importance of viral antigens as markers for human breast carcinomas. Breast cancer is one of the leading death causes in women. For the last 50 years investigators have searched for animal models to elucidate the biology of this human neoplasia. The most widely studied one has been the mouse mammary tumor.

The viral etiology of mouse mammary carcinomas has long been established dating back to the observations of Bittner (1) who demonstrated that the particles present in mother's milk are responsible for the mediation of the mouse malignancy. Since then, numerous studies have described the route of transmission, the structure of the virus and its prevalence among different inbred strains of mice (2). Until recently, however, no clues were provided as to how the mouse mammary tumor virus (MMTV) could transform mammary gland cells into malignant tumors (3).

The similarity between the mouse mammary disease and the human one focused attention on the possible involvement of human mammary tumor retroviruses like particles in human neoplasia.

Over the past few years, more and more evidence for the association of viral proteins and viral nucleic acid sequences with human mammary carcinomas has accumulated (4). Furthermore, a DNA recombinant clone has been isolated from a library of normal human cellular DNA that hybridizes with molecularly cloned MMTV genomic DNA, providing the evidence for the existence of MMTV related sequences within the human genome (5).

The fact that some homology has been established between the RNA's of the human and the murine tumor particles suggested that an antigenic relation might conceivable exist (4). We and others, have identified in paraffin sections of human breast cancer biopsies an antigen related to the group-specific antigen gp52 (a 52000 dalton envelope glycoprotein of MMTV) by the indirect immunoperoxidase technique (6). Moreover, in a study of breast cancer patients in the Israeli population we have recently shown that there is a correlation between the ability to detect the gp52 cross-reacting antigen in breast tumor sections and the severity of disease (7).

The fact that the polypeptide rather than the polysaccharide portion of the pg52 is responsible for the immunological reactivity with human breast cancer antigens adds additional significance to the biological similarities between the human and the murine mammary neoplasias (8).

A cell line (T47D) has been established in our laboratory from the pleural effusion of a patient with infiltrating duct carcinoma of the breast (9). The pg52 cross-reacting antigen has been detected on the surface of T47D cells (10) as well as in the MCF-7 breast cancer line (11). In addition, these cell lines have been shown to possess DNA sequences with partial homology to the genome of MMTV (12). The T47D cells release retrovirus-like particles (T47D-HuMTV), whose production can be increased by addition of estrogen and progesterone to the tissue culture growth media. The particles, isolated from the culture supernatants of T47D cells have the biochemical characteristics of a retrovirus; they band at a density of 1.16-1.21 g/cc in an equilibrium sucrose gradient; RNA extracted from these particles consistently gave a peak of about 70S in sedimentation gradients; reverse transcriptase activity, giving rise to 70S RNA/DNA complexes, was identified in these particles (13). The presence of gp52-MMTV crossreacting antigen(s) was recently demonstrated in both retrovirus-like particles and soluble proteins released by the T47D cells (14). These particles do not react with sera directed against gp36 and p27 of MMTV. Furthermore, it has been demonstrated by radioimmuno-competition assays that the homology between MMTV and HuMTV is quite limited, as judged from the different slopes and the displacement of the curves exhibited by the human particles (13).

Recent studies, accomplished in our laboratory, demonstrated that the cross-reacting antigenic determinant is found on glycosylated polypeptides with apparent molecular weights of about 68,000 and 60,000 daltons. These glycoproteins differ from gp52 not only in size, but also by their charge heterogeneity and polypeptide profiles after partial proteolysis. The antigenic relationship is very limited, since only a small fraction of the entire anti gp52 antibody population recognizes T47D-HuMTV soluble proteins (14).

Mouse monoclonal antibodies reactive with human breast carcinoma were produced against several immunogens such as breast cancer cell lines, fat globules-membranes, mammary carcinoma metastatic cells and lymph nodes of human breast carcinoma patients. These antibodies react with normal human breast tissue and with some tumors of non-breast origin. Some of them recognize also normal tissues (15).

Several human monoclonal antibodies against breast cancer associated antigens are available to date. One of them, produced by a mouse-human hybridoma reacts with breast tumors, as well as other tumors. This antibody has been shown to detect breast tumors and metastatic tumor cells in nude mice (16).

Another human monoclonal antibody is produced by a human-human hybridoma and recognizes MMTV structural proteins (17).

The study reported here was aimed at establishing a human monoclonal antibody (MoAb) reacting with the putative human mammary retroviruses produced by a human breast carcinoma cell line.

RESULTS AND DISCUSSION

T47D subline Cl 11 cells were grown in suspension in RPMI-1640 medium containing 10 % fetal calf serum, insulin (0.2 IU/ml), 2 mM glutamine and antibiotics. When grown in suspension for several days, the cells form aggregates that can reach 0.5-2.0 cm in diameter (Fig. 1).

Paraffin blocks were prepared from these aggregates. Five micron-thick sections were cut and stained by haematoxylin-eosin. As can be seen from Fig. 2, the cells mimic an intraductal adenocarcinoma of the breast with a central core of necrotic tumor cells.

Viral production by the T47D cells can be enhanced by addition of steroid hormones to the growth medium. Therefore, the cells were grown in medium lacking serum, to which oestrogen and progesterone were added. In parallel, cells were grown in the presence of serum without hormones. For virus detection ^{3}H-uridine was added to the medium. As shown in Fig. 3, the virus was obtained in large quantities in the presence of the hormones.

The viral region of the isopycnic sucrose gradient corresponding to a density of 1.16-1.22 g/cc was collected, centrifuged and the pellets sectioned for electron microscopy. The morphology of the HuMTV viral particles is presented in Fig. 4. The particles resemble retroviruses type D more than type C or B viruses.

As previously reported, the viral particles contain an antigen that immunologically cross-reacts with MMTV-gp52 (13). This antigen is also found on the surface of T47D cells (10), in some human breast cancer biopsies (4), as well as on cells from metastases of breast cancer patients (6). Therefore, we have chosen the HuMTV antigens as a potential breast cancer marker against which human MoAbs should be prepared. We were, however, faced with the problem of choosing the proper source for the lymphocytes to be used as fusion partners. Since serological investigations have identified antibodies in human sera that react with MMTV-gp52 (18), we decided to investigate the possibility that breast

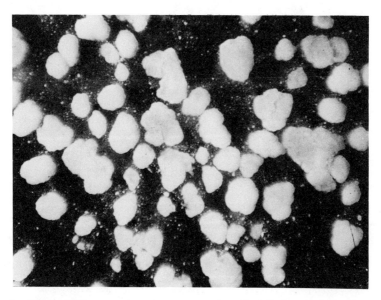

Fig. 1: T47D cells growing in suspension for 7 days, (x5; inverted microscope).

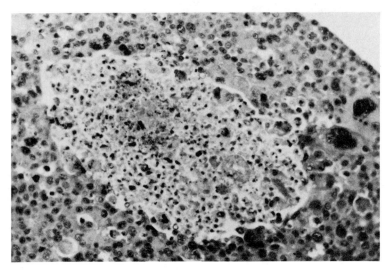

Fig. 2: Paraffin section of T47D aggregates. Haematoxylin stain (x200).

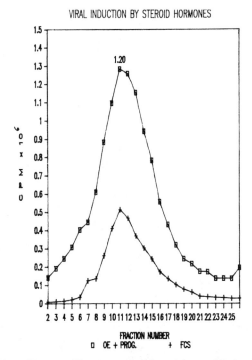

Fig. 3: Sucrose density gradient centrifugation of viral particles collected from cultures supplemented with oestrogen (10^{-8} M) and progesterone (10^{-9} M) (□). Parallel cultures were grown in medium containing only 10 % FCS (+). Twenty four hours prior to medium harvesting for virus purification, ^{3}H-uridine (50 µCi/ml) was added. Fractions were collected, TCA precipitated and counted in a Tricarb β-Counter.

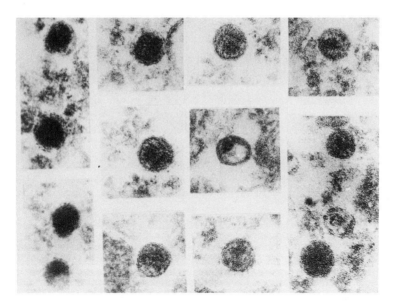

Fig. 4: Electron microscopy of HuMTV banded particles. The particles have a diameter of 100 nm. The viral pellets were fixed for 20 min. with 1 % osmium tetroxide in malonate buffer, dehydrated, and embedded in Epon. Thin sections on grids were stained overnight with a saturated solution of uranyl acetate in 30 % alcohol, post-stained for 15 min. with 0,5 % lead citrate, and observed in a Jem T-7 electron microscope.

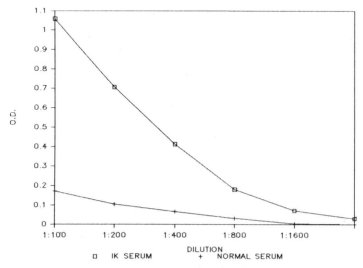

Fig. 5: Antibody titer against HuMTV was assayed by the ELISA test. IK serum (□). Normal human serum (NHS) (+).

cancer patients or laboratory investigators who have been working with the T47D cells for extended periods of time contain antibodies directed not only against MMTV, but also against HuMTV. One of our investigator's (IK) serum contained antibodies in high titer against both MMTV and HuMTV. Fig. 5 shows the level of antibodies found in IK's serum, which react with HuMTV as compared to normal human sera.

Before embarking on the preparation of monoclonal antibodies using IK's lymphocytes, we tested the ability of IK's antibodies to recognize different antigens such as MMTV, HuMTV and antigens derived from the pleural effusions (PE) of a patient with breast cancer and PE of a patient with carcinoma of the lung.

The testing was performed by the radioimmuno-precipitation assay (RIP). An example of the results obtained with the serum of the healthy investigator IK and with the serum of a breast cancer patient is given in Table 1.

The results obtained show that both the breast cancer patient and the IK sera recognize MMTV, HuMTV and an antigen present in the pleural effusion of a breast cancer patient. In contrast, the pleural effusion of a patient with carcinoma of the lung does not contain antigens recognized by the sera tested.

The presence of antibodies in IK's peripheral blood, indicated that it must contain B lymphocytes that produce these antibodies. Therefore, we have chosen to use IK as the donor of lymphocytes that will undergo hybridization with the NS0 mouse myeloma cells in order to obtain human monoclonal antibodies that will recognize HuMTV. Thirty ml of blood were drawn, lymphocytes were separated by Ficoll-Hypaque gradients and then tested for their ability to bind the HuMTV antigen, using indirect immunofluorescence detected by flow cytometry (Fig. 6).

Two populations of lymphocytes were detected: 1) small lymphocytes which comprised 90 % of the total population; 2) large lymphoblastoid cells (approximately 10 %). Eight to nine percent of the cells contained surface immunoglobulin and approximately 1 % of the total lymphocytes bound the HuMTV.

In vitro sensitization of the lymphocytes with a combination of pokeweed mitogen and HuMTV antigens showed that it is possible to increase the fraction of lymphocytes producing immunoglobulins that recognize the HuMTV antigens. These results will be reported elsewhere.

Table. 1. RIP with IK Serum and a Breast Cancer (BrCa) Patient Serum

^{125}J Antigens	CPM	
	BrCa	IK Serum
MMTV	25341	31354
HuMTV	49030	26030
PE (Breast Cancer)	141069	157722
PE (Lung Cancer)	271	7339

Net CPM = CPM of sample - CPM of normal serum

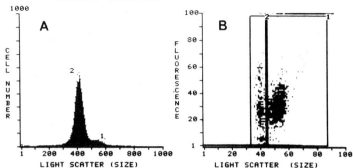

Fig. 6: Flow cytometric analysis of cell size and of the expression of immunoglobulins recognizing HuMTV antigens on the surface of IK's PBL. A: Cell size (forward light scatter) analysis. B: Fluorescent antigen binding by the cells vs. cell size. Cells were labelled with 100 µg/ml of biotinylated HuMTV antigens, washed twice and labelled with 100 µg/ml of Avidin-FITC (Sigma).

Hybridization of IK Lymphocytes with NS0 Myeloma Cells

IK lymphocytes were separated by Ficoll-Hypaque gradients, washed twice in RPMI-1640 medium containing 10 % FCS and then fused with the mouse myeloma NS0 cells as described in Fig. 7.

Cells were seeded into either 96-well microtiter plates (Falcon) or into 24-well microtiter plates. Ninety percent of the cultures produced detectable amounts of human immunoglobulin as tested by the ELISA test. As much as 14-25 % of them recognized the HuMTV antigens, depending on the inocula seeded (Table 2).

Five of the cultures, containing high titers of antibodies to the HuMTV, were chosen for subcloning (Table 3).

Table 2. Efficiency of Cloning

No. of		Colony Containing Wells		Hybridoma Secreting		
Wells	Cells/Well	No.	%	HuIg %	HuIg No	anti HuMTV %
288	2×10^5	209	73	90	30	14
20	2×10^6	20	100		5	25

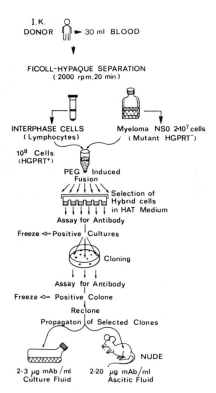

Fig. 7: Hybridization protocol

Since the donor of the cells used for fusion was immunized with influenza virus vaccine prior to the fusion with NSO, we tested the hybridoma subclones for reactivity to this virus. As can be seen from Table 3, five out of six subclones presented produced immunoglobulin that recognized HuMTV and did not recognize the Influenza virus. One clone (1.54) recognized the Influenza virus and did not react with HuMTV. In the following sections we shall describe results obtained with one monoclonal antibody - designated 4.6/6.

Table 3. Specificity of the Subclones

Subclone No.	HuIg	Anti HuMTV Activity	Anti Non Relevant Virus (Influenza Virus) Activity
1.49	+	+	-
1.52	+	+	-
1.54	+	-	+
4.6/6	+	+	-
4.11	+	+	-
5.2	+	+	-

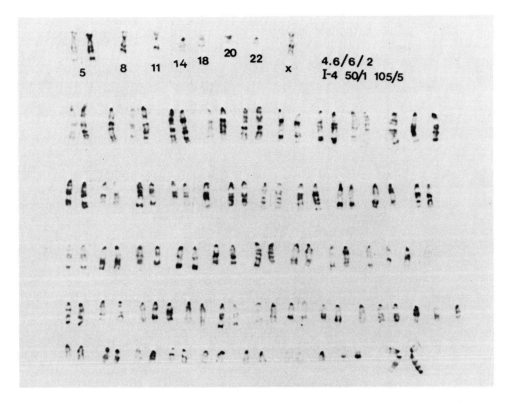

Fig. 8: Trypsin-Giemsa banded Karyotype of 4.6/6 cells. The human chromosomes are shown on top of the figure.

Characterization of the Hybridoma 4.6/6 Cells

One of the main problems encountered when fusing mouse and human cells is the instability of hybrid cells caused by loss of human chromosomes. Hybridoma 4.6/6 cells were, therefore, examined cytogenetically. All 46 metaphases tested contained human chromosomes as demonstrated in Fig. 8.

Chromosome No. 14 (responsible for human immunoglobulin heavy chain) and chromosome No. 22 (responsible for the light chain) were found in the parental hybridoma, as well as in the tested subclones.

The 4.6/6 cells were characterized by flow cytometry for:

a) The expression of human immunoglobulin on their surface;
b) Their ability to bind the HuMTV antigen.

Fig. 9 depicts the results obtained with 4.6/6 cells stained with FITC-labelled rabbit anti-human IgG. The results show that the mean fluorescence for the 4.6/6 cells was 5 times as high as that obtained with the control NS0 cells. The right hand side of the figure shows that similar results are obtained, when biotinylated HuMTV was used.

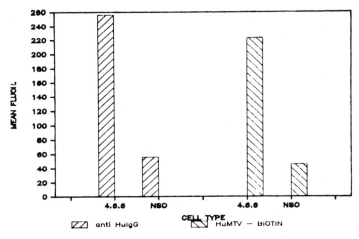

Fig. 9: The level of expression of HuIgG and of anti-HuMTV on the surface of 4.6/6 hybridoma cells, as determined by flow cytometry.

Characterization of the Human Monoclonal Antibody 4.6/6

The antibodies secreted by hybridoma 4.6/6 are IgG subclass 3. The amount of human IgG secreted by this hybridoma is 2-3 µg/ml of growth medium as measured by ELISA and RIA.

This monoclonal antibody recognizes MMTV as well as HuMTV, and does not recognize the retroviruses AMV (Avian Myeloblastosis Virus) and SSV (Simian Sarcoma Virus) (Table 4, Fig. 10). Table 4 describes Radio-immunoprecipitation tests (RIP) performed with the 4 viruses labelled with ^{125}Iodine.

As can be seen from Fig. 10, the IK serum recognizes at least 4 polypeptides of size 67, 50, 23 and 14 kilodaltons. The monoclonal antibody 4.6/6 recognizes the 67K dalton polypeptide of HuMTV. None of the other viruses were precipitated by the monoclonal 4.6/6 or by the normal Human IgG.

Table 4. Radioimmunoprecipitation of Viral Proteins with Human Serum and Human MoAb.

Antigen	IK Serum	Human MoAb 4.6/6
MMTV	28327	23682
HuMTV	14301	31153
SSV	0	0
AMV	3508	4419

Net CPM: CPM of sample - CPM of NHS; SSV - Simian Sarcoma Virus; AMV - Avian Myeloblastosis Virus.

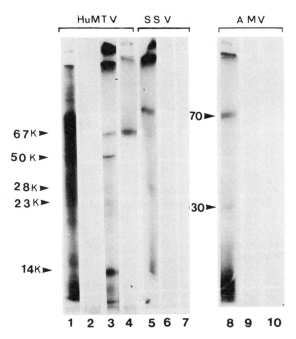

Fig. 10: Results obtained by SDS-Polyacrylamide gel electrophoresis using the precipitates described in Table 4.
Lane 1, 5 and 8 represent the ^{125}I-iodinated antigens; Lane 2 represents radioimmunoprecipitation with a normal human IgG; Lanes 3, 6 and 9 are the results obtained with IK serum. Lanes 4, 7 and 10 were obtained using antibody 4.6/6.

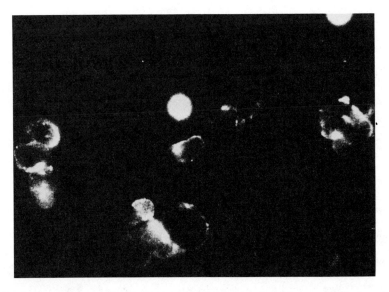

Fig. 11: T47D cells stained with 4.6/6 biotinylated monoclonal antibodies and Avidin-FITC.

The 4.6/6 antibody was biotinylated and reacted with the T47D live cells. The antibody-antigen complex was detected by Avidin conjugated to fluorescein 5-isothiocyanate. Fig. 11 demonstrates the detection of HuMTV antigen on the surface of the T47D cells. As control, HeLa cells and kidney epithelial cells were used. Neither the HeLa nor the kidney cells showed any fluorescence.

Indirect immunohistochemical staining of T47D paraffin sections showed a positive reaction in the cytoplasm of the cells (data not shown).

CONCLUSIONS

In conclusion, a human monoclonal antibody designated 4.6/6 was produced by fusion of the mouse myeloma cell line NS0 with peripheral blood lymphocytes of a healthy investigator (IK), who, for the last ten years, has been working with the T47D cell line. The donor's serum was found to contain a high titer of antibodies directed against viral particles (HuMTV) produced by the T47D cell line.

The 4.6/6 monoclonal antibody is a human IgG3. The human contribution to the hybridoma was confirmed by chromosomal analysis. These monoclonal antibodies recognize antigen(s) on the T47D cells and they react with the HuMTV secreted by the T47D cells. The results were obtained by using ELISA, RIA, radioimmunoprecipitation and by flow cytometric measurements.

It is hoped that these human monoclonal antibodies will be useful in the future for diagnosis, as well as therapy of human breast carcinomas.

REFERENCES

1. J. J. Bittner, Some possible effect of nursing on the mammary gland tumor incidence of mice, Science, 184:162 (1936).
2. P. Bentvelzen, and J. Hilgers, The murine mammary tumor virus, in: "Viral Oncology", G. Klein, ed., Raven Press, New York, pp. 311-355 (1980).
3. N. E. Hynes, B. Groner, and R. Michalides, MMTV: Transcriptional control and involvement in tumorogenesis, Adv. in Cancer Res., 41: 155-184 (1984).
4. S. Spiegelman, I. Keydar, R. Mesa-Tejada, T. Ohno, M. Ramanarayanan, R. Nayak, J. Baush, and C. Fenoglio, The presence and clinical implications of a viral-related protein in human breast cancer, Conference on cell proliferation, Cold Spring Harbor, 7: 1149-1167 (1980).
5. R. Callahan, W. Drohan, S. Tronick, and J. Schlom, Detection and cloning of human DNA sequences related to the mouse mammary tumor virus genome, Proc. Natl. Acad. Sci. USA, 79: 5503-5507 (1982).
6. R. Mesa-Tejada, I. Keydar, M. Ramanarayanan, T. Ohno, C. Fenoglio, and S. Spiegelman, Immunohistochemical detection of a cross-reacting virus antigen in mouse mammary tumors and human breast carcinomas, The Journal of Histochem. and Cytochem., 26: 532-541 (1978).
7. I. Keydar, G. Selzer, S. Chaitchik, M. Hareuveni, S. Karby, and A. Hizi, A viral antigen as a marker for the prognosis of human breast cancer, Eur. J. Cancer Clin. Oncol., 18: 1321-1328 (1982).
8. T. Ohno, R. Mesa-Tejada, I. Keydar, M. Ramanarayanan, J. Bausch, and S. Spiegelman, Human breast carcinoma antigen is immunologically related to the polypeptide of the group-specific glycoprotein of MMTV, Proc. Natl. Acad. Sci. USA, 76: 2460-2464 (1979).

9. I. Keydar, L. Chen, S. Karby, F. R. Weiss, J. Delarea, M. Radu, S. Chaitchik, and H. J. Brenner, Establishment and characterization of a cell line of human breast carcinoma origin, Europ. J. Cancer, 15: 659-670 (1979).
10. I. Keydar, Y. Ben-Shaul, and A. Hizi, Expression of an MMTV cross-reacting antigen on the cell surface of T47D, a human mammary cell line, Cancer letters, 17: 37-44 (1982).
11. N. S. Lang, C. M. McGrath, and P. Furmanski, Presence of an MMTV-related antigen in human breast carcinoma cells and its absence from normal mammary epithelial cells, J. Natl. Cancer Inst., 61: 1205-1208 (1978).
12. F. E. B. May, B. R. Westley, H. Rochefort, E. Buetti, and H. Diggelmann, MMTV-related sequences are present in human DNA, Nucleic Acids Res., 11: 4127-4139 (1983).
13. I. Keydar, T. Ohno, R. Nayak, R. Sweet, F. Simoni, F. Weiss, S. Karby, and S. Spiegelman, Properties of retrovirus-like particles produced by a cell line established from a human breast carcinoma: Immunological relationship to MMTV viral proteins, Proc. Natl. Acad. Sci. USA, 81: 4188-4192 (1984).
14. N. Segev, A. Hizi, F. Kirenberg, and I. Keydar, Characterization of a protein released by the T47D cell line immunologically related to the major envelope protein of the MMTV, Proc. Natl. Acad Sci. USA, 82: 1531-1535 (1985).
15. J. Schlom, D. Colcher, P. H. Hand, F. Greiner, D. Wunderlich, M. Weeks, P. B. Fisher, P. Noguchi, S. Pestka, and D. Kufe, Monoclonal antibodies reactive with breast tumor-associated antigens, Adv. in Cancer Res., 43: 143-173 (1985).
16. J. Schlom, D. Wunderlich, and Y. A. Teramoto, Generation of human monoclonal antibodies reactive with human mammary carcinoma cells, Proc. Natl. Acad. Sci. USA, 77: 6841-6845 (1980).
17. Y. Shoenfeld, A. Hizi, R. Tal, G. Lavie, C. Mor, S. Shteren, Z. Mamman, F. Pinkhas, and I. Keydar, Human monoclonal antibodies derived from lymph nodes of a patient with breast carcinoma react with MuMTV polypeptides, Submitted for publication.
18. M. C. Poon, M. Tomana, and W. Nedermeier, Serum antibodies against MMTV-associated antigen detected nine months before appearance of a breast carcinoma, Annals of Int. Medicine, 98: 937-938 (1984).

RECENT ADVANCES IN THYMIC HORMONE RESEARCH

Marcelo B. Sztein and Allan L. Goldstein

Departments of Medicine and Biochemistry, The George
Washington University School of Medicine
Washington DC, U.S.A.

INTRODUCTION

The essential role of the thymus gland in the development of a competent immune system was established by experiments in which it was found that the surgical removal of the thymus within 24 hours after birth, resulted in a dramatic impairment of the immune function (1,2). Neonatal thymectomy was also accompanied by a high incidence of early death due primarily to overwhelming infections. Additionally, recent evidence suggests that the thymus gland might be critical not only in mediating lymphocyte maturation and differentiation, but also in maintaining a normal immune balance by modulating normal immune response (3-12).

It is now widely accepted that the thymus exerts these activities by at least two mechanisms: a) by providing a microenvironment in which the lymphoid cells derived from the bone marrow mature and differentiate (13) and b) by secreting a number of polypeptides, collectively known as thymic hormones (TH), which have been shown to be involved in both, induction of lymphocyte differentiation and modulation of normal immune responses (3-5,14-17).

Reports from several laboratories (13,18-21) have indicated that TH are produced by thymic epithelial cells, and that at least some of them are released into the blood stream were they can be measured, indicating that they might be active not only in thymic microenvironment, but also in other areas of the economy as well.

In spite of the large amount of data available, the understanding of the role of TH in the modulation of the immune responses, and its relationship with other systems is in its early stages. For example, the production of peptides similar in structure and function to TH has been described in other organs, such as brain, kidney, liver and spleen (22-24). This suggests that the endocrine, nervous and immune system might share a number of mediators, which have recently been called immunotransmitters (25).

In this paper we will not attempt to review the extensive literature of the thymus and TH, and will refer the reader to a number of recently published extensive reviews (3-5,14-17,26-28).

We will focus on the actual status of the biochemistry and biological activities of the well characterized thymic hormones and discuss the evidence supporting the concept that they might play a key role in the interactions between immune, endocrine and central nervous systems. Finally, we will discuss the evidence suggesting that TH might be of clinical relevance in the treatment of a number of diseases associated with abnormalities of the immune system and in the control of neoplasia, which is the theme of this monograph.

BIOCHEMICAL PROPERTIES OF WELL DEFINED THYMIC HORMONES

Although some of the biological activities of thymic extracts had been described as early as the 1930's (29), to date only a limited number of thymic preparations with thymic hormone-like activity have been well characterized. For detailed information, the reader may consult a number of recent reviews (4,5,14,15,17).

Some of the TH are being used in clinical settings. Within this group, some are partially purified preparations, including thymosin fraction 5 (TF5) (30,31), thymostimulin (TS) (TP1) (32) and thymic factor X (TFX) (27,33), while some are homogeneous preparations, like thymosin alpha 1 (Tα1) (34), thymopoietin (35) and thymulin (FTS) (36,37).

Some polypeptides derived from the partially purified TH preparations have been isolated, sequenced, and subsequently synthesized by solution or solid phase procedures and/or produced by recombinant DNA technology and are now available in large quantities. A number of them have been isolated from TF5, the most extensively studied preparation. TF5, which is prepared from calf thymus tissue, contains 30-40 acidic, heat-stable polypeptides with molecular weights (MW) in the 1,000-15,000 range. Based on an isoelectric pattern exhibited by TF5 in the pH range of 3.5 to 9.5 and for nomenclature purposes, the individual peptides are divided into 3 groups: those exhibiting isoelectric points (pI's) lower than 5 are named alpha (α) thymosins, those in the pI range of 5-7 beta (β) thymosins and those above pI 7 are termed gamma (γ) thymosins (Fig. 1). Peptides which have been purified from TF5 include:

a) <u>Thymosin alpha 1</u> (Tα1), the first thymosin polypeptide isolated from TF5 (34). It consists of 28 aminoacid residues (AA), with a blocked N-terminal and a MW of 3108 and a pI of 4.2 (Fig. 2). Tα1 has been chemically synthetized by solution and solid-phase peptide procedures (38) and produced by recombinant DNA technology (39). As will be described later, Tα1 has been shown to amplify helper T cell functions and to modulate lymphokine production and neuroendocrine functions.

b) <u>Thymosin alpha 7</u> (Tα7), which has a pI of about 3.7 and a molecular weight of approximately 2,500, has not yet been purified to homogeneity. This preparation appears to induce suppressor cell activity (40) and has been localized in the thymic epithelial cells surrounding Hassall's corpuscules (14).

c) <u>Thymosin alpha 11</u> (Tα11), which consists of 35 AA and has been shown to be homologous to Tα1 through its 28 amino-terminal AA, with 7 additional AA at the carboxy-terminus (41) (Fig. 2). This polypeptide appears to be as active as Tα1, and about 30 times more potent than TF5, in protecting susceptible strains of mice from opportunistic infections (41).

d) <u>Thymosin beta 4</u> (Tβ4), composed of 43 AA, with a MW of 4982, and a pI of 5.1 (42) (Fig.2). This molecule has recently been synthesized (43) and its cDNA sequenced (44). It has been demonstrated to stimulate luteinizing hormone releasing factor in the hypothalamus (45).

e) <u>Polypeptide beta 1</u> (Pβ1), which is the predominant component on the isoelectric focusing gel of TF5. It has a pI of 6.7 and a MW of 8451 (46). This protein has been found to be identical to ubiquitin and to a portion of a nuclear chromosomal protein (A24). To date, it has not shown any consistent activity that can be related to a specific function leading to T cell maturation, or any of the biological activities described for other thymic factors.

f) <u>Thymosin beta 3</u> (Tβ3), which shares a significant internal homology with Tβ4 in most of its amino-terminal portion and exerts similar biological activities (47) (Fig. 2).

g) <u>Thymosins beta 8</u> (Tβ8) (39 AA), <u>beta 9</u> (Tβ9) (41 AA), <u>beta 10</u> (Tβ10) (42 AA), and <u>beta 11</u> (Tβ11) (41 AA). These polypeptides exhibit 67 % (29 AA) or higher homology to Tβ4 (48,49,50). Biological activity similar to Tβ4 has been described only for Tβ10.

Very recently, 2 polypeptides closely related to Tα1 and Tα11 have been found. Haritos et al. (51) described the isolation from rat thymus of a polypeptide of 113 AA, named prothymosin alpha, which has the entire Tα1 sequence on its amino-terminal (Fig. 2). It is postulated that prothymosin alpha may represent the native protein (precursor ?) from which Tα1 and other fragments are generated during the purification of TF5 bioactive peptides (51). The same authors have just reported (52) the finding of a new polypeptide: parathymosin alpha, which has been also isolated from rat thymus. Preliminary analysis of the 30 residues located at the amino-terminus of this moiety reveals a 43 % structural identity with Tα1, Tα11 and prothymosin alpha in the N-terminal region.

Preliminary observations have indicated that parathymosin alpha blocks the <u>in vivo</u> protection against opportunistic infections conferred to sensitive strains of mice by prothymosin alpha, suggesting that parathymosin alpha might modulate prothymosin alpha function (52).

Another TH preparation, thymopoietin or TP5, has been shown to be composed of at least 2 isopeptides: thymopoietin I and II (53), which differ in only 2 AA and have identical biological properties. Based on these observations, it was hypothetized that these 2 molecules probably represent iso-hormonal variations (35). Thymopoietin II has been synthetized (54), and it was demonstrated that its biological activity resides in a pentapeptide corresponding to the residues 32 to 36 of the 49 AA sequence of thymopoietin (55). This pentapeptide, called TP5 or thymopentin, has been synthetized (55).

Another TH preparation, from which the active moiety appears to have been purified to apparent homogeneity, has been named thymic humoral factor (THF). A partially purified THF preparation is currently being tested in clinical trials (56-58).

Other potentially interesting thymic factors and thymosin fraction 5-like preparations, including homeostatic thymic hormone whose active components appear to be histones (59), thymic epithelial supernatants (60), hypocalcemic and lymphocytopoietic substances (61,62), thymosterin (63), TFX (27), TP1 (32), etc are beyond the scope of this review and will not be discussed here.

In addition to the preparations previously described, which have been isolated from thymus glands, other thymic hormone-like substances were isolated directly from serum. The best characterized of these serum factors is facteur thymique serique (FTS) or thymulin. This factor was originally isolated from pig serum and has been purified, sequenced and chemically synthesized (36,37). The biologically active moiety is a nonapeptide with a MW of 847. The presence of Zn coupled to FTS appears to be necessary for its biological activity (64).

As a final note to the analysis of the biochemical properties of TH, it is important to mention that 4 of the TH which have been sequenced to date (i.e. Tα1, Tβ4, FTS and TP5) show no homology (4,17), and appear to be chemically different from other TH (i.e. TFX, THF, etc.) from which just the AA composition has been determined. However, homologies have been described between peptides obtained from the alpha or beta regions of TF5 (Fig. 2).

Further studies will be necessary to assess the biological significance of the production of this spectrum of different TH by the thymus gland.

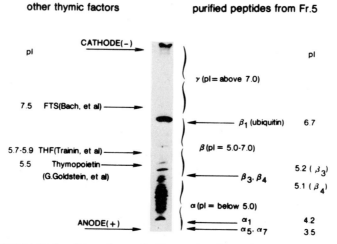

Figure 1. Isoelectric focusing of thymosin fraction 5 in LKB PAG plate (pH 3.5-9.5). Purified thymosin peptides from the alpha, beta and gamma regions are identified. The isoelectric points of several other well characterized thymic factors are illustrated for comparison.

BIOLOGICAL ACTIVITIES OF WELL CHARACTERIZED THYMIC HORMONES

The majority of TH preparations exhibit a variety of similar biological activities (4). The effects of TH can be generalized as the

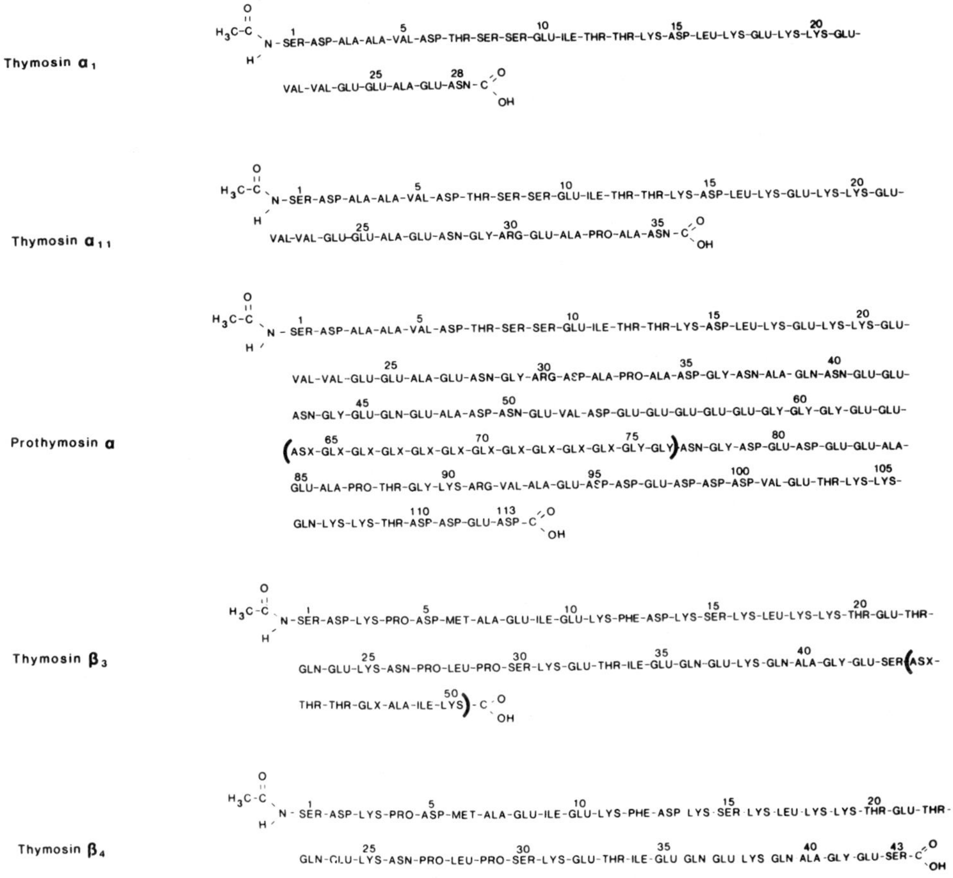

Figure 2. Amino acid sequence analysis of alpha and beta thymosins.

capability of promoting T cell differentiation, maturation and function. Examples of these activities in vivo are the immunorestorative properties of TH on T cell-dependent immunity in immunodeficient old animals (65), or the ability to protect susceptible mice against opportunistic infections (41), ect. These and other in vivo effects will be described later in relation to the potential therapeutic uses of TH.

The in vitro biological activities attributed to TH can be divided into 4 major areas:

1) Modulation of T-cell differentiation antigens

Several TH are able to modulate the expression of a variety of T cell antigen markers, including TL (T cell leukemia markers), Thy 1 and the Lyt 1,2,3 phenotype, on bone marrow and spleen lymphocyte subpopulations derived from athymic and normal mice (16,40). Most of the classical bioassays employed to measure TH activity and the active fractions obtained after the purification procedures, are based on this property of TH.

TH can either increase or decrease the expression of a particular phenotypic marker. For example, high concentration of T1 have been shown to increase (40) the expression of terminal deoxynucleotidyl transferase (TdT), a DNA polymerase found primarily in thymocytes (66), on bone marrow lymphocytes, whereas low concentrations reduced the TdT contents of murine thymocytes (4,16). This phenomenon has been interpreted as a shift toward a more "mature" population, and the same explanation has been applied to the TH-induced decrease in the percent of thymocytes which are agglutinated by peanut lectin agglutinin (PNA), another marker of differentiation exhibited in lesser amounts by medullary ("mature") than by cortical ("immature") thymocytes (16). Since the sequence of appearance and disappearance of phenotypic markers along the lymphocyte differentiation pathway has not been completely elucidated (16,28), it is very difficult to assess the precise role that TH play in lymphocyte maturation. Additionally, it should be mentioned that the modulation of phenotypic expression is subject to perturbation by a number of extrinsic factors (reviewed in 28), complicating the interpretation of the results obtained and leading sometimes to unpredictable and even contradictory results (28).

However, it is reasonable to hypothesize that the production of TH by several types of thymic epithelial cells might play a role in T cell differentiation and maturation. This hypothesis is supported by recent reports (67) that showed the differentiation of a thymocyte population expressing dim Ly 1 antigen that appears to differentiate into Lyt 2,3+; L3T4 thymocytes, which in turn lead to the appearance of Lyt 2+ and L3T4+ cells, the mature T lymphocyte subpopulations in the mouse. It was demonstrated that the expression of Lyt 2,3 antigens, and possibly L3T4, requires the presence of "cortical epithelial cells", since these cells appearance occurs at 15-17 days of gestation, immediately before the appearance of thymocytes bearing the markers of the mature phenotype (67).

In humans, similar studies have yielded the same general conclusions. For example, <u>in vitro</u> incubation of TH with human bone marrow cells leads to the expression of E-rosette receptors and human T lymphocyte antigen (HTLA), both expressed in 95-99 % of peripheral blood lymphocytes (PBL), but absent in bone marrow cells. In this system, as it is also the case in murine systems, more than one TH is able to induce the same phenotypic changes. For example, the induction of E-rosettes and human T lymphocyte antigen (HTLA) were observed after incubation with thymulin, thymopoietin and TP5 (reviewed in 14). Additionally, thymulin has recently been described to induce formation of E-rosettes and OKT3 antigen expression (present in 95-99 % of normal PBL) in a PBL subpopulation which exhibit an "immature phenotype" (OKT3-, E-rosettes-) (68).

Since to date no well characterized TH has been shown to induce all the changes which characterize T lymphocyte differentiation, it has been proposed that several TH may act sequencially to induce all the changes that occur in the normal maturational process (14).

Further studies, both in the murine and human systems, will be necessary to accurately assess the relevance that the modulation by TH on the phenotypic expression of the different lymphocyte subpopulations have on T cell differentiation.

2) <u>Modulation of T lymphocyte biological activities</u>

The regulation of a broad spectrum of T cell functions have been ascribed to TH (4,5,16,17). As we previously described with regard to induction of T cell markers, a number of well defined TH possess similar biological activities (69).

A cautionary remark should be made in analyzing the effects of TH in biological systems. It is well known that assays that measure T cell function are readily susceptible to extrinsic perturbations (5). This makes it difficult at times to separate the so called "pleiotropic effects" of TH from their true inductive capabilities, resulting in confusing and sometimes controversial findings (48). However, well controlled studies and reproducible observations made in several laboratories have established a number of currently accepted TH activities, including:

a) Regulation of proliferation. Proliferative responses to mitogens such as phytohemagglutinin (PHA) and concanavalin A (Con A), can be either enhanced or depressed by incubation with TH, depending on the preparation tested, concentration and culture conditions (14). In general terms it can be stated that TH appear to act as modulatory agents, exerting positive effects when the immune responses of normal individuals are spontaneously depressed or when in vitro culture conditions are made suboptimal, for example, by suboptimizing mitogen or antigen concentrations.

Additionally, TH have also been shown to modulate the lymphoproliferative responses of T cells to foreign histocompatibility antigens in mixed leukocyte reactions (MLR) and mixed lymphocyte tumor cultures (MLTC). It is generally accepted that TH enhance the MLR and MLTC responses in mouse (9,10,70) and human (12,71-73) systems. Current studies in our laboratories are aimed at identifying the moiety responsible for the TF5-induced increase in human MLR responses. The results indicate that neither Tα1 nor Tβ4 by themselves are able to exert this activity in vitro (Sztein et al., manuscript in preparation). These data suggest that the MLR enhancing effect is either mediated by a different peptide or by Tα1 and/or Tβ4 in the presence of other factor(s) present in TF5.

A biological activity which has been observed in the last few years is the direct mitogenic effect of TF5 on mouse (10) and human (Sztein et al., manuscript in preparation) lymphocytes. These observations might provide a clue to the mechanisms involved in TH actions. Interestingly, preliminary experiments in our laboratory appear to indicate that this activity may not be mediated by either Tα1 nor Tβ4, again leading to the suggestion that still another polypeptide not yet isolated from TF5 can directly induce lymphocytes to proliferate. However, these experiments do not rule out the possibility that either Tα1 or Tβ4 are required, but not sufficient, to induce proliferative responses.

b) Increased development of cytotoxic effector cells. A number of TH, including TF5, TP1 and TP5, have been shown to enhance specific cytotoxic T cell activity (10,74-76). For example, incubation of normal human PBL with TP1 resulted in an increase in the cytotoxic activity against allogeneic tumor cells (75). Similar results were observed in murine systems, in which TF5-mediated immunostimulation of T-cell activity in MLR and in the development of cytotoxic effector cells in MLTC assays was described (10,74).

c) Enhancement of helper T-cell activity. TH have been reported to enhance specific antibody responses in vivo (77,78) and in vitro (79,80). Injection of immunodeficient old mice with Tα1, shortly before antigen priming, enhanced the helper cell activity of their spleen cells

(77). Additionally, injection of Tα1 has been shown to augment antitetanous antibody production in response to tetanous toxoid immunization in young and old mice (78). In another study, TF5 and Tα1 were shown to enhance specific anti-influenza antibody production <u>in vitro</u> by PBL obtained from elderly volunteers (79). TF5 and THF-mediated enhancement of anti-sheep red blood cell antibody production by spleen cells from nude mice has been also described (80).

d) <u>Induction of T cell differentiation</u>. Although the phenotypic expression of specific antigens is largely employed as a measurement of how "mature" a lymphocyte population is (see above "modulation of T-cell differentiation antigens"), the response to immunoregulatory molecules can also be used a marker of T-cell maturation. For example, the resistance to corticosteroids is one of the functional characteristics associated with a more mature thymocyte cell subpopulation (16). It has recently been shown that TF5 is able to reduce the steroid binding activity and increase the resistance to the cytolytic effect of dexamethasone of human infant thymocytes (81). These findings are compatible with a role of TH in the induction of changes associated with T-cell differentiation of human thymocytes.

e) <u>Induction of suppressor T cells</u>: Enhancement of suppressor T-cell activity by TH has been observed in MLTC (82), in T cell-mediated cytotoxicity (83) and in T helper-mediated plaque forming cell (40) systems. Tα7 appears to be one of the TH mediating these activities (40).

Although most of these effects have been known for almost a decade, only recently have the mechanisms by which TH may modulate this T cell functions been studied. Central to this investigations is the role of TH in the modulation of lymphokine production. Lymphokines are products of T cells produced during mitogen or antigenic stimulation (including MLR) which play a pivotal role in the generation of normal immune responses (84). It has been demonstrated that TH are able to increase the production of a number of lymphokines, including migration inhibition factor (MIF) (85), interferon (86), colony stimulating factor (9), and interleukin 2 (IL2) (7,11,87). The latter lymphokine has been extensively studied since it plays a central role in T cell proliferation (88). We have recently demonstrated that IL2 production in response to PHA by normal PBL can be increased by TH (7). This activity could explain a number of the effects induced by TH on T cell lymphoproliferative responses, including stimulation of PHA, MLR and MLTC responses. Additionally, stimulation of lymphokine release might be one of the mechanisms involved in the <u>in vivo</u> immunorestorative effects of TH in immunodeficiencies. It is interesting to note that the molecule responsible for the enhancement of PHA-induced IL2 production by TF5 <u>in vitro</u> does not appear to be any of the well defined TH (89), indicating that this very important biological effect resides in a still undefined polypeptide in TF5. Efforts to purify the active moiety in this system are being actively pursued in our laboratory.

One of the critical steps leading to maturation and subsequent differentiation of mature T lymphocytes is the expression of receptors for interleukin 2 (IL2R). In the thymus, the presence of IL 2R has been described on murine thymocytes expressing an immature phenotype (Lyt2-/L3T4-) (90,91), suggesting that the expression of IL2R may play a role in T cell development. If TH convey a signal to thymocytes leading to their maturation, and IL2 production and IL2R expression play a role in this process, it should be then possible that TH could also induce receptors for IL2 in T cells. This indeed appears to be the case, since with the use of the anti-Tac monoclonal antibody, that reacts with human

IL2R (92,93), and flow cytometry, we have demonstrated that TF5 is able to increase in vitro IL2R expression in PHA-stimulated lymphocytes from normal individuals (6) (Fig. 3). Additionally, we have also found that TF5 is able to increase the IL2R expression of human normal lymphocytes stimulated with OKT3 monoclonal antibodies. These effects appear to be the direct effect of TF5, since abrogation of IL2 production by cyclosporin A did not affect the response (6). These data, taken together with the fact that TF5 increases IL2 production by normal human lymphocytes (7), point to a physiological role of thymic hormones in the maintenance of a competent immune system. Studies to determine if this newly described activity of TF5 is mediated by any of the well characterized TH, or a new entity are currently under way.

3) <u>Modulation of Natural Killer (NK) cell activity</u>: Very recent evidence (94) appears to support the notion of a regulatory role of the thymus in NK activity. Furthermore, in vivo administration of TP1 or Tα1 has been shown to enhance NK activity in mice (95), while no effect has been observed with TF5 (10). In vitro, both, enhancement and inhibition of NK activity by TH have been reported, depending on the concentrations of TH used. These effects have been described for TP5 (96) and FTS (97).

4) <u>Effects on Neuroendocrine System</u>: TF5 has been found to increase in vitro luteinizing hormone releasing factor (LRF) and luteinizing hormone (LH) production by the hypothalamus and pituitary (45) and in vivo increase in adrenocorticotropin (ACTH), cortisone and β-endorphin production in monkeys (98). Recently, evidence has been presented suggesting that the thymus is involved in the regulation of estrogen-induced suppression of bone marrow colony formation (CFU) leading to bone marrow hypocellularity, since thymectomy abolished the

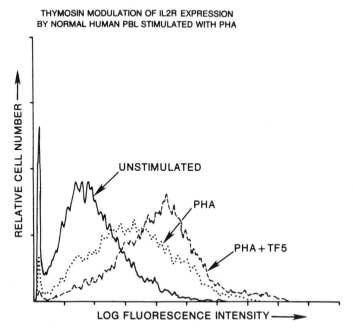

Figure 3. Normal human PBL were incubated with media (unstimulated), PHA, or PHA + TF5 for 3 days. Cells were harvested, stained for IL2R with the anti-Tac monoclonal antibody and goat F(ab')2 anti-mouse IgG-FITC, and analysed by flow cytometry.

ability of estrogens to suppress CFU proliferation (99). It was hypothesized that these changes in CFU kinetics are due, at least in part, to abnormalities of regulatory factors produced by thymic epithelial cells in response to a specific estrogen stimulus (99). On the other hand, it has also been shown that TF5 administration <u>in vivo</u> is able to advance vaginal opening and elevate estrogen levels (100), while estradiol injections decreased thymus weight and cause a transient decrease in Tα1 levels in plasma (100).

All this information, together with the fact that certain molecules produced by the immune system (including Tα1 and Tβ4) are able to act on the central nervous system (25), strongly suggest a close relationship between the endocrine thymus and the neuroendocrine systems. A model describing the network of communications between the nervous and immune systems is presented in Fig. 4. Ongoing studies with the thymosins indicate that these peptides play an important, although largely undefined role in modulating the pituitary-adrenal stress axis and the pituitary-reproductive endocrine axis. We have recently proposed that the soluble products of the immune system that have the ability to modulate and signal the brain be termed immunotransmitters (25). These molecules would include thymosins, interferons and cytokines such as interleukin 1 (endogenous pyrogen).

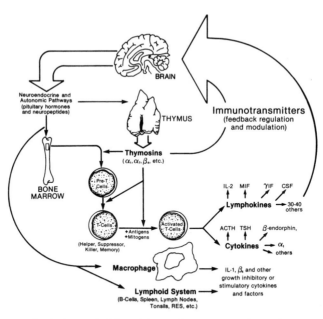

Figure 4. Proposed interrelationship of neuroactive immunotransmitters with the brain and immune system.

MEASUREMENT AND PHARMACOKINETICS OF THYMIC HORMONE ADMINISTRATION IN VIVO

a. <u>Measurement</u>. In order to study the physiology of thymic function in health and the potential use of TH in the treatment of disease, as well as to allow <u>in vitro</u> studies dealing with TH biological activities, it is of paramount importance to develop reliable assays to measure TH. The levels of TH can be determined by means of bioassays, radio-immunoassays/ELISA assays and target-cell radiobinding assays (4,101).

The original bioassay described to determine the presence of TH in human serum is based on the ability of TH to regenerate in vitro the presence of an azathioprine-sensitive splenic T-cell population obtained from mice thymectomized 2 weeks before sacrifice (102). Although difficult to set up in most laboratories, this assay when functioning properly has yielded reproducible results (4), and is able to detect levels of FTS as low as 0.01 to 0,1 ng/ml. Although it has been used during the purification of FTS, other TH have also been found to be active in this assay, including TF5, Tα1 and thymopoietin (4). Therefore, although it can detect TH-like activity in serum, this assay can not be used to identify a specific TH. Another bioassay (103) detects the ability of thymopoietin to induce the expression of the Thy 1.2 markers on lymphocytes obtained from spleens of nude mice. This assay detects levels of 0.2 ng/ml thymopoietin. With the use of these assays, it has been found that TH-like activity is decreased in serum from aging individuals, following therapeutic thymectomy in myasthenia gravis patients, in patients with immunodeficiencies, including severe combined immunodeficieny (SCID) and DiGeorge syndrome, and in 50 % of patients with systemic lupus erythematosus (SLE) (4). In contrast, with the use of the azathioprine-rosette assay, an increased TH-like activity was detected in the sera of 15 of 73 patients with mycosis fungoides (MF). This activity was demonstrated to be due to the presence of elevated FTS, since it was absorbed specifically with anti-FTS antibodies (14).

A number of radioimmunoassays (RIA) have been developed for the measurement of well characterized TH, such as Tα1 (104) and Tβ4 (19), thymulin (FTS) (20) and thymopoietin (21).

These assays are generally based in the use of heteroantisera or monoclonal antibodies against specific TH and 125 I-radiolabelled tyrosine-containing synthetic TH analogs as tracers. They are able to detect as little as 40 pg of Tα1, 5 ng of Tβ4, 1 pg of FTS and 20 pg of thymopoietin.

It is important to note that the RIA for Tα1 does not cross-react with either FTS, thymopoietin or other serum proteins, while the RIA for FTS does not cross-react with Tα1, indicating their high specificity, as opposed to the similar biological activities of several TH in the bioassays previously described.

Most recently, a microELISA assay able to measure Tα1 has been developed (105). In this assay, Tα1 in the liquid phase competes with a solid-phase bound Tα1 for a highly specific anti-Tα1 antibody, and is able to detect as little as 100 pg/ml. ELISA assays have also been developed for the measurement of FTS (14).

The development of these assays have allowed the study of the serum levels of TH in health and disease. For example, normal levels in humans have been described for Tα1 (400-1000 pg/ml), Tβ4 (450-1100 ng/ml) and FTS (20-44 pg/ml) (19,20,104,106). It has also been possible to determine circadian rhythms for Tα1, which appears to be inversely correlated to serum corticoid levels (107).

Studies of TH levels at different ages have given results which are difficult to interpret. For example, levels of Tα1 in serum have been found to be highest in newborns, decreasing significantly before puberty and becoming stable by age 20, remaining at about the same levels until passed the 6th decade. However, the bioassays of serum TH show a gradual decrease beginning at puberty, reaching background levels by the 4th to 5th decade (14). It is possible that the RIA is measuring cross-reactive polypeptides which are either biologically inactive and/or are not produced by the thymus. These hypothesis appears to be supported by the facts that a Tα1-like molecule, as measured by RIA, has been observed in other tissues, including the central nervous system (25), spleen, and liver (14) and the very recent studies showing the existence of several TH or precursor peptides with identical N-terminal sequences to Tα1 (i.e. Tα11 and prothymosin alpha), as previously discussed in the biochemistry section. Additionally, thse newly described TH have also been observed in organs other than the thymus (see sites of production section).

This tissue distribution (and production?) of Tα1 and related peptides can also provide a reasonable explanation for the lack of a consistent decrease in serum Tα1 levels observed after thymectomy in rodents (107), in spite of a sharp decrease of TH-like activity in serum as measured by bioassays, as well as a decrease in FTS levels as measured by RIA (14).

RIA have also been employed to measure TH levels in several disease states. For example, Tα1 has been shown to be decreased in children with primary immunodeficiencies, including ataxia-telangiectasia, Wiskott-Aldrich syndrome and combined immunodeficiency, which are similar to the results obtained by bioassay determinations (14). However, the results obtained with the RIA and the bioassays do not always coincide. Discrepancies have been observed in children with DiGeorge syndrome (3 out of 6 children showed decreased levels of Tα1 as determined by RIA, as compared to 6 out of 6 as measured by bioassay) (14).

Other procedures to measure polypeptide hormones include the binding to specific receptors on target cells. Receptors for FTS (108), Tα1 (109) and thymopoietin (110) have been found, but no reliable assays for the use of this methodology to measure TH levels have yet been developed.

The development of new bioassays, based in recently described biological activities of TH, such as modulation of IL2 production, IL2R expression, etc., will play a key role in the study of the biology and biochemistry of TH in the years to come.

b. <u>Pharmacokinetics</u>. One of the many issues to be addressed for the study of the therapeutic properties of TH, is the establishment of a dose schedule for the administration of these polypeptides. Central to this problem, is the determination of the pharmacokinetics of TH administration. Preliminary observations employing the azothioprine-rosette bioassay (14) suggested that FTS-like activity disappears from serum with a half life of 15 minutes. The most detailed pharmacological studies have been done during Tα1 clinical trials, in which the kinetics of Tα1 disappearance was determined on serially obtained sera samples by means of a RIA. Results indicate a sharp rise after Tα1 administration, reaching a peak at 6 hs post-injection, and returning to almost baseline levels by 24 hrs (4). Two important observations derived from these studies are 1) with the currently administered doses, the peak levels achievable are 20-50 times higher than the levels observed in newborns, which indicates that the doses being used are in the pharmacological range, and 2) the baseline levels of Tα1 during treatment appears to rise, and persist significantly elevated, as compared to placebo controls, over a 10 week study period (111). The clinical implications of these findings in trials under way are still to be evaluated. The studies todate would suggest that daily injections may not be necessary to maintain persistently elevated circulating levels of Tα1.

PHYLOGENY AND SITES OF PRODUCTION OF THYMIC HORMONES

Studies performed in the least few years seem to indicate that thymic hormones appear very early in evolution and have remained highly conserved.

The development of a specific RIA to measure Tα1 has allowed the detection of Tα1-like immunoreactive proteins in the circulation of a number of highly evolved species including man, monkey, dog, cat, baboon, cattle, rat, mouse, frog, toad and bony fish (112). In addition, recent studies have now shown that certain Tα1 cross-reactive material can be found in earthworms, tunicates, protozoan and even in prokaryotic organisms like mycobacteria tuberculosis and phlei (112). It has been suggested that the presence of this polypeptide may act as an immunomodulator of the rudimentary immune system present in organisms like fish, earthworms and tunicates (112). However, the role of Tα1-cross reactive peptides in bacteria and protozoan is not understood. A similar species distribution has been observed for Tβ4 (24), since one or more of the members of a family of peptides sharing a high degree of homology to Tβ4 (i.e. Tβ3, Tβ8, Tβ9, Tβ10 and Tβ11), have been found in mammals (including man, cat, rat, mouse, cow and rabbit), aves, reptilia, amphibia and fish (24,49,52).

The production of TH by the non-lymphoid cell component of the thymus is now very well established (13,18). Ultrastructural studies have demonstrated that many of the epithelial cells in the thymus have the characteristics of secretory cells (113). The presence of membrane-bound granules, which are present in other endocrine organs, has been described in both, cortical and medullar thymic epithelial cells (14). Additionally, a number of immunocytochemical studies, using heteroantisera raised against crude, purified and/or synthetic TH (including thymulin, Tα1, Tα7, Tβ3, Tβ4 and thymopoietin) have determined the intrathymic location of these polypeptides (13,18).

Thymulin and thymopoietin have been localized in epithelial cells in the thymic cortex and medulla. Tα1-containing human thymic epithelial cells were demonstrated in the subcapsular cortex and the medulla, while Tβ3 and Tβ4 have been found only in the subcapsular cortex area. On the other hand, the site of production of Tα7 appears to be restricted to the medullary region in human thymuses, particularly in the epithelial cells surrounding the Hassall's corpuscles (14). However, differences in cortical versus medullary locations of TH-containing epithelial cells in human, rat and mouse thymuses have been described (18), suggesting a species specificity in the distribution of those polypeptides.

Very recent studies have related the presence of TH-containing cells with the expression of surface markers specific for human thymic stroma able to be recognized by monoclonal antibodies (114). For example, cells positive for TE-4 (a marker of thymic endocrine epithelium) have been shown to contain Tα1, keratin, and to strongly express class I and II MHC antigens, while the thymic fibrous stroma cells, defined by the TE-7 monoclonal antibody, do not contain Tα1. Additionally, TE-4+ cells have been identified in the basal layer of squamous epithelium in tonsil, conjunctiva, skin and esophagous, but are absent from a number of tissues, including spleen, lymph nodes, liver, kidney, etc. (114). Another interesting study indicated that the same epithelial cells contain three of the well defined TH, i.e. thymulin, thymopoietin and Tα1, suggesting that the production of several TH is accomplished by the same epithelial cells (115).

Studies have also shown that thymic epithelial cells are not the only ones containing TH. In fact, the presence of Tα1, Tβ4 and thymopoietin have been described in human epidermal cells (18). In contrast, Tα7 (18) and thymulin (116) have only been found in the thymus.

The presence of TH in other tissues has been confirmed by a number of other immunochemical, as well as biochemical studies, in which TH-like peptides were extracted, purified and sequenced from tissues other than the thymus, conclusively showing that TH are also produced in other organs. For example, Tα1 cross-reactivity has been shown in discrete regions of the brain, with the highest content observed in the median eminence and arcuate nucleus in the hypothalamus, somewhat less in the ventromedial nucleus, and the lowest cross-reactivity was observed in the cortex and spinal cord (22). Tα1 cross-reactive material was also found in pituitary glands (22). Furthermore, the recently described prothymosin alpha (51) has been found to be present in the thymus in the highest concentrations, but cross-reactive material in amounts ranging from 15 to 65 % of the quantities found in the thymus were obtained from brain, liver, kidney, lung and spleen (23). On the contrary, parathymosin alpha, which shares a 43 % homology to prothymosin alpha in the N-terminal 30 AA, has been found to be highly concentrated in rat liver, followed by kidney, lung, brain, thymus and spleen (52). It was postulated that there is a reciprocal relationship between the concentrations of prothymosin alpha and parathymosin alpha in the different tissues, and that this could result in constant values for the sum of the concentrations of the 2 peptides. This fact may be biologically important since parathymosin alpha appears to block the immunoenhancing effects of prothymosin alpha (52).

In regard to the beta thymosins, it has been shown that Tβ4 is present not only in thymus, but also in brain and spleen, and in smaller quantities in kidney, liver and lung (24). Of interest is the fact that tissues from nude mice contained higher concentrations than tissues from normal mice (24), providing a direct evidence that tissues other than the thymus are able to produce Tβ4.

Finally, it is important to mention that TH were shown to be produced by several different cell types. Tβ4 has been shown to be produced by interdigitating cells of the thymus, Langerhans cells in the skin, peritoneal macrophages, adherent spleen cells, and in a subset of oligodendrocytes in the brain and spinal cord (24,117), and by a number of cell lines, including myoblasts, fibroblasts and glioma cells, but not by erythroleukemia, hepatoma, neuroblastoma or myeloma cells (118).

Tα1-like material has been detected in dialysates of leukocytes extracts (119) and high levels of Tα1-like material have been found in the sera of mice and patients with T-cell leukemias (120). In addition, mitogen stimulation appears to induce Tα1 production by human PBL (112). Furthermore, the presence of subsets of human PBL reactive with antibodies against Tα1, Tα7 and Tβ4 has recently been reported (121).

Further studies will be necessary to evaluate the biological significance of the ubiquitous distribution of TH in different tissues and their production by such a variety of different cell types.

THYMIC HORMONES AND THEIR ROLE IN THE TREATMENT OF DISEASE

A number of extensive reviews covering the possible clinical applications of TH in the treatment of diseases have been recently published (3-5,14,122). We will here summarize the current status and future perspectives for a therapeutic use of TH in immunodeficiencies, autoimmune and infectious diseases and cancer.

Primary Immunodeficiency Diseases

Studies in animal models and human cells isolated from patients with diseases associated with a compromised immune system have indicated that TH may be of benefit in the clinical management of such individuals. For example, in vitro increases in the percentage of E rosette-forming cells, MLR responses, etc. have been documented in PBL obtained from patients with primary immunodeficiencies by incubation with TH (3,14, 123). Additionally, results from clinical trials have indicated that at least certain TH preparations, including TF5, TP1, TP5 and thymulin are able to consistently improve the immune cell function of patients with DiGeorge syndrome (4,14,26,123,124). Improvement of T cell function have also been observed, although not as consistently, in patients with Ataxia-Teleangectasia syndrome, Wiskott-Aldrich syndrome and occasionally in patients with combined immunodeficiency (4,26,123,125). Due to the low incidence of these syndromes, the number of patients treated to date is too low to definitely establish the therapeutic relevance of TH in primary immunodeficiencies. However, the data obtained so far clearly indicate that they might be effective in improving the clinical status and immune function in these patients.

Infectious Diseases

Another group of Immunocompromised patients which may benefit from the therapeutic use of TH are those individuals with secondary immunodeficiencies associated with autoimmune and infectious diseases, cancer, burns, etc. It has been reported that TH prevent fatal infections in a burned guinea-pig model (126) and increase T cell numbers in patients with severe burns, uremia, tuberculosis, etc. (14,127).

TH may also be of value in the treatment of patients with immunodeficiencies associated with infectious agents, including individuals with acquired immune deficiency syndrome (AIDS). AIDS is an infectious disease that results in increased susceptibility to a spectrum of opportunistic infections, Kaposi's sarcoma and other malignancies (128). It has been determined that homosexuals, hemophiliacs and intravenous drug addicts are at high risk of developing AIDS (128) and several clinical syndromes (i.e. lymphadenopathy syndrome (LAS) and AIDS-related complex (ARC)) have been associated with such at risk populations (129). A lymphocytotrophic retrovirus, termed HTLV-III/LAV has been linked to the pathogenesis of AIDS and its prodromal syndromes (128,129). The array of immunological defects exhibited by these patients include a marked decrease of helper T cells with reversion of the helper/suppressor ratio, depressed proliferative responses to alloantigens and autoantigens, depressed lymphokine production, etc. (128-130). We have studied the in vivo and in vitro effects of TF5 and Tα1 on the immune status of HTLV-III seropositive subjects with depressed absolute helper T cell numbers (111,131). We found that in vitro incubation of TF5 with PBL obtained from these individuals resulted in enhanced PHA-induced IL2 production and increased their ability to proliferate in response to alloantigens (Sztein et al., manuscript in preparation). Furthermore, we observed that the daily administration for 10 weeks of 60 mg of TF5 induced a significant improvement of lymphoproliferative responses to alloantigenic stimulation (MLR) in individuals who were abnormal before treatment (111,131). Additionally, TP5 treatment also resulted in a transient increase in PHA-induced IL2 production. However, no effects were observed in absolute helper T cell number, NK activity, antibody titers to HTLV-III or in the presence of a variety of surrogate markers for AIDS (111,131). Treatment with other doses of TF5 or Tα1 were not effective in altering any of the parameters described. We concluded that although TF5 was able to improve T-cell function, the most rational strategy for the treatment of HTLV-III viremic subjects should include the prolonged administration of immunomodulatory agents (such as TF5), to immunologically reconstitute the individual, in conjunction with drugs capable of suppressing viral replication (such as suramin, HPA-23 or ribavirin) (111). Furthermore, Murray et al. (132) have reported *in vitro* effects of TF5 on induction of maturational markers in ARC patients.

A similar situation exists in regard to a possible therapeutic use of TH in patients with other infectious diseases (3,4). Studies in animals have shown that *in vivo* administration of TF5 results in enhanced immunity and increased survival of immunosuppressed mice infected with BCG (133), candida, and enhanced production of interferon in mice infected with Newcastle disease virus (3,4). Additionally, the injection of TF5 increased the resistance to candida by susceptible strains of mice (134), and the *in vivo* release of MIF and γ-Interferon (135). In humans, the administration of a number of TH preparations, including THF, TFX and TS, have resulted in a shortening in the course of Cytomegalovirus, Herpes zoster and Herpes simplex viral infections and acceleration in the restoration of T-cell immunity as compared to the expected clinical course of the disease, but unfortunately, the studies were not randomized and did not include placebo groups (4).

Only recently, the results of 2 randomized trials including the use of control groups, showing a decreased number of recurrences of herpes labialis and some improvement in the score for respiratory infections of children with recurrent respiratory infections as a results of TS treatment have been reported (136,137). Thus, as we discuss in the case of autoimmune diseases, randomized clinical trials, including placebo control groups, are necessary to firmly establish the therapeutic value of TH in the treatment of infectious diseases.

Autoimmune Diseases

A number of experimental animal models have provided evidence indicating that TH may also be of benefit in the treatment of certain autoimmune diseases (3). For example, it has been demonstrated that TF5 enhanced the ability of lymph node cells from NZB mice, used as a model of human lupus erythematosus, to respond to PHA, ConA and alloantigens and reduces the production of anti-nucleic acid antibodies (138,139). TF5 also suppressed the development of experimental autoimmune thyroiditis in a guinea pig strain which is a high responder to thyroglobulin immunization (140). However, TF5 had no suppressive effect in the incidence of experimental allergic encephalomyelitis, a model of human multiple sclerosis (141). In humans, *in vitro* experiments using PBL obtained from patients with a variety of autoimmune diseases have shown that certain TH modulate suppressor cell activity, increase the percentage of autologous rosette-forming cells, E-rosette forming cells, etc. (3,4,142,143). Preliminary reports from clinical trials (3,4,27) appear to indicate that TFX, TP5 and thymulin can be of benefit in rheumatoid arthritis (26,27,143), while no effects were observed in other autoimmune diseases, such as multiple sclerosis (3,4). Clinical trials involving more patients, randomization and the use of placebo control groups will be necessary to provide definitive answers. However, all the experimental evidence appears to indicate that TH may exert an homeostatic role in certain diseases associated with an imbalance of the immune system.

Cancer

From the discussion of the use of TH in the treatment of infectious diseases and immunodeficiencies, it becomes clear that TH may be of benefit to cancer patients in which infections and perturbations of the immune system are usually present. Thus, TH may enhance or restore the patient's immune system rendering it capable to mount an adequate immune response against the tumor. Additionally, one of the areas in which TH might be most helpful is as adjuncts to conventional chemotherapy and radiotherapy, which are accompanied by immunosuppressive side effects. Experiments in animal tumor systems have shown that TF5, FTS and THF accelerated the rejection of syngeneic tumors in immunosuppressed mice (4). Additionally, recent experiments have examined the effects of TH in tumor-bearing mice which were treated with a cytostatic agent or radiation alone or the same agent plus TF5 or Tα1. Results indicated that the combined therapy usually resulted in increased survival and/or restoration of immune functions (i.e. NK activity, lymphokine production, etc.) which were depressed as a result of chemotherapy or radiotherapy (144-147). In vitro, TH have been shown to increase E-rosette forming cells, proliferation in response to PHA and alloantigens and lymphokine production by PBL obtained from cancer patients (14). Phase I and II clinical trials have been performed in cancer patients, constituting the largest group of patients in which TH have been studied therapeutically. Extensive reviews concerning these trials, which included the use of TF5, Tα1, TS, THF and TFX, have recently been published (4,14). Patients with a variety of malignancies, including lung, head and neck, melanoma, gliomas, renal, Hodgin's, leukemia, colon and gastrointestinal cancers, have entered these trials (14). In general terms, results indicated: a) TH can be administered safely without toxicity other than occasional allergic reactions, b) normalization of some of the immune parameters usually depressed in the patient population prior to treatment have been observed with a number of TH although a consistent pattern of immunoreconstitution could not be defined and c) TH may have or have not an impact in survival depending of the particular TH preparation used, dose and schedule of administration, type of tumor, spreading of the tumor at the time of the study, etc. (14).

As an adjunct to intensive remission-induction chemotherapy TF5 was found to significantly prolong survival in patients with small cell (oat cell) bronchogenic carcinoma (148). However, these results have not been confirmed in trials with advanced oat and non-oat cell lung cancer using different chemotherapeutic and/or radiotherapeutic protocols (14).

Encouraging results have also been obtained with the use of TS in stage I melanoma patients, in which a significant improvement in the metastasis-free interval as compared to patients receiving DTIC or surgery alone has been reported (149). In spite of these somewhat positive responses most of the recent trials in which TH were employed as adjuncts of chemotherapy (including head and neck cancers, malignant melanoma at advanced stages, etc.) showed them to be ineffective or to exhibit minimal effects (14).

The results of the first phase II clinical trial with a synthetic TH, Tα1, have just been reported and the results are encouraging (150). In this study of patients with cancer of the lung treated with radiation therapy, Tα1 was administered and significantly increased the disease-free interval and survival as compared to patients receiving radiotherapy and placebo (Figure 5). The immune parameters of patients receiving Tα1 were also significantly improved (150).

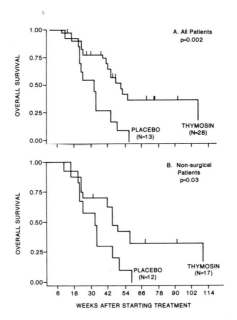

Figure 5. Synthetic Thymosin alpha 1 prolongs survival of lung cancer patients. All patients received radiotherapy and were then randomized to receive either thymosin alpha 1 or placebo. Patients were treated for 54 weeks or until relapse. Kaplan-Meier actuarial overall survival estimates for pooled patients from thymosin treatment groups and patients treated with placebo. A. All patients. B. Nonsurgical patients. (For details see Schulof et al., 150).

On the basis of these positive results, the National Cancer Institute is sponsoring two broad-based confirmatory clinical trials with Tα1. These trials are being conducted by the Radiation Therapy Oncology Group (RTOG) and the Mid-Atlantic Oncology Program (MAOP) at 30 medical centers in the United States. Tα1 is the first biological response modifier to enter the final stage of clinical testing in lung cancer.

If the RTOG trial is confirmatory, it would establish Tα1 as an effective adjunct to radiotherapy, substantially increasing the survival of patients being treated for the most common of lung cancers - those that are localized such that they cannot be removed surgically. This would represent a major advance in the treatment of the disease and would be the first significant improvement in the survival rate since the introduction of radiotherapy fifty years ago.

In addition to the potential benefit of thymosins in the treatment of lung cancer, the thymosins and other TH may be useful in treating a broad spectrum of other cancers by:

- enhancing a cancer patient's immune system so it can fight the growth of cancer cells;

- eliminating or suppressing body responses that permit cancer growth;

- enhancing a cancer patient's ability to reconstitute the immune cells damaged by other forms of cancer treatment, e.g. chemotherapy or radiation;

- stimulating overall host resistance and preventing secondary infections.

FUTURE PERSPECTIVES

The potential clinical importance of the thymus and its hormones has only recently been recognized by the scientific and medical community. We are entering what has been termed the "Age of Immunopharmacology" (151). The ongoing intensive exploration of these natural "agents of immunity" is showing the first evidence of clinical efficacy. While the TH clearly are extremely broad in potential application, additional basic research and further clinical trials are necessary to fully evaluate and refine these compounds. If the future trials with TH are as successful as the recently completed trial in lung cancer with synthetic Tα1 (150), TH may become to disorders of the immune and neuroendocrine systems a ready arsenal of specific weapons for maintaining health and increasing the quality of life.

ACKNOWLEDGEMENTS

These studies were supported in part by grants and/or gifts from the NCI (CA 24974), Alpha 1 Biomedicals, Inc., and Hoffmann La Roche, Inc.

REFERENCES

1. O. Archer, and J. C. Pierce, Role of thymus in development of the immune response, Fed. Proc. 20:26 (1961).
2. J. F. A. P. Miller, Immunological functions of the thymus, The Lancet 2:748 (1961).
3. M. B. Sztein, and A. L. Goldstein, Thymic Hormones - A Clinical Update, Springer Seminar Immunopathol. in press (1985).
4. R. S. Schulof, and A. L. Goldstein, Clinical applications of thymosin and other thymic hormones, in: "Recent Advances in Clinical Immunology", R. A. Thompson, and N. R. Rose, ed., Churchill Livingstone, New York, p. 243 (1983).
5. R. Schulof, P. H. Naylor, M. M. Zatz, and A. L. Goldstein, Thymic physiology and biochemistry, in: "Molecular and physiological basis of systemic function", H. E. Spiegel, ed., Academic Press, New York, in press (1985).
6. M. B. Sztein, S. A. Serrate, and A. L. Goldstein, Modulation of interleukin 2 receptor expression in human lymphocytes by thymic

hormones, submitted for publication.
7. M. M. Zatz, J. Oliver, C. Samuels, A. B. Skotnicki, M. B. Sztein, and A. L. Goldstein, Thymosin increases production of T-cell growth factor by normal peripheral blood lymphocytes, Proc. Nat. Acad. Sci. USA 81:2882 (1984).
8. M. M. Zatz, A. B. Skotnicki, J. M. Bailey, J. Oliver, and A. L. Goldstein, Mechanisms of action of Thymosin. II. Effects of aspirin and thymosin on enhancement of IL-2 production, Immunopharmac. 9:189 (1985).
9. M. M. Zatz, and A. L. Goldstein, Mechanisms of action of Thymosin: I. Thymosin Fraction 5 increases lymphokine production by mature murine T cells responding in a mixed lymphocyte reaction, J. Immunol. 134:1032 (1985).
10. J. E. Taldmadge, K. A. Uithoven, B. F. Lenz, and M. A. Chirigos, Immunomodulation and therapeutic characterization of thymosin fraction 5, Cancer Immunol. Immunother. 18:185 (1984).
11. T. Umiel, M. Pecht, and N. Trainin, THF, a thymic hormone, promotes interleukin-2 production in intact and thymus-deprived mice, J. Biol. Resp. Mod. 3:423 (1984).
12. J. Shoham, and I. Eshel, Thymic hormonal activity on human peripheral blood lymphocytes, in vitro. IV. Proliferative response to allogeneic tumor cells in healthy adults and cancer patients, Int. J. Immunopharmac. 5:515 (1983).
13. B. F. Haynes, The human thymic microenvironment, Adv. Immunol. 36:87 (1984).
14. R. S. Schulof, Thymic peptide hormones: Basic properties and clinical applications in cancer, in: "CRC Critical Reviews in Oncology/Hematology", S. Davis, ed., CRC Press, Boca Raton, Vol. 3, p.309 (1985).
15. T. L. K. Low, and A. L. Goldstein, Thymosin, peptidic moieties and related agents, in: "Immune modulation agents and their mechanisms", R. L. Fenickel and M. A. Chirigos, eds., Marcel Decker Inc., New York, p. 135 (1984).
16. M. M. Zatz, T. L. K. Low, and A. L. Goldstein, Role of Thymosin and other thymic hormones in T-cell differentiation, in: "Biological responses in cancer", E. Mihick, ed., Plenum Press, New York, Vol. 1, p. 219 (1982).
17. A. L. Goldstein, T. L. K. Low, G. B. Thurman, M. M. Zatz, N. R. Hall, J. E. McClure, S. Hu, and R. S. Schulof, Thymosin and other hormone-like factors of the thymus gland, in: "Immunological approaches to cancer therapeutics", E. Mihick, ed. J. Wiley and Sons, New York, p. 137 (1982).
18. H. R. Higley, and G. Rowden, Immunocytochemical localization of Thymosin and Thymopoietin in human, rat and murine thymus, in: "Thymic Hormones and Lymphokines", A. L. Goldstein, ed., Plenum Press, New York, p. 135 (1984).
19. P. H. Naylor, J. E. McClure, B. L. Spangelo, T. L. K. Low, and A. L. Goldstein, Immunochemical studies in thymosin: Radioimmunoassay of thymosin beta 4, Immunopharmac. 7:9 (1984).
20. K. Ogha, G. F. Incefy, K. F. Folk, B. W. Erickson, and R. A. Good, Radioimmunoassay for the thymic hormone serum thymic factor (FTS), J. Immunol. Methods 57:171 (1983).
21. P. J. Lisi, J. W. Teipel, G. Goldstein, and M. Schiffman, Improved radioimmunoassay technique for measuring serum thymopoietin, Clinica Chimica Acta 107:111 (1980).
22. N. R. Hall, and A. L. Goldstein, Endocrine regulation of host immunity. The role of steroids and thymosin, in: "Immune modulation agents and their mechanisms", R. L. Fenickel and M. A Chirigos, eds., Marcel Decker Inc., New York, p. 533 (1984).
23. A. A. Haritos, O. Tsolas, and B. L. Horecker, Distribution of pro-

thymosin in rat tissues, Proc. Nat. Acad. Sci. USA 81:1391 (1984)
24. B. L. Horecker, Thymosin beta 4. Distribution and biosynthesis in invertebrate cells and tissues, in: "Thymic Hormones and Lymphokines", A. L. Goldstein, ed., Plenum Press, New York, p. 77 (1984).
25. N. R. Hall, J. P. McGillis, B. L. Spangelo, and A. L. Goldstein, Evidence that thymosin and other biological response modifiers can function as neuroactive immunotransmitters, J. Immunol. 135:806s (1985).
26. J. F. Bach, and M. Dardenne, Clinical aspects of thymulin (FTS), in: "Thymic Hormones and Lymphokines", A. L. Goldstein, ed., Plenum Press, New York, p. 593 (1984).
27. A. B. Skotnicki, B. K. Dabrowska-Bernstein, M. P. Dabroski, A. J. Gorsky, J. Czarnecki, and J. Aleksandrowicz, Biological properties and clinical use of calf thymus extracts TFX-Polfa, in: "Thymic Hormones and Lymphokines", A. L. Goldstein, ed., Plenum Press, New York, p. 545 (1984).
28. O. Stutman, Role of thymic hormones in T-cell differentiation, in: "Clinics in Immunology and Allergy", J. F. Bach, guest ed., W. B. Saunders, Philadelphia, Vol. 3, p. 9 (1983).
29. W. W. Nowinsky, Fortgesetzte Beiträge zur Funktion des Thymus. Die Wirkungen des Thymocresins auf das Wachstum, Biochem. Z. 226:415 (1930).
30. A. L. Goldstein, F. D. Slatter, and A. White, Preparation, assay and partial purification of a thymic lymphocytopoietic factor (thymosin), Proc. Nat. Acad. Sci. USA 56:1010 (1966).
31. J. A. Hooper, M. C. McDaniel, G. B. Thurman, G. H. Cohen, R. S. Schulof, and A. L. Goldstein, The purification and properties of bovine thymosin, Ann. N. Y. Acad. Sci. 249:125 (1975).
32. R. Falchetti, G. Bergrsi, A. Eshkol, C. Cafiero, L. Adorini and L. Caprino, Pharmacological and biological properties of a calf thymus extract (TP1), Drugs Exptl. Clin. Res. 3:39 (1977).
33. A. B. Skotnicki, Biologiczna okthwnosc i wlasciwosci fizykochemiczne wyciagu grasiczego TFX, Pol. Tyg. Lek. 28:1119 (1978).
34. A. L. Goldstein, T. L. K. Low, M. McAdoo, J. McClure, G. B. Thurman, J. L. Rossio, C. Y. Lai, D. Chang, S. S. Wang, C. Harvey, A. H. Ramel, and J. Meienhofer, Thymosin alpha 1: isolation and sequence analysis of an immunologically active thymic polypeptide Proc. Natl. Acad. Sci. USA 74:725 (1977).
35. G. Goldstein, Isolation of bovine thymin: A polypeptide hormone of the thymus, Nature 246:11 (1974).
36. M. Dardenne, J. M. Pleau, N. K. Man, and J. F. Bach, Structural study of circulating thymic factor, a peptide isolated from pig serum. I. Isolation and purification, J. Biol. Chem. 252:8040 (1977).
37. E. Bricas, T. Martinez, D. Blanot, G. Auger, M. Dardenne, J. M. Pleau, and J. F. Bach, The serum thymic factor and its synthesis, in: "Proceedings of the fifth international peptide symposium", M. Goodman and J. Meienhofer, eds., John Wiley and Sons, New York, p. 564 (1982).
38. S. S. Wang, I. D. Kulesha, and D. P. Winter, Synthesis of thymosin alpha 1, J. Amer. Chem. Soc. 101:253 (1979).
39. R. Wetzel, H. L. Heinecker, D. V. Goeddel, P. Jhurani, J. Shapiro, R. Crea, T. L. K. Low, J. E. McClure, G. B. Thurman, and A. L. Goldstein, Production of biologically active N-desacetyl thymosin alpha 1 in Escherichia coli through expression of a chemically synthetized gene, Biochemistry 19:6096 (1980).
40. A. Ahmed, D. M. Wong, G. B. Thurman, T. L. K. Low, A. L. Goldstein, S. J. Sharkis, and I. Goldschneider, T Lymphocyte maturation: cell surface markers and immune function induced by T lymphocyte

cell-free product and by thymosin polypeptides, Ann. N. Y. Acad. Sci. 332:81 (1979).
41. J. Caldarella, G. J. Goodall, A. M. Felix, E. P. Heimer, S.B. Salvin, and B. L. Horecker, Thymosin alpha 11: a peptide related to thymosin alpha 1 isolated from calf thymosin fraction 5, Proc. Nat. Acad. Sci. USA 80:7424 (1983).
42. T. L. K. Low, and A. L. Goldstein, Chemical characterization of thymosin beta 4, J. Biol. Chem. 257:1000 (1982).
43. T. L. K. Low, S. S. Wang, and A. L. Goldstein, Solid phase synthesis of thymosin beta 4: Chemical and biological characterization of the synthetic peptide, Biochemistry 22:733 (1983).
44. A. Wodnar-Filipowitz, U. Gubler, Y. Furuichi, M. Richardson, E. F. Nowoswiat, M. S. Poonian, and B. L. Horecker, Cloning and sequence analysis of cDNA for rat spleen thymosin beta 4, Proc. Nat. Acad. Sci. USA 81:2295 (1984).
45. R. W. Rebar, A. Miyake, T. L. K. Low, and A. L. Goldstein, Thymosin stimulates secretion of luteinizing hormone-releasing factor, Science 214:669 (1981).
46. T. L. K. Low, and A. L. Goldstein, The chemistry and biology of thymosin. II. Aminoacid sequence analysis of thymosin alpha 1 and polypeptide beta 1, J. Biol. Chem. 254:987 (1979).
47. N. H. Pazmino, J. H. Ihle, R. N. McEwan, and A. L. Goldstein, Control of differentiation of thymosin precursors in the bone marrow by thymic hormones, Cancer Treat. Rep. 62:1749 (1978).
48. E. Hanapell, S. Davoust, and B. L. Horecker, Thymosin beta 8 and beta 9: Two new peptides isolated from calf thymus homologous to beta 4, Proc. Nat. Acad. Sci. USA 79:1708 (1982).
49. S. Erickson-Viitanen, S. Ruggieri, P. Natalini, and B. L. Horecker, Thymosin beta 10, a new analog of thymosin beta 4 in mammalian tissues, Arch. Biochem. Biophys. 225:407 (1983).
50. S. Erickson-Viitanen, and B. L. Horecker, Thymosin beta 11: a peptide from trout liver homologous to thymosin beta 4, Arch. Biochem. Biophys. 233:815 (1984).
51. A. A. Haritos, G. J. Goodall, and B. L. Horecker, Prothymosin alpha: Isolation and properties of the major immunoreactive form of thymosin alpha 1 in rat thymus, Proc. Nat. Acad. Sci. USA 81:1008 (1984).
52. A. A. Haritos, S. B. Salvin, R. Blacherr, S. Stein, and B. L. Horecker, Parathymosin alpha: a peptide from rat tissues with structural homology to prothymosin alpha, Proc. Nat. Acad. Sci. USA 82:1050 (1985).
53. D. H. Schlesinger, and G. Goldstein, The aminoacid sequence of thymopoietin II, Cell 5:361 (1975).
54. M. Fujino, T. Fukada, S. Kawaji, S. Shinagawa, Y. Sugino, and M. Takaoki, Synthesis of the nonatetracontapeptide corresponding to the sequence proposed for thymopoietin II, Chem. Pharm. Bull. 25:1486 (1977).
55. G. Goldstein, M. P. Scheid, E. A. Bois, D. H. Schlesinger, and J. Van Waunue, A synthetic pentapeptide with biological activity characteristic of the thymic hormone thymopoietin, Science 204:1399 (1979).
56. N. Trainin, A. Begerano, M. Strahilevitch, D. Goldring, and N. Small, A thymic factor preventing wasting and influencing lymphopoiesis in mice, Israel J. Med. Sci. 2:549 (1966).
57. Y. Yakir, and N. Trainin, Enrichment of in vitro and in vivo immunological activity of purified fractions of calf thymic hormone, J. Exp. Med. 148:71 (1978).
58. N. Trainin, V. Rotter, Y. Yakir, R. Leve, Z. Handzel, B. Shohat, and R. Zaizov, Biochemical and biological properties of THF in animal and human models, Ann. N. Y. Acad. Sci. 332:9 (1979).
59. G. Bernardi, and J. Comsa, Purification de l'hormone thymique par

chromatographie sur colonne, Experientia 21:416 (1965).
60. L. Kater, R. Oosteron, J. E. McClure, and A. L. Goldstein, Presence of thymosin in human thymic epithelium conditioned medium, Int. J. Immunopharmacol. 1:273 (1979).
61. A. Mizutani, A thymic hypocalcemic component, in: "Thymic hormones", T. D. Luckey, ed., University Park Press, Baltimore, p. 193, (1973).
62. T. D. Luckey, W. G. Robey, and B. J. Campbell, LSH, a lymphocyte-stimulating hormone, in: "Thymic Hormones", T. D. Luckey, ed., University Park Press, Baltimore, p. 167 (1973).
63. S. M. Milcu, I. Potop, R. Petersku, and E. Ghinea, Effect of thymosterin on lymphocytes in vitro, Endocrinology 14:283 (1976).
64. M. Dardenne, J. M. Pleau, B. Nabarra, P. Lefrancier, M. Derrien, M. Choay, and J. F. Bach, Contribution of Zn and other metals to the biological activity of the serum thymic factor, Proc. Nat. Acad. Sci. USA 79:5370 (1982).
65. A. L. Goldstein, T. L. K. Low, N. R. Hall, P. H. Naylor, and M. M. Zatz, Thymosin: can it retard aging by boosting immune capacity?, in: "Intervention in the aging process, part A: Quantitation, Epidemiology and Clinical Research", W. Regelson and F. M. Sinex, eds., Alan R. Liss, Inc., New York, p. 169 (1984).
66. N. H. Pazmino, J. N. Ihle, and A. L. Goldstein, Induction in vivo and in vitro of terminal deoxynucleotidyl transferase by thymosin in bone marrow cells from athymic mice, J. Exp. Med. 147:708 (1978).
67. B. J. Mathieson, and B. J. Fowlkes, Cell surface antigen expression on thymocytes: Differentiation of intrathymic subsets, Immunol. Reviews 82:141 (1985).
68. J. S. Levai, and V. Utermohlen, The effect of a human plasma thymic factor on human peripheral blood mononuclear cells subpopulations Clin. Immunol. Immunopathol. 27:433 (1983).
69. A. M. Kruisbeck, Summary of the results of the workshop, in: "The biological activity of the thymic hormones", D. W. Van Bekkum, ed., Kooker Scientific Publications, Rotterdam, p. 209 (1975).
70. G. H. Cohen, J. A. Hooper, and A. L. Goldstein, Thymosin-induced differentiation of murine thymocytes in allogeneic mixed lymphocyte cultures, Ann. N.Y. Acad. Sci. 249:145 (1975).
71. L. A. Schafer, A. L. Goldstein, J. U. Gutterman, and E. M. Hersh, In vitro and in vivo studies with thymosin in cancer patients, Ann. N.Y. Acad. Sci. 277:609 (1976).
72. D. W. Wara, D. J. Barrett, A. J. Ammann, and M. J. Cowan, In vitro and in vivo enhancement of mixed lymphocyte culture reactivity by thymosin in patients with primary immunodeficiency disease, Ann. N. Y. Acad. Sci. 332:128 (1979).
73. J. Shohan, and I. Eshel, Thymic hormonal effects on human peripheral blood lymphocytes in vitro. III. Conditions for mixed lymphocyte-tumor culture assay, J. Immunol. Methods 37:261 (1980).
74. M. M. Zatz, and A. L. Goldstein, Enhancement of murine thymocyte cytotoxic T cell responses by thymosin, Immunopharmacol. 6:65 (1983).
75. J. Shoham, and M. Cohen, Thymic hormonal activity on human peripheral blood lymphocytes in vitro. V. Effect on induction of lymphocytotoxicity, Int. J. Immunopharmacol. 5:523 (1983).
76. C. Lau, and G. Goldstein, Functional effects of thymopoietin 32-36 (TP-5) on cytotoxic lymphocyte precursor units (CLP-U). I. Enhancement of splenic CLP-U in vitro and in vivo after suboptimal antigenic stimulation, J. Immunol. 124:1861 (1980).
77. D. Frasca, M. Garavini, and G. Doria, Recovery of T-cell functions in aged mice injected with synthetic thymosin alpha 1, Cell Immunol. 72:384 (1982).
78. W. B. Ershler, J. C. Herbert, A. J. Blow, S. R. Granter, and J.

Lynch, Effect of thymosin alpha 1 on specific antibody response and susceptibility to infection in young and aged mice, Int. J. Immunopharmacol. in press (1985).
79. W. B. Ershler, A. L. Moore, and M. A. Socinski, Influenza and aging: age-related changes and the effects of thymosin on the antibody response to influenza vaccine, J. Clin. Immunol. 4:445 (1984).
80. M. J. Blankwater, L. A. Levert, A. C. W. Swart, and D. W. van Bekkum, Effect of various thymic and non-thymic factors on in vitro antibody formation by spleen cells from nude mice, Cell. Immunol. 35:242 (1978).
81. D. D. F. Ma, A. H. Ho, and A. V. Hoffbrand, Effect of thymosin on glucocorticoid receptor activity and glucocorticoid sensitivity of human thymocytes, Clin. Exp. Immunol. 55:273 (1984).
82. G. D. Marshall, G. B. Thurman, J. L. Rossio, and A. L. Goldstein, In vivo generation of suppressor T cells by thymosin in congenically athymic nude mice, J. Immunol. 126:741 (1981).
83. G. D. Marshall, G. B. Thurman, and A. L. Goldstein, Regulation of in vitro generation of cell-mediated cytotoxicity. I. In vitro induction of suppressor T lymphocytes by thymosin, J. Reticuloendothel. Soc. 28:141 (1980).
84. J. J. Oppenheim, and S. Cohen, "Interleukins, Lymphokines and Cytokines", Academic Press, San Francisco (1983).
85. G. B. Thurman, C. Seals, T. L. K. Low, and A. L. Goldstein, Restorative effects of thymosin polypeptides on purified protein derivative-dependent migration inhibition factor production by the peripheral blood lymphocytes of adult thymectomized guinea pigs, J. Biol. Resp. Modif. 3:160 (1984).
86. J. Shoham, I. Eshel, M. Aboud, and S. Salzberg, Thymic hormonal activity on human peripheral blood lymphocytes in vitro. II. Enhancement of the production of immune interferon by activated T cells, J. Immunol. 125:54 (1980).
87. D. Frasca, L. Adorini, and G. Doria, Production of and response to interleukin 2 in aging mice. Modulation by thymosin alpha 1, in: "Symposium on Lymphokines", A. de Weck, ed., Interlaken, June 14-15 (1984).
88. K. A. Smith, Interleukin 2, Ann. Rev. Immunol. 2:319 (1984).
89. M. M. Zatz, J. Oliver, M. B. Sztein, A. B. Skotnicki, and A. L. Goldstein, Comparison of the effects of thymosin and other thymic factors on modulation of interleukin-2 production, J. Biol. Resp. Modif. 4:365 (1985).
90. R. Ceredig, J. W. Lowenthal, M. Nabholtz, and R. MacDonald, Expression of interleukin D receptors as a differentiation marker on intrathymic stem cells, Nature 314:98 (1985).
91. D. H. Raulet, Expression and function of interleukin 2 receptors on immature thymocytes, Nature 314:101 (1985).
92. T. Uchiyama, S. Broder, and T. A. Waldmann, A monoclonal antibody (anti-Tac) reactive with activated and functionally mature human T cells. I. Production of anti-Tac monoclonal antibody and distribution of Tac positive cells, J. Immunol. 126:1393 (1981).
93. D. A. Cantrell, and K. A. Smith, Transient expression of interleukin 2 receptors. Consequences for T cell growth, J. Exp. Med. 158:1895 (1983).
94. J. P. Flexman, P. G. Holt, G. Mayrhofer, B. I. Latham, and G. R. Shellam, The role of the thymus in the maintainance of natural killer cells in vivo, Cell. Immunol. 90:366 (1985).
95. F. Bistoni, M. Baccarini, P. Puccetti, P. Marconi, and E. Garaci, Enhancement of natural killer cell activity in mice by treatment with a thymic factor, Cancer Immunol. Immunotherap. 17:51 (1984).
96. M. Fiorilli, M. C. Sirianni, V. Sorrentino, R. Testi, F. Aiuti, In vitro enhancement of bone marrow natural killer cells after incubation with thymopoietin 32-36 (TP-5), Thymus 5:375 (1983).

97. M. C. Dokhelar, T. Tursz, M. Dardenne, and J. F. Bach, Effect of a synthetic thymic factor (Facteur Thymique Serique) on natural killer cell activity in humans, Int. J. Immunopharmacol. 5:277 (1983).
98. D. L. Healy, G. D. Hodgen, H. M. Shulte, G. P. Chrousos, D. L. Loriaux, N. R. Hall, and A. L. Goldstein, The thymus-adrenal connection: Thymosin has corticotropin-releasing activity in primates, Science 222:1353 (1983).
99. M. I. Luster, G. A. Boorman, K. S. Korach, M. P. Dieter, and L. Hong, Mechanisms of estrogen-induced myelocytotoxicity: Evidence of thymic regulation, Int. J. Immunopharmacol. 6:287 (1984).
100. L. S. Allen, J. E. McClure, A. L. Goldstein, M. S. Barkley, and S. D. Michael, Estrogen and thymic hormone interactions in the female mouse, J. Reprod. Immunol. 6:25 (1984).
101. M. Dardenne, Evaluation of blood levels of thymic hormones in health and disease, in: "Thymic factor therapy", N. A. Byrom, and J. R. Hobbs, eds., Serono Symposia Publications for Raven Press, New York, Vol. 16 (1984).
102. M. Dardenne, and J. F. Bach, Studies on the thymic products. I. Modification of rosette-forming cells by thymic extract determination of the target RFC subpopulation, Immunol. 25:343 (1973).
103. J. J. Twomey, G. Goldstein, V. M. Lewis, A. C. Bealmear, and R.A. Good, Bioassay determinations of thymopoietin and thymic hormone levels in human plasma, Proc. Nat. Acad. Sci. USA 74:2541 (1971).
104. J. E. McClure, N. Lameris, D. W. Wara, and A. L. Goldstein, Immunochemical studies on thymosin: Radioimmunoassay of thymosin alpha 1, J. Immunol. 128:368 (1982).
105. M. G. Mutchnick, D. F. Keren, F. E. Weller, J. E. McClure, and A. L. Goldstein, Measurement of thymosin alpha 1 by dissociation microELISA, J. Immunol. Methods 60:53 (1983).
106. P. H. Naylor, A. Friedman-Kien, E. Hersh, M. Erdos, and A. L. Goldstein, Thymosin alpha 1 and thymosin beta 4 in serum: Comparison of normal, cord, homosexual, and AIDS serum, Immunopharmacol. in press (1985).
107. J. P. McGillis, N. R. Hall, and A. L. Goldstein, Circadian rhythm of thymosin alpha 1 in normal and thymectomized mice, J. Immunol. 131:148 (1983).
108. J. M. Pleau, V. Fuentes, J. L. Morgat, and J. F. Bach, Specific receptors for serum thymic factor (FTS) in lymphoblastoid cultured cell lines, Proc. Nat. Acad. Sci. USA 77:2861 (1980).
109. C. R. Garaci, M. R. Torrisi, T. Jessi, L. Frati, A. L. Goldstein, and E. Garaci, Receptors for thymosin alpha 1 on mouse thymocytes Cell Immunol. 91:298 (1985).
110. T. Audhya, M. A. Talle, and G. Goldstein, Thymopoietin radioreceptor assay utilizing lectin-purified glycoprotein from a biologically responsive T-cell line, Arch. Biochem. Biophys. 234:167 (1984).
111. R. S. Schulof, G. L. Simon, M. B. Sztein, C. R. Kessler, J. K. Courtless, J. M. Orenstein, P. D. Kind, S. Schlesselman, M. Robert-Guroff, P. H. Naylor, and A. L. Goldstein, Phase I/II trial of thymosin fraction 5 and thymosin alpha 1 in HTLV-III seropositive subjects, submitted for publication (1985).
112. P. H. Naylor, K. K. Oates, E. L. Cooper, P. Deschaux, L.F. Affronti, and A. L. Goldstein, Non-thymic sources of thymosin alpha 1, Proceedings of the 67th Endocrine Society Annual Meeting, June 19-21, Baltimore (1985).
113. J. Singh, The ultrastructure of epithelial reticular cells, in: "The thymus gland", M. F. Kendall, ed., Academic Press, London, p. 133 (1981).
114. B. F. Haynes, R. M. Scearce, D. F. Lobach, and L. L. Hensley, Phenotypic characterization and ontogeny of mesodermal-derived and endocrine epithelial components of the human thymic micro-

environment, J. Exp. Med. 159:1149 (1984).
115. W. Savini, and M. Dardenne, Thymic hormone-containing cells. VI. Immunohistological evidence for the simultaneous presence of thymulin, thymopoietin and thymosin alpha 1 in normal and pathological human thymuses, Eur. J. Immunol. 14:987 (1984).
116. W. Savino, P. C. Huang, A. Corrigan, S. Berrih, and M. Dardenne, Thymic hormone-containing cells. V. Immunohistological detection of metallothionein within the cells bearing thymulin (a zinc containing hormone) in human thymuses, J. Histochem. Cytochem. 32:942 (1984).
117. M. C. Dalakas, R. Hubbard, G. Cunningham, B. Trapp, J. L. Sever, and A. L. Goldstein, Thymosin beta 4 is present in a subset of oligodendrocytes in the normal human brain, in: "Thymic hormones and lymphokines", A. L. Goldstein, ed., Plenum Press, New York, p. 119 (1984).
118. G. J. Goodall, J. I. Morgan, and B. L. Horecker, Thymosin beta 4 in cultured mammalian cell lines, Arch. Biochem. Biophys. 221:598 (1983).
119. C. H. Kirkpatrick, A. Khan, J. E. McClure, and A. L. Goldstein, Thymosin alpha 1-like material in dialysates of leukocyte extracts, in: "Immunobiology of Transfer Factor", Academic Press, New York, p. 413 (1983).
120. M. M. Zatz, J. E. McClure, and A. L. Goldstein, Immunoreactive thymosin alpha 1 is associated with murine T-cell lymphomas, Leukem. Res. 8:1003 (1984).
121. M. C. Dalakas, D. L. Madden, A. Krezlewiccz, J. L. Sever, and A.L. Goldstein, Human peripheral blood lymphocytes bear markers for thymosins (alpha 1, alpha 7, beta 4), in: "Thymic hormones and lymphokines", A. L. Goldstein, ed., Plenum Press, New York, p. 111 (1984).
122. A. L. Goldstein, and R. S. Schulof, Thymosins in the treatment of cancer, in: "Immunity to cancer", A. Reif, and M. Mitchell, eds. in press (1985).
123. D. W. Wara, M. J. Cowan, A. J. Ammann, Thymosin fraction 5 therapy in patients with primary immunodeficiency disorders, in: "Thymic Factor Therapy", N. A. Byrom, and J. R. Hobbs, eds., Serono Symposia Publications for Raven Press, New York, Vol. 16, p.123 (1984).
124. F. Aiuti, and L. Businco, Effects of thymic hormones in Immunodeficiency, in: "Clinics in Immunology and Allergy", J. F. Bach, guest ed., Saunders Co., Philadelphia, Vol. 3, p. 187 (1983).
125. Z. T. Handzel, Z. Dolfin, S. Levin, Y. Altman, T. Hahn, N. Trainin, and N. Gadot, Effect of thymic humoral factor on cellular immune functions of normal children and of pediatric patients with Ataxia-Telangiectasia and Down's syndrome, Pediatrics Res. 13:803 (1979).
126. J. D. Stinnett, L. D. Loose, P. Miskell, C. L. Tenney, S. J. Gonce, and J. W. Alexander, Synthetic immunomodulators for prevention of fatal infections in a burned guinea pig model, Ann. Sur. 198:53 (1983).
127. T. Abiko, and H. Sekino, Deacetyl-thymosin beta 4: Synthesis and effect on the impaired peripheral T-cell subsets in patients with chronic renal failure, Chem. Pharm. Bull. 32:4497 (1984).
128. A. S. Fauci, A. M. Macher, D. L. Longo, H. C. Lane, A. H. Rook, H. Masur, and E. P. Gelmann, Acquired immunodeficiency syndrome: Epidemiologic, clinical, immunological and therapeutic considerations, Ann. Int. Med. 100:92 (1984).
129. E. M. Hersh, P. W. A. Mansell, J. M. Reuben, A. Rios, and G.R. Newel, Immunological characterization of patients with acquired immune deficiency syndrome-related symptom complex, and a related life-style, Cancer Res. 44:5894 (1984).

130. J. M. Dwayer, J. G. McNamara, L. H. Sigal, and C. C. Wood, Immunological abnormalities in patients with the acquired immune deficiency syndrome (AIDS) - A review, Clin. Immunol. Rev. 3:25 (1984).
131. P. H. Naylor, R. S. Schulof, M. B. Sztein, T. J. Spira, P. R. McCurdy, F. Darr, C. M. Kessler, G. Simon, and A. L. Goldstein, Thymosin in the early diagnosis and treatment of high risk homosexuals and hemophiliacs with AIDS-like immune dysfunction, Ann. N.Y. Acad. Sci. 437:88 (1984).
132. J. L. Murray, J. M. Reuben, C. G. Munn, G. Newell, P. W. A. Mansell, and E. M. Hersh, In vitro modulation of purine enzyme metabolism and lymphocyte marker expression by thymosin fraction 5 in homosexual males, Int. J. Immunopharmacol. 7:661 (1985).
133. F. M. Collins, and N. E. Morrison, Restoration of T-cell responsiveness by thymosin: Expression of anti-tuberculous immunity in mouse lungs, Infect. Immun. 23:330 (1979).
134. S. B. Salvin, and R. Neta, Resistance and susceptibility to infection in inbred murine strains. I. Variations in the response to thymic hormones in mice infected with Candida albicans, Cell Immunol. 75:160 (1983).
135. R. Neta, and S. B. Salvin, Resistance and susceptibility to infections in inbred murine strains. II. Variations in the effect of treatment with thymosin fraction 5 on the release of lymphokines in vivo, Cell. Immunol. 75:173 (1983).
136. F. Aiuti, M. C. Sirianni, M. Fiorilli, R. Paganelli, A. Stella, and G. Turbessi, A placebo-controlled trial of thymic hormone treatment of recurrent herpes simplex labialis infection in immunodeficient host: Results after 1-year follow-up, Clin. Immunol. Immunopathol. 30:11 (1984).
137. M. Demartino, M. E. Rossi, A. T. Muccioli, and A. Vierucci, T lymphocytes in children with respiratory infections: Effect of the use of thymostimulin on the alterations of T-cell subsets, Int. J. Tiss. Reac. VI:223 (1984).
138. M. E. Gershwin, E. Ahmed, A. D. Steinberg, G. B. Thurman, and A.L. Goldstein, Correction of T cell function by thymosin in New Zealand mice, J. Immunol. 113:1068 (1974).
139. N. Talal, M. Dauphinee, R. Pillarisetty, and R. Goldblum, Effects of thymosin on thymocyte proliferation and autoimmunity in NZB mice, Ann. N.Y. Acad. Sci. 249:438 (1975).
140. V. Tomazik, C. M. Suter, and P. B. Chretien, Experimental autoimmune thyroiditis: Modulation of the disease in high and low responser mice by thymosin, Clin. Exp. Immunol. 58:83 (1984).
141. J. Woyciechowska, A. L. Goldstein, and B. Driscoll, Experimental allergic encephalomyelitis in guinea pigs. Influence of thymosin fraction 5 in the disease, J. Neuroimmunol. 7:215 (1985).
142. S. D. Horowitz, W. Borcherding, A. Vishnu Moorthy, R. Chesney, H. Schulte-Wisserman, and R. Hong, Induction of suppressor T cells in systemic lupus erythematosus by thymosin and cultured thymic epithelium, Science 197:999 (1977).
143. E. M. Veys, H. Mielants, G. Verbruggen, T. Spiro, E. Nedweck, D. Power, and G. Goldstein, Thymopoietin pentapeptide (thymopentin, TP5) in the treatment of rheumatoid arthritis. A compilation of several short and long-term clinical studies, J. Rheumatol. 11:462 (1984).
144. M. A. Chirigos, In vivo and in vitro studies with thymosin, in: "Control of neoplasia by modulation of the immune system", M.A. Chirigos, ed., Raven Press, New York, p. 241 (1977).
145. Y. Umeda, A. Sakamoto, J. Nakamura, H. Ishitsuka, and Y. Yagi, Thymosin alpha 1 restores NK cell activity and prevents tumor progression in mice immunosuppressed by cytostatics or X-rays, Cancer Immunol. Immunother. 15:78 (1983).

146. Y. Ohta, E. Tezuka, S. Tamura, and Y. Yagi, Protection of 5-fluorouracil induced bone marrow toxicity by thymosin alpha 1, Int. J. Immunopharmac. in press (1985).
147. N. Takeichi, Y. Koga, T. Fujii, and H. Kobayashi, Restoration of T cell function and induction of anti-tumor immune response in T-cell depressed spontaneously hypertensive rats by treatment with thymosin fraction 5, Cancer Res. 45:487 (1985).
148. M. H. Cohen, P. B. Chretien, B. C. Ihed, B. E. Fossieck, R. Makuch, P. A. Bunn, A. V. Johnston, S. E. Shackney, M. J. Matthews, S. D. Lipson, D. E. Kenady, and J. D. Minna, Thymosin fraction 5 and intensive combination chemotherapy. Prolonging the survival of patients with small cell lung cancer, JAMA 241:1813 (1979).
149. E. Azizi, H. J. Brenner, and J. Shoham, Postsurgical adjuvant treatment of malignant melanoma patients by the thymic factor thymostimulin, Drug Res. 9:1043 (1984).
150. R. S. Schulof, M. J. Lloyd, P. A. Cleary, S. R. Palaszinski, D.A. Mai, J. W. Cox, O. Alabaster, and A. L. Goldstein, A randomized trial to evaluate the immunorestorative properties of synthetic thymosin alpha 1 in patients with lung cancer, J. Biol. Resp. Mod. 4:147 (1985).
151. A. L. Goldstein, The history of the development of thymosin: Chemistry, biology and clinical applications, in: "Transactions of the American Clinical and Climatological Association", 89th Annual Meeting, Ponte Vedra, Florida, October 25-27 (1976), Waverly Press, Inc., p. 79 (1977).

T-CELL GROWTH FACTOR (INTERLEUKIN-2)

Suresh K. Arya and M. G. Sarngadharan

Laboratory of Tumor Cell Biology, National Cancer Institute
National Institutes of Health, Bethesda MD 20892, and
Department of Cell Biology, Bionetics Research, Inc.
Rockville, MD 20850

Introduction

Human peripheral blood lymphocytes (PBL) undergo a blast transformation when cultured with the plant lectin phytohemagglutinin (PHA) (1). These lymphocytes go through one or two cycles of cell division but could not be maintained in culture. The conditioned media from these lymphocyte cultures were found to have a variety of protein factors that affected growth and differentiation of human cells of many lineages (2). One of these factors was identified to specifically support the growth of activated human T-lymphocytes and was termed T-lymphocyte growth factor or T-cell growth factor (TCGF) (3,4). It is now more commonly referred to as interleukin-2 (IL-2). Detailed studies have been reported on the characteristics of the cells that produce IL-2, accessory cells and factors involved in its induction, and on the target cell that responds to IL-2 (5-9). IL-2 has been purified to homogeneity from normal peripheral blood lymphocytes and also from gibbon and human leukemic cell lines (10-13). cDNAs encoding this lymphokine have been isolated and sequenced (10,14,15) and the IL-2 gene characterized (16-17). The IL-2 gene has been localized to chromosome 4q (18).

Normal resting lymphocytes do not produce IL-2, nor are they capable of responding to IL-2. When activated by antigens or lectins they are induced to produce and respond to this lymphokine. Like other polypeptide hormones, IL-2 exerts its effect through binding to high affinity-receptors on the surface of activated T-lymphocytes. These receptors are not present on resting lymphocytes, but are induced following activation by antigens or mitogens (19,20). Interaction of IL-2 with these activated T cells displaying high affinity binding sites leads to T-cell proliferation culminating in the emergence of effector T cells mediating helper, suppressor, and cytotoxic T cell function.

The lymphocyte conditioned media contain a large number of biologically active molecules that interact with cells in culture and produce a multitude of effects, some directly and others indirectly in a secondary fashion. Using crude conditioned media it was possible to establish long term cultures of human and murine T-cell clones (7,21). Such conditioned media were used in the initial establishment of T-cell lines from adult T-cell leukemia (ATL) patients (22), which led to the isolation of the first human retrovirus (HTLV-I) (23). Use of crude IL-2 preparations, however, made it difficult to define the precise functions of IL-2 and misled investigators into believing, for instance, that a B-cell growth factor (24,25) and a T-cell replacing factor (26,27) are coactivities of IL-2. These activities have subsequently been shown to be distinct from IL-2 (24,28,29). The need to purify IL-2 was recognized early, but initial attempts were thwarted by the extreme hydrophobicity of IL-2 and the consequent losses of the protein during purification steps. We will review the purificiation of IL-2 with some emphasis on procedures that have been found to be successful in our laboratories. We review also some of the major applications of the factor especially in the field of human retrovirology. A description of the characteristics of the IL-2 gene, and its organization and expression in human cell is included.

Production and Purification of IL-2

Lymphocytes from peripheral blood, spleen, tonsil, thoracic duct exudates, lymph nodes, and bone marrow have been used in the production of IL-2. However, peripheral blood mononuclear cells have been the most commonly used natural source of IL-2. Some established human and primate cell lines have been recognized as useful alternative sources. For instance Jurkat, a line derived from an acute lymphoblastic leukemia patient, makes almost 100-fold more IL-2 than normal peripheral blood lymphocytes when induced with PHA and the tumor promoter, phorbol myristate acetate (PMA) (30). Procedures to induce IL-2 production from cultures of normal human lymphocytes or from established human T-cell lines (e.g., Jurkat) have been described in detail (31). Procedures to purify IL-2 from the conditioned media from cultures of induced normal lymphocytes and Jurkat cells are similar. The following protocol was employed in our laboratory to obtain homogeneous IL-2 from cultures of normal PBL (10,11) and is equally applicable to purification of Jurkat IL-2.

Human IL-2 is present normally in small amounts in the conditioned media of PHA-stimulated lymphocyte cultures. Because of the enormous volumes of media that need to be processed to obtain significant amounts of pure IL-2, the initial step is a mjaor concentration step. This is routinely achieved by diafiltration using a Pellicon Cassette system (Millipore) employing the polysulfone membrane filter RTGC having a 10,000-dalton cut off limit.

If the medium used for the IL-2 induction contained albumin (or serum), the concentrated fraction may need to be chromatographed on a DEAE-cellulose column that effectively fractionates IL-2 from these serum proteins (31). If the factor is induced under serum-free (and albumin-free) conditions, this ion-exchange step is omitted. A major purification of IL-2 is achieved by adsorption onto controlled pore glass (CPG) and subsequent elution. The Pellicon cassette concentrate is transferred to a roller bottle and is mixed with CPG (75 Å pore size, 80 - 100 mesh) at a ratio of approximately 3 g of glass per liter of unconcentrated conditioned media. The adsorption is allowed to continue overnight at 4 °C using a roller apparatus. The medium is then removed, the glass washed well and IL-2 is eluted using 1 M tetramethylammonium chloride (TMAC) in 10 mM Tris-HCl, pH 8.2. The eluate is thoroughly dialyzed to remove TMAC which interferes in the assay for IL-2.

The CPG-purified IL-2 is then subjected to two successive high performance liquid chromatography steps using colums of C_{18}-silica. The material is acidified to pH 2 with 0.1 % trifluoroacetic acid (TFA) and pumped through the first C_{18}-silica column (2.5 x 30 cm) at a flow rate of 5 ml per minute. The column is eluted successively with 10, 45, 50, and 65 % aqueous acetonitrile containing 0.1 % TFA. The bulk of IL-2 is eluted in the 50 - 60 % acetonitrile fraction. This fraction is diluted with an equal volume of 0.1 % TFA containing 0.2 % polyethylene glycol and loaded on to a 0.38 x 30 cm column of µBondapak C_{18} (Waters Assoc.) at a flow rate of 1 ml per minute. After washing the column successively with 32.5 and 45 % acetonitrile, a 2-hr gradient between 45 and 65 % acetonitrile is initiated. The fractions containing IL-2 are pooled and, if necessary, rechromatographed on the same column, but this time using a linear 0 - 60 % acetonitrile gradient to elute IL-2. Homogeneous IL-2 elutes at about 60 % acetonitrile (Fig. 1) (11).

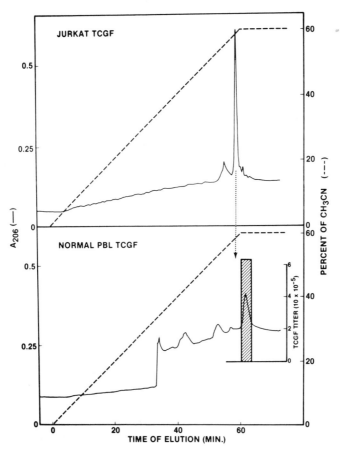

Fig. 1. Reverse phase high performance chromatography of human IL-2. The procedure is described in the text. The inset in the lower panel describes the IL-2 activity as measured using the [^{3}H]thymidine incorporation assay (31). All the activity was located in the absorbance peak. The chromatographic profile obtained with a sample of Jurkat IL-2 is given in the upper panel for comparison (Adapted from Ref. 11).

Recently, monoclonal antibodies specific for human IL-2 have been prepared and they have been useful as immuno-affinity supports for the purification of the lymphokine (32,33).

Properties of IL-2

The IL-2 molecule has been completely sequenced and it consists of 133 amino acids (10,32,33). The results confirmed the primary structure deduced from the gene sequence (10,14). As noted earlier, IL-2 is a highly hydrophobic protein and is known to appear as molecular aggregates. The same property contributed to the early findings of difficulty in purification and the enormous absorptive losses during processing of relatively pure IL-2. The molecule contains a single disulfide bond between cysteines at 58 and 105 and this appears to be important for the biological activity. Inappropriate disulfide bonding between one of the above cysteines and the cysteine at position 125 was a problem in the purification of recombinant bacterial IL-2 because of aggregation and lack of solubility. This was overcome by a site specific mutagenesis to change cysteine 125 to a serine (34). This change did not affect the activity of the IL-2. One post-translational modification that has been identified involves a threonine residue at position 3. Fast atom bombardment and mass spectroscopy of the tryptic peptide of amino acids 1-8 indicated that the threonine was either not modified or was modified initially by O-linked glycosylation with N-acetyl-galactosamine and subsequently with galactose and sialic acid (32). The differences observed in the isoelectric points and the resultant molecular heterogeneity of IL-2 from different sources are attributed to these differences in the post-translational modification of IL-2 (11,35). The fact that all forms of IL-2, including the recombinant IL-2 with no glycosylation at all are biologically active, indicate that these post-translational changes are not critical for the biological activity.

In addition to T-cell growth promoting activity, IL-2 has been implicated in a variety of other biological functions which may or may not be brought about by direct effect of IL-2. For instance, IL-2 reportedly activates natural killer cell function, presumably through the induction of interferon-γ production (36). Large quantities of IL-2 also can activate killer cells capable of lysing solid tumor cells (LAK or lymphokine activated killer cells) (37-39). IL-2 can stimulate production of B cell growth factor and interferon-γ by T cells (40). High affinity IL-2 receptors are expressed on some activated B cells (41) and addition of IL-2 may promote both B-cell proliferation and differentiation into immunoglobulin-producing cells.

T Cell Growth and Isolation of Human Retroviruses

Normal resting T cells do not express receptors for IL-2. The receptor expression is induced upon activation with an antigen or lectin. These receptor-positive T cells proliferate in response to IL-2. The proliferative response is mediated through binding of IL-2 molecules on the cell surface receptor. The IL-2 receptor was first inferred by the ability of activated T cells to absorb IL-2 from solutions and by the fact that these absorbed IL-2 molecules could be eluted from the cell surface by acidic media (42-44). The study of IL-2 receptor was greatly facilitated by the observation that a monoclonal antibody (anti-Tac) with a specificity to an activated T-cell antigen (45) was, in fact, reacting with IL-2 receptor (46). T cells have been shown to have receptors with two different affinities for IL-2 (19,47). The high affinity receptors appear to mediate the growth promoting activity. The function of the low affinity receptors is unknown. Both types of receptors bind anti-Tac antibody equally well.

In contrast to normal human lymphocytes, cells derived from patients with adult T-cell leukemia (ATL) constitutively express receptors for

for IL-2 (48). Waldmann and coworkers have been able to distinguish this syndrome, associated with HTLV-I infection (see later), from those with Sezary syndrome, a different type of leukemia of mature T cells that may present with similar cutaneous manifestations (49). ATL cells express 5 to 10-fold more receptors per cell than maximally activated normal T cells (48). Moreover, IL-2 receptor-negative cord blood T cells express large numbers of IL-2 receptors after infection with HTLV-I (48,50).

IL-2 provided the first and basic tool to produce long-term T-cell cultures of normal and neoplastic T-cells. Moreover, neoplastic T cells could be selectively grown from mononuclear cells of ATL patients because, unlike the normal cells, the leukemic cells respond to IL-2 without prior activation. Several primary T-cell lines were established from patients with mature T-cell malignancies with the help of IL-2 (22). The first human retrovirus, named human T-leukemia (lymphotropic) retrovirus type I (HTLV-I), was isolated in 1979 from such a cell line from a patient who was originally diagnosed as having an aggressive form of mycosis fungoides, the diagnosis being later reclassified on as ATL of the type commonly seen in southern Japan and the Caribbean basin (23).

Subsequently, many other isolates of HTLV-I were obtained by co-culturing normal human T cells with lymphocytes from ATL patients (50-51). A distinct but related retrovirus was isolated from lymphocytes of a patient with hairy cell leukemia (56), which has been termed HTLV-II. The experience with maintaining T cells in culture and the isolation of retroviruses played a crucial role in the more recent isolation and detailed characterization of the retrovirus associated with human acquired immune deficiency syndrome (AIDS) (57-60). Since this virus has some structural and functional similarities to HTLV-I and HTLV-II (61-63), it was termed HTLV-III. (See elsewhere in this volume).

Regulation of IL-2 Gene Expression in Normal and HTLV-Infected Human Cells

Although IL-2 initially played a critical role in successful isolation of all known human retroviruses, many of the HTLV-infected T-cell lines grow in culture independent of exogeneously added IL-2. Many of these are mature or nearly mature T-cell lines. Given the backdrop of the requirement of IL-2 for normal T-cell growth, we initially hypothesized that HTLV-infected T cells produce and respond to their own IL-2 (44,64). We surmised that these cells use either autostimulation or parastimulation to sustain the growth of the population (65). These notions were buttressed by the fact that many HTLV-infected cells display abundant membrane receptors for IL-2 (50,66), and only an uncommon HTLV-infected cell line produces IL-2 (44). The fact that many of the HTLV-I infected cell lines did not elaborate detectable IL-2 could be explained by a rapid utilization of IL-2 by these cells coupled with the lack of sensitivity of assays used to measure the release of IL-2.

The availability of cloned IL-2 DNA (10,14) allowed us to evaluate these issues and to investigate the regulation of IL-2 gene expression in normal and aberrant cells. Sequence analysis of the cloned DNA shows that the functional part of the IL-2 gene contains a coding sequence of 459 base pairs (bp), and additional 5' and 3' untranslated sequences of about 50 and 250 bp, respectively. Since we had determined the N-terminal and C-terminal amino acid sequences of the purified IL-2, a comparison of the predicted amino acid sequence with the actually determined sequence allowed us to conclude that the first 60 bp of the coding sequence of the IL-2 gene corresponds to the signal peptide of 133 amino acid residues as already described. This possibility was also earlier suggested by Taniguchi et al. (14). Furthermore, this comparison also allowed us to conclude that there was no proteolytic cleavage at the C-terminus to generate mature protein.

Cloned IL-2 DNA from different cell sources allowed the examination of the basis of physicochemical differences between normal (PBL) and leukemic (Jurkat) IL-2. These differences could result from post-translational modifications of the protein or from polymorphism of the gene itself. It was possible that IL-2 was derived from a gene family with multiple members, only one of which was expressed in a given cell type. However, comparison of the DNA sequence of our PBL cDNA clone (10) with that of Jurkat cDNA clone (14) showed them to be identical, except for one nucleotide difference which did not change the amino acid sequence of the protein (10). We therefore concluded that the differences in physicochemical properties of the two proteins were due to post-translational modifications (10,11). Als already discussed earlier, this has now been confirmed (32).

The production of IL-2 is regulated at the level of the mRNA synthesis (10). The induction of IL-2 is primarily the induction of IL-2 gene transcription (10,67). Using Northern blot analysis, we showed that the IL-2 mRNA could be detected only in those human cells which were known to produce IL-2 (Fig. 2 and Table 1). To Understand further the induction of IL-2 mRNA synthesis and hence IL-2 production, we examined whether the regulation occurred at the level of initiation of gene transcription or other post-transcriptional events such as mRNA stability were also contributory factors. Our nuclear transcription or "run off" experiments showed that the primary induction event is the initiation of IL-2 gene transcription (Fig. 3) and the degree of transcriptional activation of the IL-2 gene in Jurkat cells correlates with the abundance of mRNA in cells induced with PHA, PMA, and PHA plus PMA (67). PHA alone induces some gene transcription which is markedly enhanced by the inclusion of PMA, suggesting synergy in the effect of these two inducers. The results of experiments using RNA synthesis inhibitors to block further transcription of the IL-2 gene in induced cells suggest that PMA may additionally affect the half-life of IL-2 mRNAs possibly by increasing the specific degradation of IL-2 mRNA or alternatively, by increasing the rat of its translation (67).

Since the two inductive signals for PBL are provided by PHA and IL-1 and those for Jurkat cells by PHA and PMA, we examined if PMA acted in Jurkat cells in a manner analogous to the action of IL-1 in PBL. If this is the case, we would expect that Jurkat cells maximally induced with PMA will not display any further induction with IL-1. However, we found that IL-1 was able to further enhance the abundance of IL-2 mRNA in Jurkat cells maximally induced with PMA and PHA (Fig. 4) and thus IL-1 was synergistic with PMA (67). This implies that PMA and IL-1 act by different mechanisms in enhancing the transcriptional activity of the IL-2 gene. It is likely that both of these inducers act by binding to specific membrane receptors and generate secondary mediators. The maximal inducing concentration of a given inducer may pertain to its capacity to bind the specific membrane receptors and not necessarily to the capacity of the IL-2 gene to be induced by secondary mediators activated by that inducer. Thus, PMA and IL-1 could individually generate different or the same secondary mediators, which could affect IL-2 gene transcription by acting in concert on the same basic process such as the rate of gene transcription.

The IL-2 specific mRNA is readily detected in Jurkat cells at 2 hr

Table 1. Expression of the IL-2 Gene in Human Cells.

Cell	Inducer	Relative mRNA abundance*
T-cells:		
PBL	-	-
	PHA	+++
	PMA	NT
	PHA + PMA	+++
Jurkat	-	-
	PHA	+
	PMA	+
	PHA + PMA	+++
H4	-	+
	PHA + PMA	++
H9	-	+
	PHA + PMA	+++
HUT78	-	+
CEM	-	-
	PHA + PMA	-
Molt-4	-	-
	PHA + PMA	-
HSB-20	-	-
	PHA + PMA	-
B-cells:		
Daudi	-	-
	PHA + PMA	-
Raji	-	-
HTLV-I infected T-cells:		
HUT102	-	+
MI	-	-
	PHA + PMA	-
MJ	-	-
MT2	-	-
C2/MJ	-	-
	PHA + PMA	-
C5/MJ	-	-
1C/UK	-	-
	PHA + PMA	-
HTLV-II infected T-cells:		
MO	-	+
HTLV-III infected T-cells:		
H4/HTLV-III	-	+
	PHA	++
	PMA	++
	PHA + PMA	+++
H9/HTLV-III	-	+
	PHA + PMA	+++
Other cells:		
3A (trophoblasts)	-	-
SD (trophoblasts)	-	-
HL60 (myeloid)	-	-

* Number of (+) symbols is meant to provide approximate description; NT, not tested.

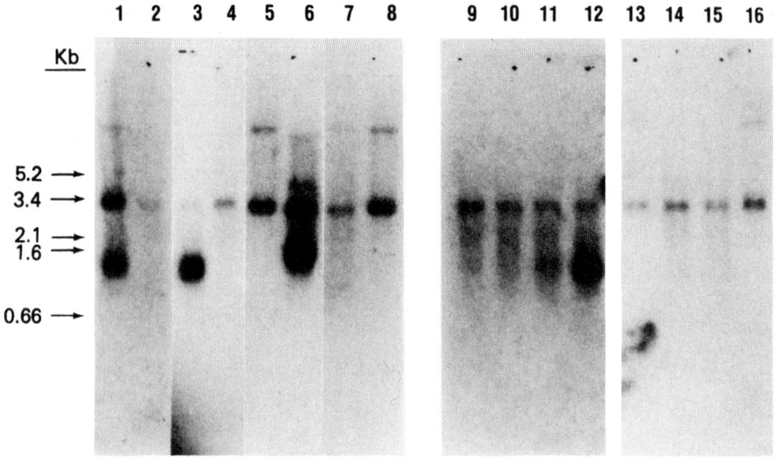

Fig. 2. Expression of IL-2 gene in human cells analyzed by hybridization of [^{32}P]-labeled IL-2 cDNA with cellular RNA. Lanes 1-16 are for: 1, PHA-stimulated lymphocytes; 2, unstimulated lymphocytes; 3, PHA + PMA stimulated Jurkat cells; 4, unstimulated Jurkat cells; 5, 6G1 cells; 6, MLA 144 cells; 7, Molt-4 cells; 8, HL-60 cells; 9, unstimulated Jurkat cells; 10, PMA-stimulated Jurkat cells; 11, PHA-stimulated Jurkat cells; 12, PHA + PMA-stimulated cells; 13, Daudi cells; 14, Raji cells; 15, trophoblast 3A cells; 16, SD cells. IL-2-specific RNA appears as a 900-nucleotide band (1100 nucleotides for MLA 144 cells). The 2300-nucleotide band is for an abundant mRNA species constitutively expressed in many cell types. (Adapted from Ref. 10.).

post-induction, rises to maximal levels at 4-6 hr and is maintained at these levels for 8-10 hr. These levels steadily decline during the next 24 hr or more. These kinetics suggest a tight control of IL-2 gene expression. The rather rapid decline suggests either a cessation of IL-2 gene transcription or increasing rate of disappearance of IL-2 mRNA by its utilization for translation and/or degradation.

An understanding of the regulation of IL-2 gene expression allows us to quantitatively evaluate the basis for the autonomous growth of the HTLV-I infected T-cell line. We should be able to detect IL-2 mRNA in these cells with sensitive Northern blot analysis utilizing a cloned IL-2 probe, if the growth regulation of these cells involves IL-2 gene expression. This would be true regardless of whether or not these cells release into the medium the IL-2 that they might synthesize. Our careful analysis of HTLV-I infected cells has not provided any evidence for

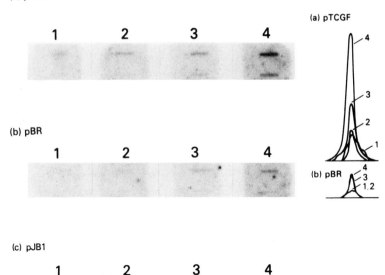

Fig. 3. Analysis of nuclear transcripts of IL-2 gene from induced and uninduced Jurkat cells. Lane 1, control cells; lane 2, cells treated with PMA; lane 3, cells treated with PHA; lane 4, cells treated with PHA plus PMA. pBR and pJB1 are negative and positive controls, respectively. Curves on the right are the respective densitometer tracings of the top band shown on left. (Adapted from Ref. 67).

active IL-2 gene transcription either constitutively or subsequent to induction with PHA and PMA in many of these cells (Fig. 5, Table I) (64). The only exceptions thus far have been the HTLV-I infected HUT102 cells which often, but not always, synthesize barely detectable IL-2 mRNA and the HTLV-II infectet MO cell line which we have not investigated in detail. We conclude that a majority of the HTLV-I infected cell lines are truly independent of IL-2 for growth. Curiously, HTLV-I infected cells invariably display abundant membrane receptors for IL-2 (50,66). But since these cells do not produce the ligand, IL-2, they bypass the IL-2 pathway of growth control. Viewed from a different perspective, the IL-2-independent growth of these cells may precisely be among the reasons underlying their immortalization.

The more recent isolation of the AIDS virus (HTLV-III/LAV) (57,58, 68,69) has allowed us to extend our studies on the impact of retrovirus infection on lymphokine gene expression in human T cells. HTLV-III, like HTLV-I and -II, is clearly T-lymphotropic and shares some structural and biological similarities with HTLV-I/BLV group of retroviruses (61, 62,70-74). However, HTLV-III also differs from other HTLVs in possessing at least two novel genes, sor (short open reading frame) and 3'-orf (3'-open reading frame), which may impart biological attributes to

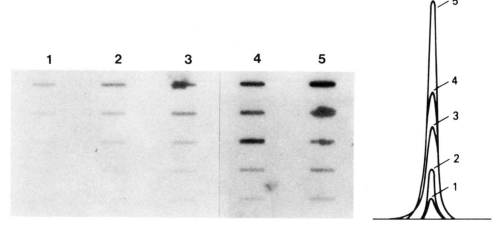

Fig. 4. Induction of IL-2 mRNA in Jurkat cells by PHA, PMA, and IL-1. Lane 1, control cells; lane 2, cells treated with PHA; lane 3, cells treated with PHA plus IL-1; lane 4, cells treated with PHA plus PMA; lane 5, cells treated with PHA plus PMA plus IL-1. Curves on the right are the respective densitometer tracings of the top bands shown on left. (Adapted from Ref. 67).

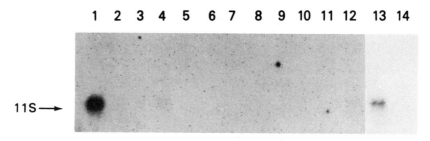

Fig. 5. Expression of IL-2 gene in HTLV-I infected cells. Lanes 1 to 14 are for: 1, PHA plus PMA-stimulated Jurkat cells; 2, IL-2-independent HUT 78 cells; 3, IL-2-independent HUT 102 cells; 4, another preparation of IL-2-independent HUT 102 cells; 5, IL-2-independent C5/MJ cells; 6, IL-2-dependent C5/MJ cells; 7, IL-2-independent C10/MJ cells; 8, IL-2-independent C10/MJ cells treated with PMA plus PHA; 9, IL-2-independent B2/UK cells; 10, IL-2-independent B2/UK cells treated with PMA plus PHA; 11, IL-2-independent MT-2 cells; 12, IL-2-dependent M1 cells; 13, IL-2-independent MO cells; and 14, IL-2-independent Molt 4 cells (immature T). (Adapted from Ref. 64).

HTLV-III that differ from HTLV-I (75,76). In fact, HTLV-III and HTLV-I have dramatically different effects on human T cells. Whereas HTLV-I is generally transforming, HTLV-III is usually cytopathic. It is this cytopathology of HTLV-III which largely underlies immune deficiency in AIDS. We thought that T-cell depletion in AIDS may be due to the diminished production of IL-2 by infected cells resulting in the starvation of T cells for their growth factor. The analysis of cultures of HTLV-III infected cells has, however, shown that the IL-2 gene expression and IL-2 production are not impaired in these cells (Fig. 6) (77). The induction of IL-2 gene expression by PHA and PMA in these cells is again a consequence of the transcriptional activation of the IL-2 gene as demonstrated by nuclear transcription experiments (77). Perhaps significantly, the induced levels of IL-2 mRNA in HTLV-III infected T cells were measurably higher than in uninfected cells, while the reverse may be the case for uninduced mRNA levels. This aspect requires further close scrutiny as it may indicate subtle differences in the regulation of IL-2 gene expression in HTLV-III infected cells. For the present, it appears that HTLV-III infection per se does not abrogate IL-2 gene expression and induction. Similar observations have been made for HTLV-III positive T cells cultured directly from the lymphocytes of AIDS patients (78). It is thus likely that T-cell depletion in AIDS is due to factors other than the unavailability of IL-2. On the other hand, unregulated production of IL-2 by HTLV-III infected cells may directly or indirectly affect T-cell proliferation and functions.

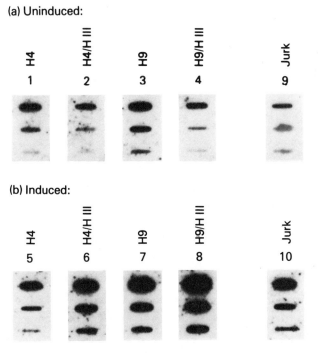

Fig. 6. Expression of IL-2 gene in HTLV-III infected cells. H4 and H9 denote uninfected cells and H4/HIII and H9/HIII are the corresponding HTLV-III-infected cells. Jurk, Jurkat cells. (Adapted from Ref. 77).

Organization and Control of IL-2 Gene

The lack of expression and inducibility of the IL-2 gene in HTLV-I infected cells poses the question of how is this gene turned off in these cells. This is not due to some general consequence of the retrovirus infection of T cells as this gene is expressed and inducible in HTLV-III infected cells. It is also unlikely to be due to a direct and in cis down-regulation of the IL-2 gene by HTLV-I genome, since HTLV-I integrates into the cell genome randomly at multiple and different sites in different infected cells. As a guide to further exploration, we postulate two broad possible mechanisms: 1) the IL-2 gene is structured and organized differently in HTLV-I infected cells than in normal and other inducible cells; 2) the transduced intracellular events that control the inducibility of IL-2 gene are impaired in HTLV-I infected cells. Viewed from a different perspective, we can postulate that lymphokine expression in T cells is a differentiation state-specific and/or cell cycle-specific phenomenon. We can further postulate that HTLV fixes the state of genetic expression of a cell at the time of the establishment of infection. We now need only to postulate that HTLV-I infection occurs when IL-2 gene is not expressed, and conversely, HTLV-III infection occurs when the IL-2 gene is in the expression mode.

We have shown that the IL-2 gene exists as a single copy gene in normal, HTLV-I- and HTLV-III-infected or uninfected cells (Ref. 10 and Arya, S.K., unpublished results). Others have made similar observations for other cell types (17). This is consistent with a single location of the IL-2 gene on human chromosome 4q (18). We have not detected any difference at the level of restriction enzyme site mapping in the organization of the IL-2 gene in HTLV-I-infected cells and in a variety of other cells, including HTLV-III-infected cells. Thus, no rearrangement or polymorphism of the IL-2 gene is detectable in producer and non-producer cells at this level. This does not rule out subtle differences in the structure and organization of the gene between HTLV-I-infected and normal cells.

The IL-2 gene in normal human cells (PBL) consists of four exons and three introns of varying lengths (16,17). The first exon contains the 5'-untranslated region (47 bp) and codes for the first 49 amino acids (147 bp) of the protein, including the 20 amino acids of the signal peptide. The second and third exons encode the next 20 and 48 amino acids (60 and 144 bp), respectively. The fourth exon codes for the last 36 amino acids (108 bp) of the carboxy terminus and also contains 216 bp of the 3'-untranslated sequence. The first, second and third introns are, respectively, 91, 2292, and 1364 bp long. Thus, the gene encoding the 133 amino acid (399 bp)-mature IL-2 protein is distributed over about 5 kb of DNA length.

Alternative to structural difference is the difference in the intracellular events that control the expression and induction of the IL-2 gene in HTLV-I infected cells. It is possible that the IL-2 gene contains both positive (enhancer-promoter) and negative (repressor-abrogator) regulatory sequence elements. The interaction of certain effectors with the positive regulatory sequence enhances gene transcription and interaction with the negative regulatory sequence results in the abrogation of transcription. The balance between these interactions governs the transcription of the gene. We assume that inducing agents operate by transducing signals through intracellular effector molecules. While in normal cells, the positive regulatory interactions are functional, these interactions may not be functional in HTLV-I infected cells or be dominated by negative regulatory interactions.

In summary, the IL-2 gene is an inducible lymphokine gene which exists as a single copy gene in all cell types examined. Its expression is primarily regulated at the level of gene transcription. Though other factors may play a role, they are secondary to the induced transcriptional activation of the gene. The molecular basis for the lack of expression and induction of the IL-2 gene in HTLV-I infected T cells remains to be explained, especially in light of the fact that this gene is transcriptionally active and inducible in HTLV-III-infected T cells. It is paradoxical that the IL-2 gene is inactive in HTLV-I-infected cells which are immortal and it is active in HTLV-III-infected cells which are prone to cell death. The molecular events in transduction of extracellular events into intracellular signals influencing lymphokine gene expression are yet a mystery. The fundamentals of regulatory circuits involving these lymphokines in immune regulation are far from understood.

References

1. P.C. Nowell, Phytohemagglutinin induces DNA synthesis in leucocyte cultures. Cancer Res., 20: 462-467 (1960).
2. J.J. Oppenheim, M.B. Mizel, and M.S. Meltzer, Biological properties of lymphocyte-derived and macrophage-derived mitogenic "amplification" factors, in: "The Biology of Lymphokines" (S. Cohen, E. Pick, and J.J. Oppenheim, Eds.), Academic Press, New York, pp. 399-402 (1979).
3. D.A. Morgan, F.W. Ruscetti, and R.C. Gallo, Selection in vitro growth of T-lymphocytes from normal human bone marrows. Science, 193: 1007-1008 (1976).
4. F.W. Ruscetti, D.A. Morgan, and R.C. Gallo, Functional and morphologic characterization of human T cells continously grown in vitro. J. Immunol., 119: 131-138 (1977).
5. F.W. Ruscetti, and R.C. Gallo, Human T-lymphocyte growth factor: Regulation of growth and function of T lymphocytes. Blood, 57: 379-394 (1981).
6. K.A. Smith, T-Cell growth factor. Immunol. Rev., 51: 337-356 (1980).
7. M.H. Schrier, N.N. Iscove, R. Tees, L. Aardon, and H. von Boehmer, Clones of killer and helper T cells: Growth factor requirements, specificity, and retention of function in long term culture. Immunol. Rev., 51: 314-336 (1980).
8. H.von Boehmer, and W. Haas, H-2 restricted cytolytic and non-cytolytic T-cell clones. Isolation, specificity and functional analysis. Immunol. Rev., 54: 27-56 (1981).
9. B.Sredni, and R.H. Schwartz, Antigen-specific T-cell clones: Methodology, specificity, MHC restriction and alloreactivity. Immunol. Rev., 54: 198-225 (1981).
10. S.C. Clark, S.K. Arya, F. Wong-Staal, M. Matsumoto-Kobayashi, R.M. Kay, R.J. Kaufman, E.L. Brown, C. Shoemaker, T. Copeland, S. Oroszlan, K. Smith, M.G. Sarngadharan, S.G. Lindner, and R.C. Gallo, Human T-cell growth factor: Partial amino acid sequence, cDNA cloning, and organization and expression in normal and leukemic cells. Proc. Natl. Acad. Sci. USA, 81: 2543-2547 (1984).
11. R.C. Gallo, S.K. Arya, S.G. Lindner, F. Wong-Staal, and M.G. Sarngadharan, Human T-cell growth factor, growth of human neoplastic T cells, and human T-cell leukemia-lymphoma virus, in: "Thymic Hormones and Lymphokines" (A.L. Goldstein, Ed.) New York, Plenum Publishing Corporation, pp. 1-17 (1984).

12. L.E. Henderson, J.F. Hewetson, R.F. Hopkins, R.C. Sowder, R.H. Neubauer, and H. Rabin, A rapid, large-scale purification procedure for gibbon interleukin-2. J. Immunol., 131: 810-815 (1983).
13. R.J. Robb, R.M. Kutny, and V. Chowdhry, Purification and partial sequence of human T-cell growth factor. Proc. Natl. Acad. Sci. USA, 80: 5990-5994 (1983).
14. T.Taniguchi, H. Matsui, T. Fujita, C. Takaoka, N. Kashima, R. Yoshimoto, and J. Hamuro, Structure and function of a cloned cDNA for human interleukin-2. Nature, 301: 306-310 (1983).
15. R.Devos, G. Placetinck, H. Cheroutre, G. Simons, W. Degrave, J. Tavernier, E. Remaut, and W. Fiers, Molecular cloning of human interleukin-2 cDNA and its expression in E. coli. Nucleic Acids Res., 11: 4307-4322 (1983).
16. T.Fujita, C. Takaoka, H. Matsui, and T. Taniguchi, Structure of the human interleukin-2 gene. Proc. Natl. Acad. Sci. USA, 80: 7437-7441 (1983).
17. N.Holbrook, K.A. Smith, A.J. Fornace, C. Comean, R.L. Wiskocil, and G.R. Crabtree, T-cell growth factor: Complete nucleotide sequence and the organization of the gene in normal and malignant cells. Proc. Natl. Acad. Sci. USA, 81: 1634-1638 (1984).
18. L.J. Siegel, M.E. Harper, F. Wong-Staal, R.C. Gallo, W.G. Nash, and S.J. O'Brien, Gene for T-cell growth factor: Localization on human chromosome 4q and feline chromosome B1. Science, 223: 175-178(1984)
19. R.J. Robb, A. Munck, and K.A. Smith, T-cell growth factor receptors: Quantitation, specificity, and biological relevance. J. Exp. Med., 154: 1455-1474 (1981).
20. W.C. Green, and R.J. Robb, Receptors for T-cell growth factor: Structure, Function, and expression on normal and neoplastic cells. in: "Contemporary Topics in Molecular Immunology" (S.Gillis, and F.P. Inman, Eds.), Plenum Publishing Corp., New York, Vol. 10, pp.1-34 (1985).
21. S.Gillis, and K.A. Smith, Long-term culture of tumor-specific cytotoxic T-cells. Nature, 268: 154 (1977).
22. B.J. Poiesz, F.W. Ruscetti, J.W. Mier, A.M. Woods, and R.C. Gallo, T-cell lines established from human T-lymphocytic neoplasias by direct response to T-cell growth factor. Proc. Natl. Acad. Sci. USA, 77: 6815-6819 (1980).
23. B.J. Poiesz, F.W. Ruscetti, A.F. Gazdar, P.A. Bunn, J.D. Minna, and R.C. Gallo, Detection and isolation of type-C retrovirus particles from fresh and cultured lymphocytes of a patient with cutaneous T-cell lymphoma. Proc. Natl. Acad. Sci. USA, 77: 7415-7419 (1980).
24. S.L. Swain, G. Dennert, J.F. Warner, and R.W. Dutton, Culture supernatants of a stimulated T-cell line have helper activity that acts synergistically with interleukin-2 in the response of B cell to antigen. Proc. Natl. Acad. Sci. USA, 71: 2517-2521 (1981).
25. H.J. Leibson, P. Marrick, and J.W. Kappler, B-cell helper factors. I. Requirement for both interleukin-2 and another 40,000 mol. wt. factor. J. Exp. Med., 154: 1681-1693 (1981).
26. J.D. Watson, S. Gillis, J. Marbrook, D. Mochizuki, and K. Smith, Biochemical characterization of lymphocyte regulatory molecules. I. Purification of a class of murine lymphokines. J. Exp. Med., 150: 849-856 (1979).
27. A.Granelli-Piperno, J.D. Vassalli, and E. Reich, Purification of murine T-cell growth factor. A lymphocyte mitogen with helper activity. J. Exp. Med., 154: 422-431 (1981).
28. K.Takatsu, K. Tanaka, A. Tominaga, Y. Kumahara, and T. Hamaoka, Antigen-induced T-cell replacing factor (TRF). III. Establishment of T-cell hybrid clone continuously releasing TRF. J. Immunol., 125: 2646-2653 (1980).
29. W.E. Paul, B. Sredni, and R.H. Schwartz, Long-term growth and cloning

of nontransformed lymphocytes. Nature, 294: 697-699 (1981).
30. S.Gillis, and J. Watson, Biochemical and biological characterization of lymphocyte regulatory molecules. V. Identification of an interleukin-2 producing human T cell line. J. Exp. Med., 152: 1709-1719 (1980).
31. M.G. Sarngadharan, R.C. Ting, and R.C. Gallo, Methods for production and purification of human T-cell growth factor, in: "Methods for Serum-Free Culture of Neuronal and Lymphoid Cells", (D.W. Barnes, D.A. Sirbasku, and G.H. Sato, Eds.) Alan R. Liss, Inc., New York, pp. 127-144 (1984).
32. R.J. Robb, R.M. Kutny, M. Panico, H.R. Morris, and V. Chowdhry, Amino acid sequence and post translational modification of human interleukin 2. Proc. Natl. Acad. Sci. USA, 81: 6486-6490 (1984).
33. T.D. Copeland, K.A. Smith, and S. Oroszlan, Characterization of immunoaffinity-purified human T-cell growth factor from Jurkat cells, in: "Thymic Hormones and Lymphokines" (A.L. Goldstein, Ed.) Plenum Publishing Corp., New York, pp. 181-189 (1984).
34. A.Wang, S.-D. Lu, and D.F. Mark, Site specific mutagenesis of the human interleukin-2 gene: Structure-function analysis of the cysteine residues. Science, 224: 1431-1433 (1984).
35. R.J. Robb, and K.A. Smith, Heterogeneity of human T cell growth factor(s) due to variable glycosylation. Mol. Immunol., 18: 1087-1094 (1981).
36. J.J. Farrar, W.R. Benjamin, M.L. Hilfiker, M. Howard, W.L. Farrar, and J. Fuller-Farrar, The biochemistry, biology and role of interleukin-2 in the induction of cytotoxic T cell and antibody forming B cell responses. Immunol. Rev., 63: 129-166 (1982).
37. M.T. Lotze, E.A. Grimm, A. Mazumdar, J.L. Strausser, and S.A. Rosenberg, In vitro growth of cytotoxic T lymphocytes: Lysis of fresh and cultured autologous tumor by lymphocytes cultured in T-cell growth factor. Cancer Res., 41: 4420-4425 (1981).
38. E.A. Grimm, D.J. Wilson, and S.A. Rosenburg, Lymphokine-activated killer cell phenomenon. II. The precursor phenotype is serologically distinct from peripheral blood T cells, memory CTL, and NK cells. J. Exp. Med., 157: 884-897 (1983).
39. E.A. Grimm, R.J. Robb, J.A. Roth, M. Neckers, L.B. Lachman, D.J. Wilson, and S.A. Rosenberg, Lymphocyte activated killer cell phenomenon: Evidence that IL-2 is sufficient for direct activation of peripheral blood lymphocytes into lymphokine-activated killer cells. J. Exp. Med., 158: 1356-1361 (1983).
40. M.Howard, L. Matis, T.R. Malek, E. Shevach, W. Kell, D. Cohen, K. Nakanishi, and W.E. Paul, Interleukin-2 induces antigen-reactive T cell lines to secrete BCGF-1. J. Exp. Med., 158: 2024-2039 (1983)
41. T.A. Waldmann, C.K. Goldman, R.J. Robb, J.M. Depper, W.J. Leonard, S.O. Sharrow, K.F. Bongiovanni, S.J. Korsmeyer, and W.C. Greene, Expression of interleukin-2 receptors on activated human B cells. J. Exp. Med., 160: 1450-1466 (1984).
42. G.D. Bonnard, D. Yosaka, and D. Jacobson, Ligand-activated T-cell growth factor induced proliferation: Absorption of T-cell growth factor by activated T cells. J. Immunol., 123: 2704-2708 (1979).
43. A.Coutinho, E.L. Larsson, K.O. Gronvik, and J. Anderson, Studies on T lymphocyte activation. II. The target cell for concanavalin A induced growth factors. Eur. J. Immunol., 9: 587-592 (1979).
44. J.E. Gootenberg, F.W. Ruscetti, J.W. Mier, A. Gazdar, and R.C. Gallo, Human T-cell lymphoma and leukemia cell lines produce and respond to T-cell growth factor. J. Exp. Med., 154: 1403-1417 (1981).
45. T.Uchiyama, S. Broder, and T.A. Waldmann, A monoclonal antibody (anti-Tac) reactive with activated and functionally mature human T cells. J. Immunol., 126: 1393-1397 (1981).

46. W.J. Leonard, J.M. Depper, T. Uchiyama, K.A. Smith, T.A. Waldmann, and W.C. Greene, A monoclonal antibody that appears to recognize the receptor for human T-cell growth factor. Partial characterization of the receptor. Nature, 300: 267-269 (1982).
47. R.J. Robb, W.C. Greene, and C.M. Rusk, Low and high affinity cellular receptors for interleukin-2. Implications for the level of Tac antigen. J. Exp. Med., 160, 1126-1146 (1984).
48. J.M. Depper, W.J. Leonard, M. Kronke, T.A. Waldmann, and W.C. Greene, Augmented T-cell growth factor receptor expression in HTLV-infected human leukemic T cells. J. Immunol., 133: 1691-1695 (1984).
49. T.A. Waldmann, W.C. Greene, P.S. Sarin, C. Saxinger, D.W. Blayney, W.A. Blattner, C.K. Goldman, K. Bongiovanni, S. Sharrow, J.M. Depper, W. Leonard, T. Uchiyama, and R.C. Gallo, Functional and phenotypic comparison of human T cell leukemia/lymphoma virus positive adult T cell leukemia with human T cell leukemia/lymphoma virus negative Sezary leukemia, and their distinction using anti-Tac. J. Clin. Invest., 73: 1711-1718 (1984).
50. M. Popovic, G. Lange-Wantzin, P.S. Sarin, D. Mann, and R.C. Gallo, Transformation of human umbilical cord blood T cells by human T-cell leukemia/lymphoma virus. Proc. Natl. Acad. Sci. USA, 80: 5402-5406 (1983).
51. I. Miyoshi, I. Kubonishi, S. Yoshimoto, T. Akagi, Y. Ohtsuki, Y. Shireishi, K. Nagato, and Y. Hinuma, Type C virus particles in a cord T-cell line derived by co-cultivating normal human cord leukocytes and human leukemic T cells. Nature, 294: 770-771 (1981).
52. M. Yoshida, I. Miyoshi, and Y. Hinuma, Isolation and characterization of retrovirus from cell lines of human adult T-cell leukemia and its implication in the disease. Proc. Natl. Acad. Sci. USA, 79: 2031-2035 (1982).
53. R.C. Gallo, Human T-cell leukaemia-lymphoma virus and T-cell malignancies in adults, in: "Cancer Surveys" (J. Wyke, and R. Weiss, Eds.), Oxford University Press, Oxford, Vol. 3, pp. 113-159 (1984).
54. M. Popovic, P.S. Sarin, M. Robert-Guroff, V.S. Kalyanaraman, D. Mann, J. Minowada, and R.C. Gallo, Isolation and transmission of human retrovirus (human T-cell leukemia virus). Science, 219: 856-859 (1983).
55. P.D. Markham, S.Z. Salahuddin, V.S. Kalyanaraman, M. Popovic, P. Sarin, and R.C. Gallo, Infection and transformation of fresh human umbilical cord blood cells by multiple sources of human T-cell leukemia-lymphoma virus (HTLV). Int. J. Cancer, 31: 413-420 (1983).
56. V.S. Kalyanaraman, M.G. Sarngadharan, M. Robert-Guroff, I. Miyoshi, D. Blayney, D. Golde, and R.C. Gallo, A new subtype of human T-cell leukemia virus (HTLV-II) associated with a T-cell variant of hairy cell leukemia. Science, 218: 571-573 (1982).
57. M. Popovic, M.G. Sarngadharan, E. Read, and R.C. Gallo, Detection, isolation, and continuous production of cytopathic retroviruses (HTLV-III) from patients with AIDS and pre-AIDS. Science, 224: 497-500 (1984).
58. R.C. Gallo, S.Z. Salahuddin, M. Popovic, G.M. Shearer, M. Kaplan, B.F. Haynes, T.J. Palker, R. Redfield, J. Oleske, B. Safai, G. White, P. Foster, and P.D. Markham, Frequent detection and isolation of cytopathic retroviruses (HTLV-III) from patients with AIDS and at risk for AIDS. Science, 224: 500-503 (1984).
59. J. Schuepbach, M. Popovic, R.V. Gilden, M.A. Gonda, M.G. Sarngadharan, and R.C. Gallo, Serological analysis of a subgroup of human T-lymphotropic retroviruses (HTLV-III) associated with AIDS. Science, 224: 503-505 (1984).

60. M.G. Sarngadharan, M. Popovic, L. Buch, J. Schuepbach, and R.C. Gallo, Antibodies reactive with human T-lymphotropic retroviruses (HTLV-III) in the serum of patients with AIDS. Science, 224: 506-508 (1984).
61. S.K. Arya, R.C. Gallo, B.H. Hahn, G.M. Shaw, M. Popovic, S.Z. Salahuddin, and F. Wong-Staal, Homology of genomes of AIDS-associated virus with genomes of human T-cell leukemia viruses. Science, 225: 927-930 (1984).
62. L. Ratner, W. Haseltine, R. Patarca, K.J. Livak, B. Starcich, S.F. Josephs, E.R. Doran, J.A. Rafalski, E.A. Whitehorn, K. Baumeister, L. Ivanoff, S.R. Petteway, M.L. Pearson, J.A. Lautenberger, T.S. Papas, J. Ghrayeb, N.T. Chang, R.C. Gallo, and F. Wong-Staal, Complete nucleotide sequence of the AIDS virus, HTLV-III, Nature, 313: 277-284 (1985).
63. B. Starcich, L. Ratner, S.F. Josephs, T. Okamoto, R.C. Gallo, and F. Wong-Staal, Characterization of long terminal repeat sequences of HTLV-III. Science, 227: 538-540 (1985).
64. S.K. Arya, F. Wong-Staal, and R.C. Gallo, T-cell growth factor gene: Lack of expression in human T-cell leukemia-lymphoma virus-infected cells. Science, 223: 1086-1087 (1984).
65. S.K. Arya, F. Wong-Staal, and R.C. Gallo, Dexamethasone-mediated inmhibition of human T cell growth factor and γ-interferon messenger RNA. J. Immunol., 133: 273-276 (1984).
66. D.L. Mann, M. Popovic, C. Murray, C. Neuland, D.M. Strong, P. Sarin, R.C. Gallo, and W.A. Blattner, Cell surface antigen expression in newborn cord blood lymphocytes infected with HTLV. J. Immunol., 131: 2021-2024 (1983).
67. S.K. Arya, and R.C. Gallo, Transcriptional modulation of human T-cell grwoth factor gene by phorbol ester and interleukin-1. Biochemistry 23: 6685-6690 (1984).
68. F. Barre-Sinoussi, J.-C. Chermann, F. Rey, M.T. Nugeyre, S. Chamaret, J. Gruest, C. Dauguet, C. Axler-Blin, F. Brun-Vezinet, C. Rozioux, W. Rozenbaum, and L. Montagnier, Isolation of a T-lymphotropic retrovirus from a patient at risk for acquired immunodeficiency syndrome (AIDS). Science, 220: 868-871 (1983).
69. S.Z. Salahuddin, P.D. Markham, M. Popovic, M.G. Sarngadharan, S. Orndorff, A. Fladagar, A. Patel, J. Gold, and R.C. Gallo, Isolation of infectious human T-cell leukemia/lymphotropic virus type III (HTLV-III) from patients with acquired immunodeficiency syndrome (AIDS) or AIDS-related complex (ARC) and from healthy carriers: A study of risk groups and tissue sources. Proc. Natl. Acad. Sci. USA 82: 5530-5534 (1985).
70. B.H. Hahn, G.M. Shaw, S.K. Arya, M. Popovic, R.C. Gallo, and F. Wong-Staal, Molecular cloning and characterization of the HTLV-III virus associated with AIDS. Nature, 312: 166-169 (1984).
71. M.A. Muesing, D.H. Smith, C.D. Cabradilla, C.V. Benton, L.A. Lasky, and D.J. Capon, Nucleic acid structure and expression of the human AIDS/lymphadenopathy retrovirus. Nature, 313: 450-458 (1985).
72. S. Wain-Hobson, P. Sonigo, O. Danos, S. Cole, and M. Alizon, Nucleotide Sequence of the AIDS virus, LAV. Cell, 40: 9-17 (1985).
73. R. Sanchez-Pescador, M.D. Power, P.J. Barr, K.S. Steimer, M.M. Stempien, S.L. Brown-Shimer, W.W. Gee, A. Renard, A. Randolph, J.A. Levy, D. Dina, and P.A. Luciw, Nucleotide sequence and expression of an AIDS-associated retrovirus (ARV-2). Science, 227: 484-492 (1985).
74. J. Sodroski, C. Rosen, F. Wong-Staal, S.Z. Salahuddin, M. Popovic, S. Arya, R.C. Gallo, and W.A. Haseltine, Trans-acting transcriptional regulation of human T-cell leukemia virus type III long terminal repeat. Science, 227: 171-173 (1985).

75. S.K. Arya, G. Chan, S.F. Josephs, and F. Wong-Staal, Trans-activator gene of human T-lymphotropic virus type III (HTLV-III). Science, 229: 69-73 (1985).
76. S.K. Arya, and R.C. Gallo, Three novel genes of human T-lymphotropic virus type III: Immune reactivity of their products with sera from acquired immune deficiency syndrome patients. Proc. Natl. Acad. Sci. USA, 83: 2209-2213 (1986).
77. S.K. Arya, and R.C. Gallo, Human T-cell growth factor (interleukin-2) and γ-interferon genes: Expression in human T-lymphotropic virus type III- and type I-infected cells. Proc. Natl. Acad. Sci. USA, 82: 8691-8695 (1985).
78. D.Zagury, M. Fouchard, J.-C. Vol, A. Cattan, J. Leibowitch, M. Feldman, S. Sarin, and R.C. Gallo, Detection of infectious HTLV-III/LAV virus in cell-free plasma from AIDS patients. Lancet, ii: 505-506 (1985).

DRUG RESISTANCE: NEW APPROACHES TO TREATMENT

J.R. Bertino[1], S. Srimatkandada, M.D. Carman, M. Jastreboff
L. Mehlman, W.D. Medina, E. Mini, B.A. Moroson,
A.R. Cashmore, and S.K. Dube

Departments of Medicine and Pharmacology
Yale University School of Medicine
New Haven, Connecticut

ABSTRACT

Mechanisms by which malignant cells may become resistant to chemotherapeutic agents are reviewed, with emphasis on methotrexate resistance. At least four mechanisms of resistance have been described in experimental systems, including human tumor cells propagated in vitro: impaired uptake of methotrexate, an altered target enzyme (dihydrofolate reductase), and an elevated level of dihydrofolate reductase, or decreased methotrexate polyglutamylation. Combinations of these changes have been noted to occur in cells acquiring resistance to methotrexate. In the clinic, examples of resistance due to alteration of dihydrofolate reductase or elevated levels of this enzyme due to gene amplification have been reported. A strategy for selectively eradicating these resistant cells with second generation antifolates that are cytotoxic to resistant cells is discussed.

INTRODUCTION

The development of effective drugs and combinations of these drugs has markedly improved the therapy of certain malignancies, in particular acute leukemia, diffuse lymphoma and choriocarcinoma. However, drug resistance remains a major obstacle to cure of patients with these diseases. For example, 90 % of patients with acute lymphocytic leukemia (ALL) and 70 % of patients with acute non-lymphocytic leukemia (ANLL) achieve a complete remission with drug combinations. In contrast, 5 year disease free survivals in these diseases are 50 % and 15 %, respectively (1). Retreatment with the same agents is less effective, almost certainly because of development of drug resistance. The understanding of the detailed mechanisms whereby cells become resistant to these drugs may allow development of therapeutic programs that prevent development of drug resistance, or allow development of drugs directed toward selectively eradicating resistant cells. In this manuscript, the current status of resistance to anticancer drugs is reviewed, with emphasis on methotrexate (MTX) resistance. New therapeutic strategies directed toward eradication of resistant cells are suggested.

---

[1] American Cancer Society Professor

MECHANISM OF RESISTANCE TO ANTICANCER DRUGS

Table 1 lists chemotherapeutic agents in common use for the treatment of malignant diseases, and some mechanisms by which mammalian cells have been reported to develop resistance to these agents.

Table 1. Mechanisms of cellular resistance to anticancer drugs

Drug	Mechanism of Resistance	Reference
Methotrexate	Impaired transport	(2)
	Altered DHFR	(3)
	Elevated DHFR	(4)
Prednisone	Altered cytoplasmic receptor	(5)
Vincristine	Impaired transport	(6)
6-mercaptopurine	Decreased activation to nucleotide	(7)
Daunomycin	Impaired transport	(8)
Cytosine arabinoside	Decreased deoxycytidine kinase	(9)

In most instances, drug resistance is a result of emergence of pre-existent mutants that survive as a consequence of the selective pressure exerted by these agents. Resistance to certain drugs may occur by diverse mechanisms; some cells selected for high levels of resistance have even been noted to have more than two causes of resistance, e.g. impaired transport and elevated levels of enzyme, or elevated and altered levels of the target enzyme (10-12). In this circumstance, the second presumed mutation has been noted to occur after development of partial resistance by the first mechanism, rather than simultaneously.

Cell culture systems and transplanted experimental tumors have been valuable models for studies of drug resistance, since highly resistant lines may be developed, and the mechanism(s) of resistance easily elucidated. In the clinic, of course, because of the narrow therapeutic index of most drugs employed, relatively small changes (e.g. a 2-4fold elevation in a target enzyme), may result in what may be termed "clinical" resistance. In this cirumstance, resistance is more difficult to detect biochemically, unless sensitive assays that can measure these small changes are available.

MECHANISMS OF RESISTANCE TO METHOTREXATE

MTX is a useful drug in the treatment of patients with acute lymphocytic leukemia, choriocarcinoma, head and neck cancer, diffuse lymphoma, osteogenic sarcoma, bladder cancer, and breast cancer. In contrast, because of its low activity as a single agent, or in combination regimens, it is not considered to be a useful drug in the treatment of patients with certain other malignancies, including acute non-lymphocytic leukemia (ANLL) and most other solid tumors. Limited uptake of MTX by cells, presumably due to an efficient efflux of the drug, as well as rapid "induction" of DHFR may be important factors in the natural resistance to MTX (13). We have suggested that the increase in total dihydrofolate reductase (DHFR) that occurs when immature leukocytes are exposed to MTX is a consequence of drug binding and protection of DHFR from proteolytic degradation (13,14). A recent study using human KB cells, resistant to MTX came to a different conclusion, namely that the increase in DHFR noted was modulated at the translational level, or by as yet undefined post translational mechanisms (15).

Intracellular polyglutamylation of MTX is a recently discovered phenomenon that appears to be an important determinant of MTX responsiveness. This process results in accumulation of polyanionic forms of this drug (the addition of up to 3 additional glutamates in carboxyl linkage) (16-19). Thus cells capable of rapidly converting intracellular MTX to polyglutamate forms appear to be more susceptible to this drug (20,21).

Four mechanisms by which cells acquire resistance to MTX have been well documented experimentally in mammalian tumor cells (reviewed in 22): [1] an increase in DHFR activity, [2] an altered DHFR with less binding affinity to MTX, [3] impaired transport of MTX, and [4] decreased polyglutamylation of MTX.

Gene amplification as the cause of elevated DHFR levels in MTX resistant cells

In most, but not all instances (23) of acquired drug resistance to MTX associated with an increased level of DHFR, amplification has been detected. As much as several hundred fold increases in DHFR, mRNA for DHFR, and DHFR gene copies have been found in lines highly resistant to MTX (24-29). We recently described a subline of the human leukemia blast cell line (K562), that is markedly resistant to MTX, which has a 240fold increase in DHFR activity (28). This subline, obtained by exposure to stepwise, gradual increases in MTX concentrations, has corresponding elevated levels of DHFR mRNA(s) and DHFR gene copies. Of interest is the presence of three chromosomes with "homogeneous staining regions", not found in the parent line which we identified as chromosomes 5, 6 and 19. In studies with mouse and hamster lines resistant to MTX, as well as in this subline (30), the homogeneous staining regions were found to be the site of the amplified DHFR genes. These lines containing an HSR are associated with a stable karyotype (26,27). An increase in overall paired chromosomes, lacking a centromere, called "double minute" chromosomes has also been associated with unstable methotrexate resistance (31).

The availability of cDNA DHFR probes, both mouse (25) and human (32), has allowed a study of the size and organization of the DHFR gene from mouse, hamster and human sources. The human gene, like the mouse gene is very large (ca. 30 Kb) and has at least 4 intervening sequences. By using both the mouse and human cDNA probes, we have noted significant sequence diversity in the 3' end of the gene, in the non-coding region (28). The mouse, hamster and human MTX resistant lines produce multiple DHFR mRNA's (33,34,28). In the case of the Chinese Hamster lung line, two different DHFR enzymes have been noted (33). No evidence exists for more than one species of DHFR produced in the mouse and human resistant lines. In contrast to the mouse DHFR mRNA's, in which the predominant cytoplasmic mRNA is 1.6 Kb in size (34), the major human DHFR mRNA is 3.8 Kb (28,32).

Impaired Transport as a Cause of MTX Resistance

Several mouse and human lines have been described that are resistant to MTX as a consequence of impaired transport of MTX (2,22,35-37). This probable alteration in the membrane carrier also results in impaired transport of the natural cofactors, 5-methyltetrahydrofolate, and 5-formyltetrahydrofolate. A subline of the L1210 leukemia highly resistant to MTX has been found to have both an elevated DHFR (12fold) and a markedly impaired uptake for MTX (10).

Altered DHFR as a Mechanism for MTX Resistance

Although resistance to MTX as a consequence of alteration of DHFR occurs less frequently than an elevated DHFR, or impaired transport of MTX, several lines have been described with this alteration (3,11,12, 38,39).

Table 2. Resistance to MTX due to an Altered DHFR

Cell line	Inh by MTX (Compared to sensitive)	Reference
CHO	5	(11)
L1210	10	(40)
HCT-8	100	*
3T6	270	(12)
L5178Y	100,000	(39)

* unpublished observation by this laboratory.

Haber and Schimke recently described a mouse 3T6 subline with an elevated and altered DHFR; this enzyme required a concentration of MTX several logs higher for 50 % inhibition than the parent line (12). Recently, the cDNA for this enzyme has been sequenced, and the change in the amino acid sequence that is associated with decreased binding of MTX to the enzyme has been found to be due to a point muation in the coding for amino acid 22. This base substitution (40) results in an arginine instead of leucine in the enzyme.

Decreased Polyglutamylation of MTX as a Mechanism for MTX Resistance

Since polyglutamylation appears to be an important determinant for MTX cytotoxicity, it was not surprising to find that in some circumstances, MTX resistance was attributed to the lack of this activity (41,42). Of interest is the finding that when the activity of the enzyme responsible for this polyglutamylation (folyl polyglutamate synthetase) was examined in extracts of human breast cancer cells, the enzyme activity was similar to the sensitive parental line (41). The intracellular factor(s) responsible for this decreased activity in the cell have not been elucidated thus far.

MTX Resistance in Patients

Acquired MTX resistance of blast cells from patients with leukemia may be demonstrated by a simple in vitro test in which DNA synthesis (^{3}H-deoxyuridine uptake into DNA) is measured in the absence and presence of MTX (43,44). After a one hour exposure to 10^{-6} M MTX, DNA inhibition of leukemia cells that have not been previously exposed to this drug is greater than 95 % (44). Both an altered DHFR and elevated enzyme levels have been noted to occur in cells from MTX-resistant (43,44). However, since elevated levels of DHFR may occur rapidly as a consequence of exposure to the drug by virtue of "induction", genotypic evidence for drug resistance was difficult to obtain.

Recently, there have been three reports of gene amplification occuring in tumor cells of patients who were clinically resistant to MTX (Table 3).

Table 3. Examples of Gene amplification in Patients Clinically Resistant to MTX

Tumor Type	Fold Amplified	Reference
Acute myelocytic leukemia	3-4	(45)
Blast crisis, chronic myelocytic leukemia	3-4	(46)
Small cell lung cancer	2	(47)

In each of these instances gene amplification was of low magnitude (2-4 fold) and a sensitive dot blot assay was necessary to establish the increase in gene copy number observed (45-47).

DRUG RESISTANT CELLS AS TARGETS FOR CHEMOTHERAPY

A strategy for preventing or eradicating drug resistant tumor cells that survive treatment with a drug or combination of drugs is to use another drug or another drug combination with a different mechanism(s) of action. Thus cells surviving the first treatment by virtue of drug resistance might be eradicated by the second treatment. Success of this strategy depends on the relative effectiveness of the second treatment in killing the remainder of the tumor cells, and lack of cross-resistance of the cells to the second treatment (48).

Another strategy that might even be more specific and effective, is to use drugs that <u>selectively</u> kill resistant cells that emerge as the consequence of the first treatment. By using resistance to MTX as an example, the following strategies may be formulated and deserve further testing. Successful use of these strategies would require specific identification of the mechanism for drug resistance.

<u>Impaired Drug Uptake</u>

Two general approaches are worth considering to overcome resistance due to decreased uptake of a drug. In the case of MTX, encapsulation into liposomes (49,50) or modification of its structure so that the newly designed compound is transported differently are both approaches that have been explored using experimental tumor systems.

MTX is transported in most tumor cells by an active transport system that functions to transport reduced folates (51,52). By modifying the MTX molecule by attaching substituents to the α and γ carboxyls of the glutamate, compounds may be fabricated that are transported by alternate mechanisms. For example, when macromolecules such as albumin and poly-lysine are coupled to MTX via peptide linkage to the ·glutamate carboxyls, these compounds are effective <u>vs</u> MTX transport resistant tumor cell lines (53-55). In both cases, evidence exists that MTX is released intra-cellularly. Lipophilic derivatives (i.e. the mono and dibutylesters of MTX) that rapidly cross the membrane have also been shown to be effective vs MTX-transport resistant cells.

Potent inhibitors of DHFR have also been synthesized with structures not resembling folic acid, i.e. so called "non classical" antifolates. These compounds are transported by transport systems that are not folate-carrier mediated (56,57). Several of these compounds show promise in experimental testing, and have been completed in phase 1 trials (58-60).

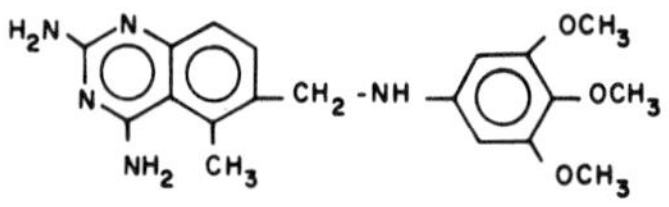

Trimetrexate
(NSC 249008, JB-II, TMQ)

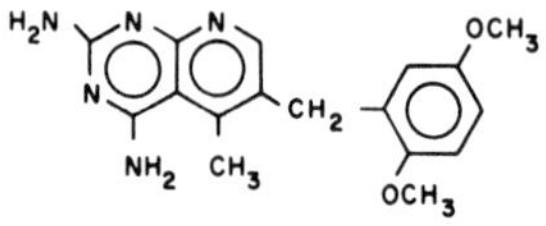

BW30IU

Figure 1. Structure of two "non-classical" antifolates

These compounds are even more effective in MTX resistant cells than MTX sensitive cells (61). An explanation for this increased sensitivity might derive from the observation that transport resistant cells may have less than normal intracellular levels of folate coenzymes as a consequence of this transport defect. In addition, the opportunity for selective rescue of normal cells, but not MTX-resistant cells with leucovorin exists; normal stem cells (bone marrow, GI mucosa) will be rescued by concurrent leucovorin while the MTX-transport resistant cells may not (62).

Increase in DHFR Activity

A strategy to selectively kill cells with an increase in DHFR was first suggested by Friedkin (63). Compounds that were substrates for DHFR and then became inhibitors of thymidylate synthetase upon reduction, were sought. The analogs, homofolate and dihydrohomofolate, are prototypes; these drugs are substrates for DHFR, and were more inhibitory to MTX resistant cells than sensitive cells, presumably as a consequence of selective inhibition of thymidylate synthetase. We have also recently demonstrated that cells with low levels of DHFR gene amplification are relatively more sensitive to trimetrexate as compared to MTX, presumably due to the higher intracellular concentrations achieved of the former drug (61).

Altered DHFR With Less Binding Affinity to MTX

Although an alteration in DHFR that leads to MTX resistance is relatively uncommon in cells made resistant to MTX in experimental systems, if the nature of this alteration is defined, selective inhibitors of the altered DHFR may be synthesized. The three-dimensional structure of prokaryotic and eukaryotic DHFR enzymes is now known, and the exact conformation and amino acids involved in binding of MTX of the normal enzymes is now almost completely understood (63). It thus may be possible to rationally design inhibitors that more selectively inhibit the altered DHFR as compared to the normal enzyme.

CONCLUDING REMARKS

Drug resistance remains a major obstacle to the chemotherapeutic cure of cancer. The understanding of the mechanisms by which cells become resistant to drugs is important, since new therapeutic approaches that utilize this knowledge may be formulated, and offer the possibility of obtaining great selectivity in cancer treatment.

REFERENCES

1. F. W. Gunz, E. S. Henderson, eds., "Leukemia", Grune and Stratton, Inc. New York (1983).
2. G. A. Fischer, Defective transport of amethopterin (methotrexate) as a mechanism of resistance to the antimetabolite in L5178Y leukemia cells, Biochem. Pharmacol., 11:1233-1234 (1962).
3. G. Blumenthal, and D. M. Greenberg, Evidence for two molecular species of dihydrofolate reductase in amethopterin resistant and sensitive cells of the mouse lekemia L4946, Oncology, 24:223-229 (1970).
4. D. K. Misra, S. P. Humphreys, M. Friedkin, A. Goldin, E. J. Crawford, Increased dihydrofolate reductase activity as a possible basis of drug resistance in leukemia, Nature, 189:39-42 (1961).
5. C. Sibley, and G. M. Tomkins, Mechanisms of steroid hormone resistance Cell, 12:221-226 (1979).
6. J. R. Riordan, and V. Ling, Purification of P-glycoprotein from plasma membrane vesicles of Chinese Hamster ovary cell mutants with Reduced Colchicine permeability, J. Biol. Chem., 254:12701-12705 (1979).
7. W. F. Brockman, Mechanism of resistance to anticancer agents, in: "Adv. Cancer Res.", A. Haddow, and S. Weinhouse, eds., Academic Press, New York (1963).
8. K. Dano, Experimental developed cellular resistance to daunomycin, Acta Path. et Micro. Scand., Sec. A., Suppl. 256 (1976).
9. M. Y. Chu, and G. A. Fischer, Comparative studies of leukemic cells sensitive and resistant to cytosine arabinoside, Biochem. Pharmacol., 14:333-341 (1965).
10. C. A. Lindquist, Characterization of a new murine leukemia line, L1210RR, and comparative studies of human dihydrofolate reductase enzyme, Ph.D. Thesis, Yale Univ. Sch. Med. (1979).
11. W. F. Flintoff, and K. Essani, Methotrexate-resistant Chinese Hamster Ovary cells contain a dihydrofolate reductase with an altered affinity for methotrexate, Biochemistry, 19:4321-4327 (1980).
12. D. A. Haber, S.M. Beverly, M.L. Kiely, and R.T. Schimke, Properties of an altered dihydrofolate reductase encoded by amplified genes in cultures of mouse fibroblasts, J. Biol. chem., 256: 9501-9510 (1981).

13. J. R. Bertino, W.L. Sawicki, A.R. Cashmore, E.C. Cadman, and R.T. Skeel, Natural resistance to methotrexate in human acute non-lymphocytic leukemia, Cancer Treat. Rep., 61:667-673 (1977).
14. B. L. Hillcoat, V. Swett, and J.R. Bertino, Increase in dihydrofolate reductase activity in cultured mammalian cells after exposure to methotrexate, Proc. Nat. Acad. Sci., 58:1632-1637 (1967).
15. B. A. Domin, S.P. Grill, K.F. Bastow, and Y.C. Cheng, Effect of methotrexate on dihydrofolate reductase activity in methotrexate-resistant human KB cells, Mol. Pharmacol., 21: 478-482 (1982).
16. C. M. Baugh, C.L. Krumdieck, and M.G. Nair, Polygammaglutamyl metabolites of methotrexate, Biochem. Biophys. Res. Commun., 52:27-34 (1973).
17. M. G. Nair, and C.M. Baugh, Synthesis and biological evaluation of polyglutamyl derivatives of methotrexate, Biochemistry, 12: 3923-3927 (1973).
18. D. S. Rosenblatt, V.M. Whitehead, M.M. Dupont, M.J. Vuchich, and N. Vera, Synthesis of polyglutamate in cultured human cells, Mol. Pharmacol., 14:210-214 (1978).
19. D. A. Gewirtz, J.C. White, J.C. Randolph, and I.D. Goldman, Transport, binding and polyglutamylation of methotrexate in freshly isolated rat hepatocytes, Cancer Res., 40:573-578 (1980).
20. J. Galivan, Evidence for the cytotoxic activity of polyglutamate derivatives of methotrexate, Mol. Pharmacol., 17:105-110 (1980)
21. R. L. Schilsky, B.D. Bailey, and B.A. Chabner, Methotrexate polyglutamate synthesis by cultured human breast cancer cells, Proc. Nat. Acad. Sci. USA, 77:2919-2922 (1980).
22. J. R. Bertino, S. Srimatakandada, M.D. Carmen, E. Mini, J. Jastreboff B.A. Moroson, and S.K. Dube, Mechanisms of drug resistance in human leukemia, in: "Modern Trends in Human Leukemia VI," Neth, Gallo, Greares, Janka, eds., Springer-Verlag,Berlin(1985)
23. S. Dedhar, D. Hartley, and J.H. Goldie, Increased dihydrofolate reductase activity in methotrexate-resistant human promyelocytic-leukemia (HL-60) cells. Lack of correlation between increased activity and overproduction, Biochem. J., 225:609-617 (1985).
24. F. W. Alt, R.E. Kellems, J.R. Bertino, and R.T. Schimke, Multiplication of dihydrofolate reductase genes in methotrexate-resistant variants of cultured murine cells, J. Biol. Chem., 253:1357-1370 (1978).
25. J. H. Nunberg, J.R. Kaufman, A.C.Y. Chang, S.N. Cohen, and R.T. Schimke, Structure and genomic organization of the mouse dihydrofolate reductase gene, Cell, 19:355-364 (1980).
26. B. J. Dolnick, R.J. Berenson, J.R. Bertino, R.J. Kaufman, J.H. Nunberg, and R.T. Schimke, Correlation of dihydrofolate reductase elevation with gene amplification in a homogeneously staining chromosomal region of L5178Y cells, J. Cell. Biol., 83:394-402 (1979).
27. J. H. Nunberg, R.J. Kaufman, R.T. Schimke, G. Urlaub, and L.A. Chasin Amplified dihydrofolate reductase genes are localized to homogeneously staining region of a single chromosome in a methotrexate-resistant Chinese hamster ovary cell line, Proc. Nat. Acad. Sci. USA, 75:5553-5556 (1978).
28. S. Srimatkandada, W.D. Medina, A.R. Cashmore, W. Whyte, D. Engle, B.A. Moroson, C.T. Franco, S. Dube, and J.R. Bertino, Amplification and organization of dihydrofolate reductase genes in a human leukemic cell line, K-562, resistant to methotrexate Biochemistry, 22:5774-5781 (1983).

29. P. W. Melera, J.A. Lewis, J.L. Biedler, and C. Hession, Antifolate-resistant Chinese hamster cells. Evidence for dihydrofolate reductase gene amplification among independently derived sublines overproducing different dihydrofolate reductases, J.Biol. Chem., 255:7024-7028 (1980).
30. L. Mehlman, M. Carman, B.A. Spengler, J. Biedler, and J.R. Bertino, Evolution of homogeneous staining regions (HSR's) in a human myelogenous leukemia cell line during development of resistance to methotrexate (MTX), Proc. Amer. Assoc. Cancer, Res., in press (1986).
31. R. J. Kaufman, P.C. Brown, and R.T. Schimke, Amplified dihydrofolate reductase genes in unstably methotrexate-resistant cells are associated with double minute chromosomes, Proc. Nat. Acad. Sci. USA, 76:5669-5673 (1979).
32. C. Morandi, and G. Attardi, Isolation and characterization of dihydrofolic acid reductase from methotrexate-sensitive and resistant human cell lines, J. Biol, Chem., 256:10169-10175 (1981).
33. B. J. Dolnick, and J.R. Bertino, Multiple messenger RNA's for dihydrofolate reductase, Arch. Biochem. Biophys., 210:691-697 (1981).
34. P. W. Melera, D. Wolgemuth, J.L. Biedler, and C. Hession, Antifolate-resistant Chinese hamster cells. Evidence from independently derived sublines for the over-production of two dihydrofolate reductases encoded by different mRNAs, J. Biol. Chem., 254:319-322 (1980).
35. B. A. Kamen, A.R. Cashmore, R.N. Dreyer, B.A. Moroson, P. Hsieh, and J.R. Bertino, Effect of [^{3}H]methotrexate impurities in apparent transport of methotrexate by a sensitive and resistant L1210 line, J. Biol. Chem., 255:3254-3257 (1980).
36. M. T. Hakala, On the role of drug penetration in amethopterin resistance of sarcoma 180 cells in vitro, Biochem. Biophys. Acta, 102:198-209 (1965).
37. F. M. Sirotnak, D.M. Moccio, L.E. Kelleher, and L.J. Goutas, Relative frequency and kinetic properties of transport defective phenotypes among methotrexate-resistant L1210 clonal cells derived in vivo, Cancer Res., 41:4447-4452 (1981).
38. J. H. Goldie, G. Krystal, D. Hartley, G. Gudauskas, and S. Dedhar, A methotrexate insensitive variant of folate reductase present in two lines of methotrexate-resistant L5178Y cells, Eur. J. Cancer, 16:1539-1546 (1980).
39. T.H. Duffy, S.B. Beckman, and F.M. Huennekens, Forms of L1210 dihydrofolate reductase differing in affinity for methotrexate, Fed. Proc., 43:3436 (abstract) (1984).
40. C. C. Simonsen, and A.D. Levinson, Isolation and expression of an altered mouse dihydrofolate reductase cDNA, Proc. Nat. Acad. Sci. USA, 80:2495-2499 (1983).
41. K.H. Cowan, and J. Jolivet, A methotrexate resistant human breast cancer cell line with multiple defects, including diminished formation of methotrexate polyglutamates, J. Biol. Chem., 259:10789-10800 (1984).
42. E. Frei, A. Rosowsky, J.E. Wright, C.A. Cucchi, J.A. Lippke, T.J. Ervin, J. Jolivet, and W.A. Haseltine, Development of methotrexate resistance in a human squamous cell carcinoma of the head and neck in culture, Proc. Nat. Acad. Sci., 81:2873-2877 (1984).
43. W. M. Hryniuk, and J.R. Bertino, Treatment of leukemia with large doses of methotrexate and folinic acid: clinical-biochemical correlates, J. Clin. Invest., 48:2140-2155 (1969).

44. J. R. Bertino, and R.T. Skeel, On natural and acquired resistance to folate antagonists in man, in: "Pharmacological Basis of Cancer Chemotherapy", Williams and Wilkins, Baltimore (1975).
45. M. D. Carman, J.H. Schornagel, R.S. Rivest, S. Srimatkandada, C.S. Portlock, and J.R. Bertino, Clinical resistance to methotrexate due to gene amplification, J. Clin. Oncol., 2:7-11 (1984).
46. G. A. Curt, D.N. Carney, K.H. Cowan, J. Jolivet, B.D. Bailey, J.C. Drake, C.S. Kao-shan, J.D. Minna, and B.A. Chabner, Unstable methotrexate resistance in human small-cell carcinoma associated with double minute chromosomes, N. Engl. J. Med., 308:199-202 (1983).
47. R. C. Horns, W.J. Dower, and R.T. Schimke, Gene amplification in a leukemia patient treated with methotrexate, J. Clin. Oncol., 2:2-7 (1984).
48. J. H. Goldie, and A.J. Coldman, A mathematic model for relating the drug sensitivity of tumors to their spontaneous mutation rate, Cancer Treat. Rep., 63:1727-1733 (1979).
49. S. B. Kaye, J.A. Boden, and B.E. Ryman, The effect of liposome (phospholipid vesicle) entrapment of actinomycin D and methotrexate on the in vivo treatment of sensitive and resistant solid murine tumours, Eur. J. Cancer, 17:279-289 (1981).
50. M. J. Kosloski, F. Rosen, R.J. Milholland, D. Papahadjopoulous, Effect of lipid vesicle (liposome) encapsulation of methotrexate on its chemotherapeutic efficacy in solid rodent tumors, Cancer Res., 38:2848-2853 (1978).
51. F. M. Huennekens, K.S. Vitols, and G.B. Henderson, Transport of folate compounds in bacterial and mammalian cells, in: "Adv. in Enzymology", A. Meister, ed., John Wiley, New York (1978).
52. I. D. Goldman, N.S. Lichenstein, and V.T. Oliverio, Carrier mediated transport of the folic acid analogue, methotrexate, in the L1210 leukemia cell, J. Biol. Chem., 243:5007-5017 (1968).
53. S. A. Jacobs, M. d'Urso-Scott, and J.R. Bertino, Some biochemical and pharmacologic properties of amethopterin-albumin, Ann. N.Y. Acad. Sci., 186:284-286 (1971).
54. B. C. Chu, C.C. Fan, and S.B. Howell, Activity of free and carrier-bound methotrexate against transport-deficient and high dihydrofolate dehydrogenase-containing methotrexate-resistant L1210 cells, J.N.C.I., 66:121-124 (1981).
55. H. J. Ryser, and W.C. Shen, Conjugation of methotrexate to poly (L-lysine) as a potential way to overcome drug resistance, Cancer, 45:1207-1211 (1980).
56. B. A. Kamen, B. Eibl, A.R. Cashmore, and J.R. Bertino, Uptake and efficacy of trimetrexate, a non-classical antifolate in methotrexate-resistant leukemia cells in vitro, Biochem. Pharmacol., 33:1697-1699 (1984).
57. B. Neuenfeldt, D. Van Hoff, J. Whitecar, and T. Williams, Comparison of activity of lipid soluble pyrido-pyrimidine BW 301U and methotrexate against human colony forming units, Proc. AACR, 38:181 (1986).
58. J. Lin, A. Cashmore, M. Baker, M. Ernstoff, J. Marsh, J.R. Bertino, R. DeLap, and A.J. Grillo-Lopez, Trimetrexate in metastatic colorectal cancer - Early phase II results in previously treated patients, in press (1986).
59. R. C. Donehower, M.L. Graham, G.E. Thompson, G.B. Dole, and D.S. Ettinger, Phase I and pharmacokinetic study of trimetrexate in patients with advanced cancer, Proc. Amer. Soc. Clin. Oncol., 4:32 (1985).
60. S. Legha, D. Tenney, D.H. Ho, and I. Krakoff, Phase I clinical and pharmacologic study of trimetrexate, Proc. Amer. Soc. Clin. Oncol., 4:48 (1985).

61. E. Mini, B.A. Moroson, C.T. Franco, and J.R. Bertino, Cytotoxic effects of folate antagonists against methotrexate-resistant human leukemia lymphoblast CCRF-CEM cell lines, Cancer Res., 45:325-330 (1985).
62. B. T. Hill, L.A. Price, S.I. Harrison, and J.H. Gredi, The difference between "selective folinic acid protection" and folinic acid rescue in L5178Y cell culture, Eur. J. Cancer, 13:861-871 (1977).
63. M. Friedkin, Enzyme studies with new analogues of folic acid and homofolic acid, J. Biol. Chem., 242:1466-1476 (1967).
64. K. W. Volz, D.A. Metthews, R.A. Alden, S.T. Freer, C. Hanech, B.T. Kaufman, and J. Kraut, Crystal structure of avian dihydrofolate reductase containing phenyltriazine and NADPH, J. Biol. Chem., 257:2527-2536 (1982).

MECHANISMS OF MEMBRANE-MEDIATED CYTOTOXICITY BY ADRIAMYCIN

Thomas Grace, Yigal H. Ehrlich and Thomas R. Tritton

Department of Pharmacology, Psychiatry and Biochemistry and Vermont Regional Cancer Center, University of Vermont Burlington, VT 05405, USA

INTRODUCTION

The principal objective of cancer chemotherapy has been to acquire chemical agents which are cytotoxic to tumor cells. To a very significant degree, this quest has been successful, at least as attested to by the existence of scores of drugs which have been useful in the management of neoplasia. A case can be made, however, that at least with respect to the choice of cellular targets for drug action, the strategy for obtaining new drugs has had a rather narrowly defined focus. When one considers a cell as a collection of organs and subcellular systems, it seems reasonable that any of these targets could be susceptible to poisoning by cytotoxic agents. Despite this apparent richness of molecular targets, one site, DNA, has held the pre-eminent position as the major target for anticancer drug action. Most of the available drugs act at this target either by direct chemical interaction (e.g. alkylating agents) or by interference with DNA biosynthesis (e.g. antimetabolites). The major limitation of this approach is lack of selectivity because normal, non-cancer cells also have DNA, and this DNA and its synthesis are only imperceptibly (to a drug) different than the DNA of tumor cells. As a result, all of the existing antineoplastics have undesirable toxic effects on normal cells. There are at least two ways around this dilemma. First, one could increase the specificity of drugs for tumor cells or second one could pick a target other than the genetic apparatus at which to aim drug action. Our hypothesis is that the cell surface or plasma membrane can be used to meet both of these objectives. Selectivity at this site is possible because of the fact that numerous differences in composition, structure, organization, dynamics and function exist between the surface of normal and tumor cells (1,2). Because of these differences a drug can in principle recognize the difference between a normal and tumor cell solely by sampling the cell surface, which of course is the first structure a drug "sees" when it encounters a cell. Drug action at the cell surface is possible because numerous important biologic functions are carried out on or in membranes, and disruption of these essential functions could lead to cytotoxicity.

So far in this essay we have provided a rationale for directing anticancer drug action to the cell surface. The question now becomes - is there any experimental evidence supporting this notion? Most chemotherapists are surprised to learn that for virtually every drug in the chemotherapeutic armamentarium, there is published evidence of action on cell membranes (reviewed in 3 and 4). Although this literature is rather extensive, it can't be said that there are cohesive or comprehensive theories explaining how the drugs could bring about their medicinal action by altering the plasma membrane, nor is there any rigorous proof for any agent that the cell surface is, in fact, the primary site for drug action. The best case for membrane action can be made for adriamycin, and our intention is to summarize some of the evidence supporting this assertion and then describe potential mechanisms by which cytotoxicity could be wrought at the cell surface.

EVIDENCE THAT ADRIAMYCIN CAN ACT ON CELL SURFACES

The paradigm for adriamycin has been that it enters cells and subsequently causes damage to DNA by one of several postulated mechanisms. The most prominent of these mechanisms are shown in Table I. In order for the first mechanism (intercalation) to account for cytotoxicity, the drug has to bind to DNA. Since there exist anthracycline derivatives (e.g. 9,10) that do not have measurable affinity for DNA, but which are cytotoxic, direct DNA interaction would not seem to be a necessary and sufficient condition for drug action. Each of the other three possibilities in Table I requires that the drug access the interior of the cell; the prediction of this requirement is that cytotoxicity should be proportional to uptake. This prediction has been shown to be invalid, at least under some circumstances. For example, in a series of Sarcoma 180 sublines selected for varying degrees of adriamycin resistance, it was shown that no relationship exists between cytotoxicity and uptake (11). Furthermore, resistant cell lines have been isolated which have <u>increased</u> drug uptake relative to sensitive lines (12), again providing argument against the idea that the amount of drug which enters the cell determines the degree of pharmacologic action. Still further evidence in this same vein is the observation (13) that adriamycin loses its ability to be cytotoxic to L1210 and HL-60 cells below a temperature of about 20 °C, despite the fact that substantial drug accumulation takes place at this temperature.

TABLE I. Intracellular or DNA Actions of Adriamycin

Action	Reference
Intercalation into DNA double helix	5
Generation of reactive oxygen	6
Generation of alkylating species	7
Stabilization of Topoisomerase II/ DNA cleavage complexes	8

These results can be used to argue against the importance of the mechanisms listed in Table I but of course do not prove the point. Nor do they prove that the cell surface is the target for drug action, even though the surface is the only target which does not require uptake to be operative. In order to support the membrane target hypothesis, the drug would have to be shown to [1] have reasonable affinity for membranes and [2] have modulatory effects on the biological properties of membranes. Table II shows that both of these criteria are met by adriamycin, and in fact this drug is an extraordinarily membrane active molecule.

The list of membrane actions of adriamycin and its relatives is

TABLE II. Membrane Actions of Adriamycin

Membrane Property Affected	Reference
lectin binding	14b
fluidity	14b,16,17,18
fusion	24,32
ion permeability	33,34
lipid composition	23
hormone receptors	19
binding of drug to phospholipids	28,29,30,31
morphology	27,35
prostaglandin formation	22
redox functions	26
glycosylation	21
capping	25

even more extensive than indicated here, but the existence of the list does not prove that any of those membrane responses are critical for cytotoxic action. The strongest evidence that the cell surface can be a primary target for adriamycin comes from our experiments using polymer-immobilized drug (36-38), and from the Tokes group using polyglutaraldehyde microspheres (39,40). In these forms adriamycin can only interact with the cell-surface since the polymers used are larger than cells. Consequently, DNA interactions are precluded. We have approached this problem by using several different types of polymeric support, and several chemical strategies for drug attachment and orientation on the polymer matrix. We have tested the biological activity of these materials against numerous cell lines and measured the cytotoxicity by survival, cloning and continuous growth techniques. Assessment of possible release of free anthracycline has been accomplished by very sensitive HPLC and RIA techniques. The essential conclusion is that adriamycin is actively cytotoxic under conditions where it can be operationally demonstrated that no free drug enters the cells. Stated differently - an agent which was heretofore postulated to have intracellular DNA as its major target, can still be active even when restricted solely to interaction at the cell surface. The significance of this work was even more sharply focused when we learned that the immobilized adriamycin is at least 1000times more potent than free adriamycin, conclusively demonstrating that the cell surface represents a very responsive site for drug action and that nonpenetrating derivatives might be a very useful new class of anticancer agents.

MEMBRANE MECHANISMS OF ACTION OF ADRIAMYCIN

We have sought experimental verification for three potential mechanisms whereby a plasma membrane interaction could initiate cytotoxicity: [1] a drug receptor; [2] modulation of membrane fluidity and [3] disruption of normal surface-directed growth control.

One could argue against the existence of a drug receptor or anthropomorphic grounds, that is, it seems unlikely that a cell would synthesize receptors for a substance which was toxic. Despite the argument, we have also devised experiments to look for receptors and two such approaches have failed to provide evidence for the existence of anthracyclines receptors. First, photoaffinity labeling experiments using daunomycin as the photoactive reagent (13) suggest that although the drug covalently labels cell surface sites, these sites are related neither to a transport system nor to a formal drug receptor for this compound. Thus, these results suggest that there is no well-defined binding site which, when occupied by drug, leads to the endpoint of drug action, namely cell death. <u>Second</u>, experiments with immobilized adriamycin attached to the polymeric support in different orientations (38) show that the polymer-bound drug is cytotoxically active under all geometries tested. Thus, no single, preferred, orientation on the drug or the surface is necessary for pharmacologic action. This is not the result one expects if a precise, well-defined, interaction with a drug receptor is required. Rather, it appears that if drug is merely present on (or in) the plasma membrane, it may be active regardless of its exact binding geometry.

These results suggesting the absence of a usual type of receptor binding site for adriamycin, lead us to wonder whether modulation of a <u>global</u> membrane property like fluidity could be involved in mediating the cytotoxic action of the drug. Using spin labeled Sarcoma 180 cells (14b) we showed that adriamycin induces a decrease in membrane fluidity of exposed cells and that the effect is dose-dependent and rapid. It is possible that this fluidity response may be a primary event in the initiation of cytotoxicity by adriamycin. However, since many compounds can modulate membrane fluidity and most of them are not effective antineoplastic agents, one is left with a dilemma of interpretation in determining whether or not the effect of a drug on membrane fluidity is a primary event in its biological mechanism of action, or is simply a result of the fact that many substances will alter fluidity non-specifically in the course of interaction with cellular membranes. To attempt to address this problem, conditions which influence the cytotoxic action of adriamycin were analyzed for their ability to influence in parallel ESR detected changes in the fluidity of S180 cell membranes.

Sarcoma 180 cells made hypoxic by exposure to 95 % N_2/5 % CO_2 and thereby more sensitive to the cytotoxic action of adriamycin (15) exhibit an increase in membrane fluidity. The enhanced inhibition of cell survival exhibited by adriamycin against the hypoxic cells was completely reversible upon reoxygenation and this reversal of the enhanced cytotoxicity followed the same time course of reoxygenation as the reversal of the order parameter changes. These findings indicate that: (a) a change in membrane fluidity in the region of the paramagnetic molecule is caused by exposure of Sarcoma 180 cells to an atmosphere of low oxygen tension, and (b) this change in the property of the membrane appears to be coupled to the enhanced cytotoxicity of the anthracycline in hypoxic cells.

Further evidence that the physical state of the membrane may be a determinant of the cytotoxicity of the anthracyclines was found in studies comparing the order parameter of anthracycline sensitive and resistant Sarcoma 180 sublines. A progressive increase in membrane fluidity was found as cells developed increasing resistance to this antibiotic (16). Likewise, Wheeler et al. (17) showed that progressively increasing resistance to adriamycin in MDAY-K2 cells was associated with decreased polarization of DPH i.e. increased fluidity. Complicating the issue, Ramu et al. (18) recently reported that P388 cells resistant to adriamycin were more rigid than sensitive cells. Apparently, the direction of the change in membrane fluidity with progressive insensitivity to the drug depends on the cell line under study, but these changes in the fluidity properties of membranes appear to be coupled to the expression of anthracycline resistance.

Since it is clear that the fluidity of membranes plays an important role in regulating biological function, the correlations found between fluidity, hypoxia and sensitivity to adriamycin are suggestive of a direct relationship between these phenomena. In a certain sense, though, fluidity is an unsatisfying explanation because fluidity itself is not a biological function, rather it is merely the milieu in which function is supported. Thus it is important to ask to what function alterations in fluidity are coupled. We have been investigating the possibility that drug disruption of normal growth regulatory processes through hormone, growth factor or mitogen systems on cell surfaces could produce unbalanced growth or cytotoxicity. To test this idea we have investigated the effects of adriamycin on the expression of receptors for epidermal growth factor (EGF). We originally showed (19) that adriamycin causes an up regulation in the number of EGF receptors in growing cells. The number of receptors in drug treated cells can increase dramatically - up to one hundred fold - and it is evident that alterations in membrane receptor levels can lead to altered cellular responses to the normal physiological regulator (EGF in this case).

Unfortunately, the up regulation phenomenon occurs too slowly (1 - 3 days) to be directly involved in the initial signals for cytotoxicity. We realized, however, that more important than the number of receptors for EGF, is the mechanism coupling EGF binding to regulation of cell growth. One of the ways that the EGF receptor can transduce signals is by acting as a protein kinase, so we have been investigating this (and other) phosphorylation reactions and their perturbation by drugs. Figure 1 shows a polyacrylamide gel autoradiogram of phosphoproteins in A-431 cells under various conditions (see legend for details). Although a variety of changes occur in drug treated cells, two bear special mention. First, the 170 kD EGF receptor itself, both in the presence and absence of EGF, has enhanced phosphorylation levels in response to the treatment of cells with drug. Since the ability of the receptor to act as a tyrosine kinase toward normal substrates is affected by its phosphorylation level, this drug induced change represents a potential way in which receptor function (i.e. control of growth) could be altered. In addition, the increased EGF receptor phosphorylation occurs within five minutes of exposure of cells to adriamycin. This is the earliest known biologic response to the drug and thus could represent a very early initiation signal in the cytotoxic cascade.

The second major change induced by adriamycin is increased phosphorylation of a 110 kD protein (Figure 1). Unlike the EGF receptor which is a membrane protein, the 110 kD protein is cytoplasmic. The phosphorylation of this protein is also controlled by other agents which damage cells from the surface (e.g. asbestos fibers - ref. 20) but not by agents which kill cells by intracellular mechanisms (e.g. methotrexate). This protein may thus be a common mediator of cell kill by surface membrane interaction.

CONCLUSION

It is clear that cytotoxic action may occur at the cell surface, and that this target represents a potentially useful site for the design of new anticancer drugs. Less clear are the mechanistic details by which surface attack brings about the biologic response of cell death. We have made progress in delineating certain possible mechanisms, however, and the phosphorylation system in particular appears to offer rich possibilities for dissecting mechanisms cells use to control normal growth and the alteration of these mechanisms by antineoplastic agents.

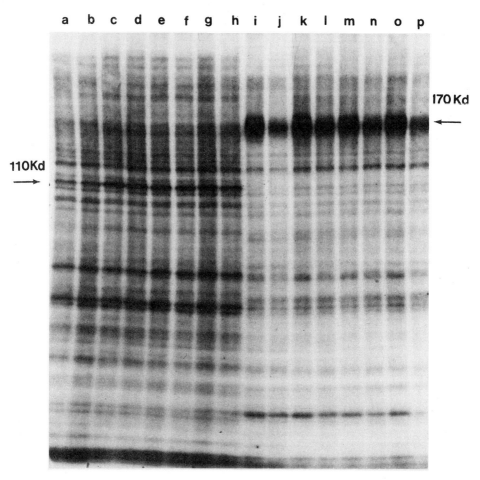

Figure 1. Endogenous protein phosphorylation in total homogenates of A431 cells.

The cells were grown to confluence, then incubated in DMEM + 10 % FBS + 1 x 10^{-7} molar adriamycin for five minutes, one hour and two hours. The cells were washed twice with PBS (Ca^{++}, Mg^{++}), harvested with a rubber policeman and washed again (Tris-sucrose/Mg^{++} buffer), placed in 1.25 ml of HEPES sucrose Mg^{++} buffer and homogenized with a 2 ml glass tissue homogenizer. The homogenates were then diluted to 1 mg/ml protein (Lowry method) with HEPES sucrose Mg^{++} buffer and assayed for endogenous protein phosphorylation (41) at 4 °C für 10 min using 1 µM γ-^{32}P-ATP (10 µCi per reaction) and 20 µl of protein sample. Lanes A through H were assayed using 10 µl of 1 mM $MgCl_2$ buffer + 10 µl of H_2O. Lanes I through P were assayed using 10 µl of 5 mM $MnCl_2$ buffer + 10 µl of 5 x 10^{-7} molar EGF. The reactions were stopped by adding sodium dodecyl sulfate (SDS) solution, the solubilized reaction products were heated at 30 °C for one hour and aliquots (20 µg of protein) were applied to 7 - 14 % exponential gradient polyacrylamide gels and electrophoresed using an SDS discontinuous buffer system, stained and dried and placed in contact with Kodak-X-Omat x-ray film for 24 hours and then developed (42).

Lanes A and B are without exposure of cells to ADR (control).
Lanes C and D are with 5 min exposure to ADR.
Lanes E and F are with 1 hr exposure to ADR.
Lanes G and H are with 2 hr exposure to ADR.

The 110 kD band has stimulated phosphorylation with the exposure of the cells to ADR and when assayed using a Mg^{++} buffer.

Lanes I and J are without exposure of cells to ADR (control).
Lanes K and L are with 5 min exposure to ADR.
Lanes M and N are with 1 hr exposure to ADR.
Lanes O and P are with 2 hr exposure to ADR.

The 170 kD band likewise has a stimulated phosphorylation with the exposure of the cell to ADR and when assayed using a Mn^{++} buffer. In each pair above, the first sample is without EGF, and the second with 5 x 10^{-7} M EGF.

REFERENCES

1. D. F. H. Wallach, Membrane Molecular Biology of Neoplastic Cells, Elsevier, New York (1975).
2. R. D. Hynes, Surface of Normal and Malignant Cells, Wiley, New York (1979).
3. J. D. Hickman, K. Scanlon, and T. R. Tritton, Trends in Pharm. Sci., 5:15 (1983).
4. T. R. Tritton, and J. A. Hickman, in: "Experimental and Clinical Progress in Cancer Chemotherapy", F. M. Mussia, and Martinus Nijoff, eds., Boston 81 (1985).
5. A. DiMarco, Cancer Chemotherapy Dep., 6:91 (1975).
6. N. R. Bachur, S. L. Gordon, M. V. Gee, and H. Kon, Proc. Nat. Acad. Sci., 76:954 (1979).
7. H. W. Moore, Science, 197:529 (1977).
8. K. M. Tewey, T. C. Rowe, L. Yang, B. P. Halligan, and L. F. Liu, Science, 226:466 (1984).
9. S. K. Sengupta, D. Seshadri, E. J. Modest, and M. Israel, Proc. Amer.

Assoc. Cancer Res., 17:109 (1976).
10. E. M. Acton, in: "Anthracyclines: Current Status and New Developments", S. T. Crooke, and S. D. Reich, eds., Academic Press, New York, 11 (1980).
11. J. M. Siegfried, A. C. Sartorelli, and T. R. Tritton, Eur. J. Cancer Clin. Oncol., 19:1133 (1983).
12. H. H. Sedlacek, personal communication.
13. P. Lane, and T. R. Tritton, manuscript in preparation.
14a. G. Yee, M. Carey, and T. R. Tritton, Cancer Res., 44:1898 (1984).
14b. S. A. Murphree, T. R. Tritton, P. L. Smith, and A. C. Sartorelli, Biochim. Biophys. Acta, 649:317 (1981).
15. J. M. Siegfried, K. A. Kennedy, A. C. Sartorelli, and T. R. Tritton, Cancer Res., 43:54 (1983).
16. J. M. Siegfried, K. A. Kennedy, A. C. Sartorelli, and T. R. Tritton, J. Biol. Chem., 258:339 (1983).
17. C. Wheeler, R. Rader, and D. Kessel, Biochem. Pharm., 31:2691 (1982).
18. A. Ramu, D. Glaubinger, I. T. Magrath, and A. Joshi, Cancer Res., 43:5533 (1983).
19. G. Zuckier, and T. R. Tritton, Exp. Cell Res., 148:155 (1983).
20. K. Hamsen, G. Cameron, Y. Ehrlich, and B. T. Mossman, Proc. Amer. Assoc. Cancer Res., 45:314 (1985).
21. D. Kessel, Mol. Pharm., 16:306 (1979).
22. K. Ohuchi, and L. Levine, Prostaglandin and Medicine, 1:433 (1978).
23. S. I. Schlager, and S. H. Ohanian, J. Nat. Cancer Inst., 63:1475 (1979).
24. S. A. Murphee, D. Murphy, A. C. Sartorelli, and T. R. Tritton, Biochim. Biophys. Acta, 691:97 (1982).
25. M. Gosalvez, L. Pezzi, and C. Vivero, Biochem. Soc. Trans., 6:659 (1978).
26. F. Crane, W. C. Mackellar, D. J. Morre, T. Ramasarma, H. Goldenberg, G. Grebing, and H. Low, Biochem. Biophys. Res. Comm., 93:746 (1980).
27. R. B. Mikkelson, P. S. Lin, and D. F. H. Wallach, J. Mol. Med., 2:73 (1977).
28. R. F. Tayler, L. A. Teague, and D. W. Yesair, Cancer Res., 41:4316 (1981).
29. E. Goormaghtigh, M. Vandernbraden, J. A. Ruysschaert, and B. DeKruijff, Biochim. Biophys. Acta, 685:137 (1982).
30. G. Karczmar, and T. R. Tritton, Biochim. Biophys. Acta, 557:306 (1979).
31. T. Burke, and T. R. Tritton, Biochemistry, 24:1768 (1985).
32. A. Necco, and M. Ferraguti, Exp. Mol. Pathol., 31:353 (1979).
33. J. R. Harper, E. P. Orringer, and J. C. Parker, Res. Comm. Chem. Path. Pharm., 26:2 (1979).
34. T. Dasdia, A. DiMarco, M. Goffredi, A. Minghetti, and A. Necco, Pharm. Res. Comm., 11:19 (1979).
35. S. Chahwala, and J. A. Hickman, Cancer Res., in press (1985).
36. T. R. Tritton, and G. Yee, Science, 217:248 (1982).
37. L. Wingard, and T. R. Tritton, in: "Affinity Chromatography and Biological Recognition", Z. M. Chaiken et al., eds., Academic Press, New York 583 (1984).
38. L. Wingard, T. R. Tritton, and K. A. Eggler, Cancer Res., 45:3529 (1985).
39. Z. A. Tokes, K. E. Rogers, and A. Rembaum, Proc. Nat. Acad. Sci., 79:2026 (1982).
40. K. E. Rogers, B. J. Carr, and Z. A. Tokes, Cancer Res., 43:2741 (1983).
41. Y. H. Ehrlich, Handbook of Neurochemistry, 2. edt. 6:541 (1984).
42. Y. H. Ehrlich, E. G. Brunngraber, P. K. Sinha, and K. N. Prasad, Nature, 265:238 (1977).

TERMINAL TRANSFERASE AND ADENOSINE DEAMINASE ACTIVITIES IN HUMAN NEOPLASIA: THEIR ROLE IN MODULATING CANCER TREATMENT

P. S. Sarin, A. Thornton and D. Sun

Laboratory of Tumor Cell Biology
National Cancer Institute
Bethesda, Maryland 20892

Terminal deoxynucleotidyl transferase (TdT) and adenosine deaminase (ADA) are two enzyme markers that have been exstensively utilized to define the stage of the disease and the cell types involved in human leukemia B cell proliferation and severe combined immune deficiency syndrome (SCID). The use of these biological markers has been very useful in selecting treatment protocols for patients with leukemia/lymphoma and SCID.

Terminal Transferase

Terminal deoxynucleotidyl transferase or terminal transferase (TdT) is an enzyme that catalyzes the polymerization of deoxyribonucleotide initiators in the absence of a DNA or RNA template. It was discovered in calf thymus tissue during early attempts to isolate and characterize DNA polymerases from thymus tissue (1,2). In contrast to other cellular DNA polymerases which require all four deoxyribonucleoside triphosphates for optimal DNA synthesis, TdT is capable of carrying out terminal addition of deoxyribonucleotides in the presence of only a single nucleotide substrate. In early studies (3,4) this enzyme was considered to be specific for thymus tissues but high levels of TdT were subsequently detected in various forms of human leukemia, including acute lymphocytic leukemia (ALL) (2,5-9) and chronic myelogenous leukemia (CML) in acute blast phase of the disease (6,10-15). High levels of this enzyme were also detected in two cell lines (4802 and Molt-4) derived from patients with acute lymphoblastic leukemia (16,17). High levels of TdT have also been detected in gibbons with acute lymphoblastic leukemia (18). The detection of TdT in blast phase of CML patients suggested the presence of an overabundance of lymphoid cells of pre T cell type as seen in the case of patients with acute lymphoblastic leukemia and hence suggested the use of vincristine and prednisone therapy, commonly used in treatment of ALL, for the treatment of CML patiens in blast phase of the disease (6,14,15).

Examination of leukocytes from various forms of leukemias and lymphomas including acute lymphoblastic leukemia, acute myelogenous leukemia (AML), chronic myelogenous leukemia in stable and blast phase of the disease, both B and T cell types of chronic lymphocytic leukemia (CLL), Hodgkins lymphoma, diffuse histiocytic lymphoma and Sezary syndrome incidate that majority (92 %) of the patients with ALL are

TdT positive, indicating the accumulation of TdT positive pre T cells. In addition, approximately 40 % of the CML patients in blast phase of the disease are TdT positive whereas only 1 % or less of the CML patients in the chronic phase of the disease show the presence of this enzyme. No case of B or T CLL, Sezary syndrome, Hodgkins lymphoma or diffuse histiocytic lymphoma were found to be TdT positive. Table 1 summarizes the results obtained with various tissues. The CML patients in blast crisis who were treated with ALL protocol of vincristine and prednisone showed a marked reduction in the circulating blasts as well as a marked reduction in the levels of terminal transferase (14,15). However, resistance to therapy has been observed in some cases after sometime and the second round of treatment with vincristine-prednisone may then be less effective.

Table 1. Distribution of Terminal Deoxynucleotidyl Transferase in Hematologic Malignances

Diagnosis	Number of patients Tested	TdT +	% TdT +
1. Acute Lymphoblastic Leukemia (ALL)	510	467	92
2. Acute Meylogenous Leukemia (AML)	350	28	8
3. Chronic Myelogenous Leukemia (CML) (Blast Phase)	280	106	38
4. Chronic Myelogenous Leukemia (CML) (Stable Phase)	120	1	less than 1
5. Chronic Lymphocytic Leukemia (CLL-T)	67	0	0
6. Chronic Lymphocytic Leukemia (CLL-B)	89	0	0
7. Sezary Syndrome	36	0	0
8. Hodgkins Lymphoma	29	0	0
9. Diffuse Histiocytic Lymphoma	28	0	0
10. Xeroderma Pigmentosa	12	0	0
11. Hairy Cell Leukemia	14	0	0

The TdT levels detected in the leukocytes of CML patients in acute blast phase of the disease and ALL patients vary considerably and are dependent on the stage of the disease. High levels of TdT, comparable to the levels observed in thymus gland, have been observed in T cell lines established from ALL patients as well as some CML patients in blast crisis. Results summarized in Table 2 show that patients with chronic phase CML, CLL, Sezary Syndrome, Xeroderma pigmentosa, systemic lupus erythematosus (SLE) and infectious mononucleosis are negative for TdT suggesting thereby a lack of TdT positive pre T cells in peripheral blood leukocytes. Low levels of TdT have also been reported in a few patients with acute myelomonocytic leukemia (AMML) (19), acute undifferentiated leukemia (AUL) (20), and myelodysplastic syndrome (21).

Table 2. Terminal Transferase Levels in Various Tissues

Tissue	Diagnosis	nmols/10^9 cells
1. Thymus gland (human or calf)	normal	400-500
2. a. Lymphocytes (PHA stimulates)	normal	< 0.5
b. bone marrow	normal	1-2
3. T cell lines		
8042	ALL	400
Molt-4	ALL	350-400
4. B cell lines		
NC37	normal	< 0.5
8392	ALL	< 0.5
SB	ALL	< 0.5
5. Leukocytes	ALL	50-200
	CML (Blast-phase)	300-600
	CML (chronic phase)	< 0.5
	AML	2-10
	CLL	< 0.5
	Infectious mononucleosis	< 0.5
	Systemic Lupus erythematosus (SLE)	< 0.5
	Xeroderma pigmentosa	< 0.5
	Sezary Syndrome	< 0.5

Efforts to increase the sensitivity of detection of terminal transferase activity have focused on methods of enriching the population of cells positive for TdT by cell separation techniques (22). Various cell separation techniques (Table 3) used to enrich TdT positive cells include separation on ficoll-hypaque, unit gravity sedimentation in sucrose gradients, centrifugal elutriation, free flow electrophoresis, affinity chromatography and fluorescence activated cell sorting (FACS). In early studies (23) TdT positive cells in human bone marrow were enriched by unit gravity sedimentation, resulting in 14 - 20 fold enrichment of TdT activity in the enriched cell population. Subsequently, the technique of free flow electrophoresis (22) was utilized to enrich TdT positive cell populations from bone marrow and leukocytes of patients with ALL and CML in blast phase of the disease. As shown in Table 4, there is a 7 to 33 fold enrichment in the TdT positive cells, thus making detection of small number of TdT positive cells simpler either by enzyme studies or by immunofluorescence technique utilizing the polyclonal or the monoclonal antibody (24) made against terminal transferase. TdT antibody has not only been valuable in the detection of TdT positive cells by immunofluorescence (25) but a radioimmunoassay has also been successfully developed for the detection of terminal transferase with an antibody developed in mice (26).

Table 3. Cell Separation Techniques for Enrichment of TdT Positive Cells

1. Ficoll-Hypaque
2. Staput Gradients. Unit Gravity Sedimentation in Sucrose Gradients
3. Centrifugal Elutriation
4. Free Flow Electrophoresis
5. Electrophoresis in Sucrose Gradients
6. Affinity Chromatography
7. Fluorescence Activated Cell Sorting (FACS)

Table 4. Enrichment of Terminal Transferase Activity After Cell Separation

Patient No.	Diagnosis	TdT Activity (nmoles/10^9 cells)		Enrichment (Fold)
		Unfractionated Cells	Peak Fraction	
1	Normal (bone marrow)	0.7	14	20
2	ALL*	18	225	12
3	ALL	54	390	7
4	ALL	5	85	17
5	ALL	2.5	75	30
6	CML (BC)	7	234	33
7	CML (BC)	29	630	22
8	CML (BC)	2	47	24

* ALL, acute lymphoblastic leukemia;
CML (BC), chronic myelogenous leukemia in blast phase

In our early studies on the detection of TdT in CML patients in blast phase of the disease (6,11) we had proposed that the blast phase in these patients represented a lymphoblastic transformation. Hence, these patients should respond to treatment with vincristin-prednisone therapy commonly used in the treatment of patients with acute lymphoblastic leukemia. Studies carried out in various laboratories (14,15) show that approximately 80 % of the TdT positive CML patients, who were also Philadelphia chromosome (Ph-1) positive, respond to treatment with vincristine-prednisone (Table 5). These results suggest that vincristine-prednisone therapy should be the treatment of choice for TdT positive CML patients in blast phase.

Table 5. Response of Chronic Myelogenous Leukemia (Blast Crisis) Patients to Vincristine/Prednisone Therapy

Reference	TdT (+)	TdT (-)	Ph_1
Oken et al. (14)	1/1	NT	+
Marks et al. (27)	8/13	0/9	+
Ross et al. (15)	1/1	NT	+
Janossy et al. (29)	6/7	0/14	+
McCaffrey et al. (28)	11/16	1/14	+
Total (number +/number Tested)	29/38	1/37	

Efforts to obtain both potent and specific inhibitors of terminal transferase have so far been unsuccessful despite the existence of several reports on the inhibition of TdT by nucleotide analogs, synthetic polymers and other compounds of unknown mechanism of action (30-35). In summary, terminal transferase is a unique biological marker (Table 6), which has been extremely useful in the classification of leukemias and the prediction of treatment protocols for TdT positive CML patients in the blast phase.

Table 6. Characteristics of Terminal Deoxynucleotidyl Transferase

1. A unique DNA polymerase which does not require a template for direction

2. Present in 90 % of the patients with acute lymphoblastic leukemia

3. Present in 40 % of the chronic myelogenous leukemia patients in blast crisis

4. A marker for predicting response of chronic myelogenous leukemia patients in blast crisis to Vincristine and Prednisone therapy

5. A possible marker for predicting the onset of relapse before any morphologic indication of relapse

6. A marker for immature T cells

7. A marker for classification of leukemia into terminal transferase positive and negative leukemia

8. A marker for selection of treatment protocols for leukemic patients

It will be of interest to develop specific inhibitors of terminal transferase for treatment of TdT positive leukemias and an understanding of the biological function of this enzyme will be extremely important in future directions of research in human leukemias and lymphomas.

Adenosine Deaminase

Adenosine deaminase (ADA) is an important enzyme in the purine salvage pathway, and is involved in the deamination of adenosine, and deoxyadenosine. The enzyme is present in all mammalian tissues and the absence of this enzyme congenitally results in a severe immunodeficiency classified as severe combined immune deficiency disease (SCID) (37-39). The immune deficiency is probably a result of excessive build-up of deoxyadenosine triphosphate which inhibits DNA synthesis by feedback inhibition (40). Variable levels of ADA activity have been observed in the lymphocytes of patients with leukemias and lymphomas. Thus decreased levels of ADA have been observed in patients with chronic lymphocytic leukemia (CLL) whereas increased levels of this enzyme have been observed in patients with acute lymphoblastic leukemia (ALL), chronic myelogenous leukemia (CML) in acute blast phase, and acute myelogenous leukemia (AML) (41). Elevated levels of ADA have also been observed in patients with carcinoma (42,43), and infectious mononucleosis (44). Although, ADA activity has been observed in both T and B cells, levels of ADA activity in T cells are considerably higher (45).

Analysis of the levels of ADA activity in ALL patients in active phase of the disease and in remission indicates that the ADA levels return to normal levels in remissions but are elevated again in relapse (46). A correlation of the elevated levels of ADA in ALL with increased levels of terminal transferase has also been observed. The mean values of ADA activity in different forms of leukemia are summarized in Table 7. Adenosine deaminase has been purified from human leukemia cells and a radioimmunoassay has also been developed (47,48). Molecular cloning of ADA has also been recently accomplished (49,50). With the availability of radioimmunoassays and cloned DNA it should be possible to quantitate more fully low levels of this enzyme.

Table 7. Adenosine Deaminase Levels in Lymphocytes of Leukemia Patients and Healthy Donors

No.	Diagnosis*	Mean ADA activity nmoles/10^8 cells/min
1.	ALL T-cell	6500
2.	ALL B-cell	< 300
3.	ALL Non T- Non B	1800
4.	ALL Remission	< 300
5.	ALL Relapse	4800
6.	CML Blast phase	2500
7.	CLL	1500
8.	Normal	< 300

*ALL, acute lymphoblastic leukemia; CML, chronic myelogenous leukemia; CLL, chronic lymphocytic leukemia. The values shown in the table are mean values obtained from 175 patients with ALL, 40 patients with CML, 70 patients with CLL and 50 healthy donors. ADA activity is expressed as nmoles of inosine producerd in one minute from 10^8 cells.

Several laboratories have been involved in the development of drugs that could selectively inhibit ADA for potential use in treatment of lymphoid malignancies. A number of inhibitors of ADA have been developed and a representative group of ADA inhibitors with Ki ranging from 10^{-4} to 10^{-11} M are summarized in Table 8. Deoxycoformycin along with coformycin and EHNA have been the most widely studied ADA inhibitors both in the laboratory and in the clinic (51,52). Studies to date indicate that ADA inhibitors either alone or in combination with other drugs may be useful in the treatment of T cell malignancies including T-ALL, mycosis fungoides and Sezary syndrome (53-60).

Table 8. Apparent Ki values of representative ADA inhibitors

Inhibitor	Ki (M)
2'-Deoxyinosine	6×10^{-5}
Inosine	1.2×10^{-4}
6-Thioguanosine	9.2×10^{-5}
6-Thioinosine	3.3×10^{-4}
Isocoformycin	8×10^{-8}
Deoxycoformycin	1.5×10^{-11}
Coformycin	1.2×10^{-10}
EHNA*	6.5×10^{-9}

* EHNA, erythro-9-(2R-hydroxy,3S-nonyl)-adenine

REFERENCES

1. P.S. Sarin, Terminal transferase as a biological marker for human leukemia, in: "Recent Advances in Cancer Research" (R.C. Gallo, edt.), CRC Press, Cleveland, Vol. I, p. 131 (1977).
2. F.J. Bollum, Terminal deoxynucleotidyl transferase as a hematopoietic cell marker, Blood, 54: 1203 (1979).
3. J.S. Krakow, S. Coustogeorgopoulos, and E.S. Cannelakis, Incorporation of deoxyribonucleotides into terminal positions of DNA, Biochem. Biophys. Res. Comm., 5: 477 (1961).
4. F.J. Bollum, Oligodeoxyribonucleotide primed reaction catalyzed by calf thymus polymerase, J. Biol. Chem., 237: 1945 (1962).
5. R.McCaffrey, D.F. Smolar, and D. Baltimore, Terminal deoxynucleotidyl transferase in a case of childhood acute lymphoblastic leukemia, Proc. Natl. Acad. Sci. USA, 70: 521 (1973).
6. P.S. Sarin, P.N. Anderson, and R.C. Gallo, Terminal deoxynucleotidyl transferase activities in human blood leukocytes and lymphoblast cell lines. High levels in lymphoblast cell lines and in blast cells of some patients with chronic myelogenous leukemia in acute phase, Blood, 47:11 (1976).
7. M.S. Coleman, M.F. Greenwood, J.J. Hutton, F.J. Bollum, B. Lampkin, and P. Holland, Serial observations on terminal deoxynucleotidyl transferase activity and lymphoblast surface markers in acute lymphoblastic leukemia, Cancer Res., 36:120 (1976).

8. G. Janossy, F.J. Bollum, K.R. Bradstock, A. McMichael, N. Rapson, and M.F. Greaves, Terminal transferase positive human bone marrow cells exhibit the antigenic phenotype of common acute lymphoblastic leukemia, J. Immunol., 123: 1525 (1979).
9. P.S. Sarin, and R.C. Gallo, Terminal deoxynucleotidyl transferase as a biological marker for human leukemia, in: "Modern Trends in Human Leukemia" (R. Noth, Ed.) J. F. Lehmann's Verlag, Munich, Vol. 2, p. 491 (1971).
10. P.H. Wiernik, L.S. Edwards, and P.S. Sarin, Marrow terminal deoxynucleotidyl transferase activity in adult acute leukemia, Hämatol. Bluttransfusion, 23: 125 (1979).
11. P.S Sarin, and R.C. Gallo, Terminal deoxynucleotidyl transferase in chronic myelogenous leukemia, J. Biol. Chem., 249: 8051 (1974).
12. J.J. Hutton, and M.S. Coleman, Terminal deoxynucleotidyl transferase measurements in the differential diagnosis of adult leukemias. Brit. J. Haematol., 34: 447 (1976).
13. R. McCaffrey, T. Harrison, R. Parkman, and D. Baltimore, Terminal deoxynucleotidyl transferase activity in human leukemia cells and in normal human thymocytes, New Engl. J. med., 292: 775 (1975).
14. M.M. Oken, P.S. Sarin, R.C. Gallo, J.G. Johnson, B.J. Gormus, R.E. Rydell, and M.E. Kaplan, Terminal transterase levels in chronic myelogenous leukemia in blast crisis and in remission, Leuk. Res., 2: 173 (1978).
15. D.D. Ross, P.H. Wiernik, P.S. Sarin, and J. Whang-Peng, Loss of terminal deoxynucleotidyl transferase (TdT) activity as a predictor of emergence ot resistance to chemotherapy in a case of chronic myelogenous leukemia in blast crisis, Cancer, 44: 1566 (1979).
16. P.S. Sarin, and R.C. Gallo, Characterization of terminal deoxynucleotidyl transferase in a cell line (8402) derived from a patient with acute lymphoblastic leukemia, Biochem. Biophys. Res. Comm., 65: 673 (1975).
17. B. Srivastava, and J. Minowada, Terminal deoxynucleotidyl transferase in a cell line (Molt-4) derived from the peripheral blood of a patient with acute lymphoblastic leukemia, Biochem. Biophys. Res. Commun., 51: 529 (1973).
18. P.S. Sarin, M. Virmani, and B. Friedman, Terminal transferase in acute lymphoblastic leukemia in gibbons, Biochim. Biophys. Acta, 608: 62 (1980).
19. M.S. Coleman, J.J. Hutton, P.D. Simone, and F.J. Bollum, Terminal deoxyribonucleotidyl transferase in human leukemia, Proc. Natl. Acad. Sci. USA, 71: 4404 (1974).
20. S.L. Marcus, S.W. smith, C.I. Jarowski, and M.J. Modak, Terminal deoxyribonucleotidyl transferase activity in acute undifferentiated leukemia, Biochem. Biophys. Res. Commun., 70: 37 (1976).
21. I.P. Brusamolino, E.P. Alessandrino, A.I. Servassi, U. Bertazzoni, and C. Bernasconi, Terminal deoxynucleotidyl transferase positive acute leukemias evolving from a myelodysplastic syndrome, Am. J. Hematol., 20: 187 (1985).
22. P.S. Sarin, M. Virmani, and R.C. Gallo, Enrichment of cell populations containing terminal transferase activity by free flow electrophoresis, Int. J. Cancer, 29: 507 (1982).
23. R.D. Barr, P.S. Sarin, and S. Perry, Terminal transferase in human bone marrow lymphocytes, Lancet, 1: 508 (1976).
24. F.J. Bollum, Antibody to terminal deoxynucleotidyl transferase, Proc. Natl. Acac. Sci. USA, 72: 4119 (1975).
25. F.J. Bollum, Terminal deoxynucleotidyl transferase as a hematopoietic cell marker, Blood, 54: 1203 (1979).
26. P.C. Kung, P.D. Gottlieb, and D. Baltimore, Terminal deoxynucleotidyl transferase. Serological studies and radioimmunoassay, J. Biol. Chem., 251: 2399 (1976).

27. S.M. Marks, D. Baltimore, and R.P. McCaffrey, Terminal transferase as a predictor of initial responsiveness to vincristine and prednisone in blastic chronic myelogenous leukemia, New Engl. J. Med., 298: 812 (1978).
28. R.McCaffrey, A. Lillquist, S. Sallan, E. Cohen, and M. Osband, Clinical utility of leukemia cell terminal transferase measurements, Cancer Res., 41: 4814 (1981).
29. G.Janossy, R.K. Woodruff, M.J. Pippara, H.G. Prentice, A.V. Hoffbrand, A. Paxton, A. Lister, T.A. Brunch, and M.F. Greaves, Relation of lymphoid phenotype and response to chemotherapy incorporating vincristine-prednisolone in the acute phase of Ph-1 positive leukemia, Cancer, 43: 426 (1979).
30. K.Ono, Y. Iwata, H. Nakamura, and A. Matsukage, Selective inhibition of terminal deoxynucleotidyl transferase by diadenosine tetraphosphate, Biochem. Biophys. Res. Commun., 95: 34 (1980).
31. W.E.G. Müller, R.K. Zahn, and J. Arendes, Differential mode of inhibition of terminal deoxynucleotidyl transferase by 3'-dATP, ATP, ß-araATP and α-araATP, FEBS Letters, 94: 47 (1978).
32. R.A. DiCioccio, and B.I.S. Srivastava, Inhibition of deoxynucleotide polymerizing enzyme activities of human cells and of simian sarcoma virus by heparin, Cancer Res., 38: 2401 (1978).
33. M.S. Coleman, and M.R. Deibel, Terminal deoxynucleotidyl transferase, in: "Enzymes of Nucleic Acid Synthesis and Modification" (S. Jacob, Ed.) Academic Press, New York, Vol. 1, p. 93 (1983).
34. S.Capitani, G. Mazzotti, S. Papa, P. Santi, and F.A. Manzoli, Effect of phospholipid vesicles on the activity of terminal transferase, Biochem. Biophys. Res. Comm., 89: 1206 (1979).
35. R.McCaffrey, R. Bell, and A. Lillquist, Selective killing of leukemia cells by inhibition of TdT, Hematol. Bluttransf., 28: 24 (1983).
36. M.J. Modak, Biochemistry of terminal deoxynucleotidyl transferase: Mechanism of inhibition by adenosine 5' triphosphate, Biochemistry, 17: 3116 (1978).
37. M.S. Coleman, J. Donofrio, J.J. Hutton, I. Hahn, A. Daoud, B.J. Lampkin, and J. Dynminski, Identification and quantification of adenine deoxynucleotides in erythrocytes of a patient with adenosine deaminase deficiency and severe combined immunodeficiency, J. Biol. Chem., 253: 1619 (1978).
38. E.R. Giblett, J.E. Anderson, F. Cohen, B. Pollara, and H.J. Mewissen, Adenosine deaminase deficiency in two patients with severely impaired cellular immunity, Lancet, 2: 1067 (1972).
39. R.Bell, R.P. Agarwal, A. Lillquist, and R. McCaffrey, Biochemical markers in neoplastic lymphoid cells, in: "Immunology of the lymphomas" (S.B. Sutcliffe, Ed.), CRC Press, Boca Raton, Florida, p. 154 (1985).
40. A.Cohen, R. Hirschhorn, S.D. Horowitz, A. Rubinstein, S.H. Polmar, R. Hong, and D.W. Martin, Deoxyadenosine triphosphate as a potentially toxic metabolite in adenosine deaminase deficiency, Proc. Natl. Acad. Sci. USA, 75: 472 (1978).
41. J.F. Smyth, and K.R. Harrap, Adenosine deaminase activity in leukemia, Brit. J. Cancer, 31: 544 (1975).
42. H.Nishihara, S. Ishikawa, K. Shinkai, and H. Akedo, Multiple forms of human adenosine deaminase, Biochim. Biophys. Acta, 302: 429 (1973).
43. C.Sufrin, G.L. Trisch, A. Mittleman, and G.P. Murphy, Adenosine deaminase activity in patients with carcinoma of the bladder, J. Urol., 119: 343 (1978).
44. L.H. Koehler, and E.J. Benz, Serum adenosine deaminase: Methodology and clinical application, Clin. Chem., 8: 133 (1962).
45. G.L. Tritsch, and J. Minowada, Differences in purinic metabolizing activities in human leukemia T cell, B cell and ALL cell lines, J. Natl. Cancer Inst., 60: 1301 (1978).

46. E. Brusamolino, P. Isernia, M. Lazzarino, I. Scovassi, U. Bertazzoni, and C. Bernasconi, Clinical utility of terminal deoxynucleotidyl transferase and adenosine deaminase determinations in adult leukemia with a lymphoid phenotype, J. Clin. Oncol., 2: 871 (1984).
47. P.E. Daddona, and W.N. Kelley, Human adenosine deaminase. Purification and subunit structure, J. Biol. Chem., 252: 110 (1977).
48. D.A. Wiginton, M.S. Coleman, and J.J. Hutton, Purification, characterization and radioimmunoassay of adenosine deaminase from human leukemic granulocytes, Biochem. J., 195: 389 (1981).
49. D.A. Wiginton, G.S. Adrian, R.L. Friedman, D.P. Suttle, and J.J. Hutton, Cloning of cDNA sequences of human adenosine deaminase, Proc. Natl. Acad. Sci. USA, 80: 7481 (1983).
50. D. Valerio, M.G. Duvestyn, P.M. Khan, A.G. Van Kessel, A.D. Waard, and A.J.V. Eb, Isolation of cDNA clones for human adenosine deaminase, Gene, 25: 231 (1983).
51. R. Agarwal, Inhibitors of adenosine deaminase, Pharmacol. Therap., 17: 399 (1982).
52. M.S. Coleman, Selective enzyme inhibitors as antileukemic agents, Bioscience, 33: 707 (1983).
53. J.F. Smythe, M.M. Chassin, K.R. Harrap, R.H. Adamson, and D.G. Johns, 2-Deoxycoformycin phase I clinical trial and clinical pharmacology, Proc. Am. Assoc. Cancer Res., 20: 47 (1979).
54. M. Grever, J. Miser, S. Balcerzak, and J. Neidhart, Ablation of adenosine deaminase activity by deoxycoformycin induces cell lysis in refractory leukemia, Proc. Am. Assoc. Cancer Res., 21: 335 (1980).
55. D.G. Poplack, S.E. Sallan, G. Rivera, J. Holcenberg, S.B. Murphy, J. Blatt, J.M. Lipton, P. Venner, D.L. Glaubiger, R. Ungerleider, and D. Johns, Phase I Study of 2'-deoxycoformycin in acute lymphoblastic leukemia, Cancer Res., 41: 3343 (1981).
56. A.L. Yu, B. Bakay, F.H. Kung, and W.L. Nyhan, Effect of 2'-deoxycoformycin on the metabolism of purines and the survival of malignant cells in patients with T cell leukemia, Cancer Res., 41: 2677 (1981).
57. P.P. Major, R.P. Agarwal, and D.W. Kufe, Clinical pharmacology of deoxycoformycin, Blood, 58: 91 (1981).
58. H.G. Prentice, K. Ganeshguru, K.F. Bradstock, A.H. Goldstone, J.F. Smyth, B. Wonke, G. Janosky, and A.V. Hoffbrand, Remission induction with adenosine deaminase inhibitor 2'-deoxycoformycin in thy-lymphoblastic leukemia, Lancet, 2: 170 (1980).
59. J.R. Kanofsky, D.G. Roth, J.F. Smyth, J.F. Bann, D.L. Sweet, and J.E. Uttman, Treatment of lymphoid malignancies with 2'-deoxycoformycin - a pilot study, Am. J. Clin. Oncol., 5: 179 (1982).
60. D.P. Gray, M.R. Grever, M.F.E. Slaw, M.S. Coleman, and S.P. Balcerzak, 2'-Deoxycoformycin and ara-A in the treatment of refractory acute myelocytic leukemia, Cancer Treatment Reports, 66: 253 (1982).

INHIBITORS OF TERMINAL TRANSFERASE: A NEW STRATEGY

FOR THE TREATMENT OF HUMAN LEUKEMIA

> Ronald McCaffrey, Amy Ahrens, Richard Bell, Robert Duff, Henry Hoppe, Anne Lillquist, and Zachary Spigelman
>
> Section of Medical Oncology, Evans Memorial Department of Clinical Research, Boston University Medical Center Boston, Massachusetts, U.S.A.

INTRODUCTION

Terminal deoxynucleotidyl transferase (TdT, EC 2.7.7.31) is a unique DNA polymerase which catalyzes the polymerization of deoxyribonucleotides on the 3'-hydroxyl ends of preformed oligo- or polydeoxynucleotide initiators, in a template-independent manner (1). Its expression is restricted, in normal animals, to subsets of primitive lymphocytes, and, in disease states, to the blast cells of certain forms of acute leukemia and diffuse lymphoma (2). For immunobiologists TdT has emerged as a useful marker for characterizing subsets of pre-B and pre-T lymphocytes (3-7). For physicians caring for patients with leukemia and lymphoma, neoplastic cell TdT status has turned out to be a useful criterion for patient assignment to therapeutically meaningful categories (8).

Although the existence of this enzyme has been known for 25 years, its physiologic function in the cells in which it is expressed, whether normal or malignant, remains unknown. To further define what the precise function of TdT might be, we have begun a series of studies aimed at the development of specific TdT inhibitors. With such probes one would be able to examine a variety of cellular processes and functions in TdT-positive cells, and thus gain insight into the role of this enzyme in both normal and malignant states. Our underlying hypothesis is that TdT plays a critical role in the biology of TdT-positive cells, and that therefore its inhibition would constitute a serious metabolic insult to such cells.

Our work to-date has focused on 6-anilinouracils, a class of compounds known to be potent, selective inhibitors of replicative DNA synthesis in bacterial and mammalian cells (9). Following the report that one 6-anilinouracil could specifically inhibit DNA polymerase α (EC 2.7.7.7) (10), we began a systematic search for structurally related compounds which might show specificity for TdT. Our preliminary work has uncovered several related 6-anilinouracils which can specifically inhibit TdT. Our data also show that under defined conditions growth inhibition of TdT-positive cell lines and normal in vivo TdT-positive cells occurs upon exposure to two of these agents; this strongly supports our hypothesis that TdT is indeed in some way critical to the viability of those cells in which it is expressed.

BIOCHEMICAL SCREENING FOR SPECIFIC INHIBITORS

Following the publication by Wright et al. (10) on the inhibition of DNA polymerase α by p-butylanilinouracil, we began a study of related compounds to determine if other anilinouracil derivatives might have specificity for TdT. The ability of one member of the 6-anilinouracil class to specifically and selectively inhibit one DNA polymerase suggested to us that a compound with a related structure might specifically inhibit TdT. Each derivative in this initial screening evaluation was employed at 200 μM final concentration in 1 % DMSO (which is required for solubility); TdT was purified from human acute lymphoblastic leukemia cells, using ion exchange and affinity chromatography (1,11). Control TdT reactions were also run in the presence of 1 % DMSO. In this initial screen two derivatives, p-methoxyanilinouracil and p-aminoanilinouracil, showed significant inhibition of TdT (Table 1).

Table 1. Effect of Uracil Analogues on TdT.

TdT activity was assayed using oligo(dA)$_{50}$ as initiator and ^{3}H-dGTP as substrate, under the conditions defined in ref. 11. Enzyme was purified from human acute lymphoblastic leukemia cells.

COMPOUND	nMoles ^{3}H-dGMP Inc.
Control	1.22
6-anilinouracil	1.21
6-(benzylamino)uracil	1.03
6-(phenetylamino)uracil	1.31
6-(p-butylanilino)uracil	1.12
6-(p-hydroxyanilino)uracil	1.21
6-(p-acetamidobenzylamino)uracil	1.43
6-(cyclohexylamino)uracil	1.31
6-(cyclohexylmethylamino)uracil	1.33
6-(n-pentylamino)uracil	1.20
6-(iso-pentylamino)uracil	1.21
6-(3',4'-trimethyleneanilino)uracil	1.23
6-(d-naphthylamino)uracil	1.24
5-(p-methoxybenzyl)-6-aminouracil	1.20
6-(p-methoxyanilino)uracil	0.51
6-(p-aminoanilino)uracil	0.69

The p-aminoanilinouracil derivative was available in larger quantity; therefore, it was used to further characterize the nature of the inhibition seen. The inhibition seen with this agent was specific for TdT. Date for representative experiments involving the constitutive mammalian cell DNA polymerases, polymerases α, β, and γ, are shown in Figure 1. As noted, only at high concentrations of the agent is there slight inhibition seen with DNA polymerases α, β and γ.

Table 2 shows that the inhibition of TdT is neither initiator- nor substrate-dependent. Although a hierarchy of inhibitable substrate-initiator combinations is obvious (with oligo(dA)$_{50}$-dATP being the most inhibitable), all initiator/substrate combinations are inhibited by p-aminoanilinouracil.

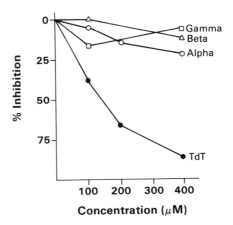

Figure 1. Inhibition of TdT and DNA Polymerases α, β and γ by Various Concentrations of p-Aminoanilinouracil

TdT was purified from human acute lymphoblastic leukemia cells; DNA polymerases α and γ from HeLa cells; and DNA polymerase β from acute myeloblastic leukemia cells. TdT purified from calf thymus gland gave identical results.

Table 2. Inhibition of TdT, assayed using various initiator/substrate combinations, by p-aminoanilino-uracil at 100 µM and 400 µM final concentrations.

100 µM Inhibitor

SUBSTRATES	INITIATORS/% INHIBITION			
	oligo(dA)	oligo(dC)	oligo(dG)	oligo(dT)
dATP	77 %	42 %	38 %	34 %
dCTP	68 %	37 %	46 %	45 %
dGTP	37 %	36 %	16 %	23 %
dTTP	42 %	51 %	22 %	34 %

400 µM Inhibitor

SUBSTRATES	INITIATORS/% INHIBITION			
	oligo(dA)	oligo(dC)	oligo(dG)	oligo(dT)
dATP	93 %	62 %	71 %	69 %
dCTP	92 %	71 %	74 %	74 %
dGTP	84 %	77 %	55 %	70 %
dTTP	83 %	72 %	71 %	84 %

Figure 2 shows a Lineweaver-Burk plot establishing an uncompetitive nature for the inhibition seen with oligo(dA)$_{50}$ and dGTP. Presumably similar uncompetitive inhibition is present for all initiator-substrate combinations. The nature of the trivial degrees of inhibition seen with DNA polymerase α, β and γ at the higher concentrations of the p-aminoanilinouracil compound has not yet been investigated. The uncompetitive nature of the inhibition of TdT by p-aminoanilinouracil is unlike the inhibition of DNA polymerase α by p-butylanilinouracil, which was shown to be due to competitive displacement of dGTP by the inhibitor (10).

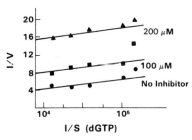

Figure 2. Double reciprocal plots of 1/V vs. 1/S for inhibition of TdT by p-aminoanilinouracil. Slopes remain constant, with intercepts varying with $\underline{i}$, indicating uncompetitive inhibition.

THE EFFECT OF p-ANILINOURACILS ON TdT(+) and TdT(-) CELLS IN CULTURE

Three murine lymphoid lines (kindly given to us by Dr. N. Rosenberg, Tufts University) were studied in detail: Line 2M3: TdT negative: Line 298-26: TdT-positive; and Line B244: TdT-positive. The TdT status of each line was established by biochemical assay, as previously described (11). The p-aminoanilinouracil was solubilized in 1 % DMSO and added, in fresh medium, to cells every 24 hours for 4 days. The outcome of these experiments is shown graphically in Figure 3. Similar data were generated using HeLa and L1210 cells (both TdT-negative and both unaffected by p-aminoanilinouracil), and the murine EL-4 TdT-positive line (which was growth-inhibited by p-aminoanilinouracil in a manner similar to that seen with the 298-26 cells). Although it was obvious to us that the growth inhibition we were seeing with these cell lines had two components (one due to the DMSO itself and one due to the DMSO/aminouracil combination), and thus represented a less than optimal model system, we nevertheless proceeded to examine growth characteristics of several additional cell lines upon exposure to p-aminoanilinouracil. These experiments were done in association with Dr. Jun Minowada, then at Roswell Park Memorial Institute. Thirty permanent human cell lines, initiated from patients with various forms of acute leukemia, were studied. Twenty-one lines were TdT-positive, 9 were TdT-negative. However, in this series of experiments we discovered that the requirement for 1 % DMSO for the solubility of the agent made the interpretation of the data impossible: the "DMSO control" cultures behaved with extreme variability in the presence of 1 % DMSO, sometimes showing no effect; sometimes showing enhanced growth, and sometimes showing growth inhibition.

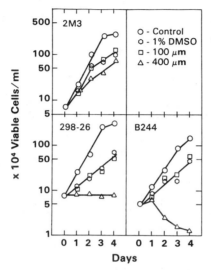

Figure 3. Effect of p-aminoanilinouracil and 1 % DMSO on growth of three murine lymphoid lines. Line 2M3 is TdT-negative; lines 298-26 and B244 are TdT-positive. Cells were seeded into T25 cm² flasks on day zero at the densities shown, in the presence of either 1 % DMSO, 100 µM, or 400 µM p-aminoanilinouracil in 1 % DMSO. Control cultures had medium only.

Likewise, in considering an exploration of the effects of p-aminoanilinouracil on intact animals, we discovered that DMSO (at 10 %), given i.p., caused variable changes in spleen, liver, thymus, and total body weight in mice given a 4-day i.p. exposure. Therefore, because it was clear that DMSO per se had the potential for significant growth effects, we turned our attention to the development of a water soluble derivative of one of our active compounds. We attempted to confer water-soluble characteristics on both the p-methoxy- and p-amino-compounds by mixing each with an equivalent amount of NaOH, followed by filtration of the resulting sodium salt, evaporation, solubilization in water, and final evaporation to yield a powder. Both compounds were rendered water soluble by this process, but were no longer inhibitors of TdT activity in the "salt" form. We have not investigated the basis for this loss of inhibitory activity following the "salting" procedure. We then sought to render yet another compound water soluble by converting it by the same process to a sodium salt. Although this compound, derived from the p-aminoanilinouracil prototype compound by the introduction of a CH_3 group in the meta-position of the aniline ring, did not display perfect selectivity for TdT inhibition, it nevertheless remained relatively specific as a TdT inhibitor (Figure 4) after becoming water soluble, and we therefore studied its effects on TdT-positive and TdT-negative cells in culture (Figure 5).

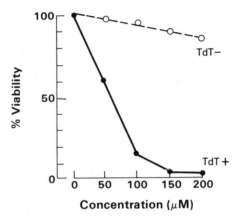

Figure 4. Inhibition of TdT and DNA polymerases α, β and γ by various concentrations of the water-soluble sodium salt of p-amino-3'-methyl-anilinouracil

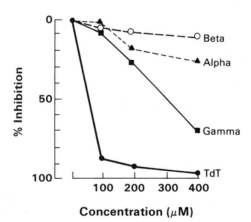

Figure 5. Effect of various concentrations of the water-soluble sodium salt of p-amino-3'-methyl-anilinouracil on viability at 24 hours of 2M3 TdT(-), and B244 TdT(+) cell lines. Viability was assessed by trypan blue dye exclusion. Cells were seeded into T-25 cm² flasks at 5×10^4 cells/ml in the presence of the compound at the concentrations shown from 0 to 200 μM. Data represent means of triplicate flasks for each concentrations.

Neither the biochemical data (Figure 4) nor the cell culture data (Figure 5) show absolute selectivity for TdT. We speculate that the low but significant degree of trypan staining seen in the TdT-negative 2M3 cells may be related to the inhibition of DNA polymerase γ activity shown in Figure 4.

STUDIES IN MICE

We next did a series of preliminary studies on intact normal mice, to determine (a) whether general toxicity with p-amino-3'-methyl-anilinouracil might be seen; and (b) how TdT-positive cells, particularly in the thymus, might be affected. The data from these experiments are summarized in Table 3 and Figure 6. At the doses used, no acute toxicity was noted. What is clear from the data, however, is that while the maximum effect of the higher dose is seen in the thymus (the major site of TdT-positive cells in normal animals), there are nevertheless significant changes seen in total body weight, and in the liver and spleen. Here again our speculation is that these effects may be due to the significant inhibition of DNA polymerase γ which this compound has (Figure 4). Work in progress is now directed at the development of 6-anilinouracil derivatives which have both specificity for TdT, and aqueous solubility.

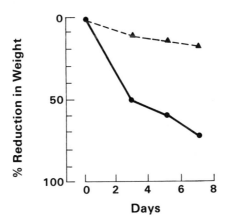

Figure 6. Effect of daily intraperitoneal injections of the water-soluble sodium salt of p-amino-3'-methyl-anilinouracil on thymic weight in 6-week old BALB-C mice. For each time point, groups of 5 mice were sacrificed, and thymus weights recorded and averaged for each group. Percentage reduction in thymus weight was calculated from the difference between average thymus weight of the controls (groups of 5 mice for each of the 4 time points), and the test groups for each time: dashed line = 17 mg/kg/day; solid line = 68 mg/kg/day.

CONCLUSION

The physiologic function of TdT in the cells in which it is expressed remains essentially unknown. The most compelling speculation comes from work exploring the molecular events involved in both immuno-

Table 3. Effect of the water soluble sodium salt of p-amino-3'-methyl-anilinouracil on thymic, liver, splenic and total body weight of 6-week old BALB-C mice after 5 daily intraperitoneal injections of compound. Experimental details are as on the legend to Figure 6.

% Loss of Weight on Day 7

Dose	Thymus	Liver	Spleen	Total Body
17 mg/kg/day	18	16	15	9
68 mg/kg/day	75	20	11	24

globulin gene rearrangement and T-cell antigen receptor gene rearrangement. Alt and Baltimore have suggested a key role for TdT in the somatic rearrangement of germ-line DNA elements to form complete variable region genes for immunoglobulin light and heavy chains in pre-B cells (12). At V/J_H and V_H/D joints of IgM heavy-chain rearranged genes, extra (non-germ line) nucleotides are found inserted between the D and J_H, and V_H and D segments. The insertion of these new elements (referred to as N regions) in rearranged heavy-chain genes has been speculatively attributed to site-directed TdT activity during the process of gene rearrangement. Similar N regions at V, D, and J junctions of the T-cell receptor β-chain gene are also now being identified in thymocytes undergoing T-cell antigen receptor gene rearrangement, suggesting that this too may be the function TdT subserves in thymocytes undergoing maturation to functional T-cells (13).

Whatever the physiological role ultimately assigned to TdT may be, the hypothesis which animates our work is that the inhibition of this enzyme would result in damage which would ultimately be inconsistent with cell viability. We believe our preliminary data partially validate this speculation. The availability of potent, water-soluble inhibitors of TdT, which our future work should provide, will allow for studies on the sequelae of TdT inhibition in both normal and malignant TdT-positive cells. Our long-range expectation is that the experimental agents which we plan to study will be further developed to provide a new class of clinically useful agents capable of selectively damaging TdT-positive cells, thus adding to our pharmacological anti-leukemic armamentarium. Our work, in addition, should provide new insight into molecular mechanisms operative in pre-B and pre-T cells.

REFERENCES

1. F. J. Bollum, in: "The Enzymes: Therminal Deoxynucleotidyl Transferase", R. D. Boyer, ed., Academic Press, New York, p. 145 (1974).
2. R. McCaffrey, A. Lillquist, S. Sallan, E. Cohen, and M. Osband, Cancer Res., 41:4814 (1981).
3. A. E. Silverstone, H. Cantor, G. Goldstein, and D. J. Baltimore, Exp. Med., 144:453 (1976).
4. G. Janossy, F. J. Bollum, K. F. Bradstock, A. McMichael, N. Rapson, and M. F. Greaves, J. Immunol., 123:1525 (1979).
5. F. J. Bollum, Blood, 54:1203 (1979).
6. J. Blatt, G. Reaman, and D. G. Poplack, New Engl. J. Med., 303:918 (1980).
7. M. F. Greaves, Cancer Res., 41:4752 (1981).
8. S. M. Marks, D. Baltimore, and R. P. McCaffrey, New Engl. J. Med., 298:812 (1978).

9. N. C. Brown, J. Gambino, and G. E. Wright, J. Med. Chem., 20:1186 (1977).
10. G. E. Wright, E. F. Baril, and N. C. Brown, Nucl. Acids Res., 1:99 (1980).
11. R. P. McCaffrey, T. A. Harrison, R. Parkman, and D. Baltimore, New Engl. J. Med., 292:775 (1975).
12. R. W. Alt, and D. Baltimore, Proc. Nat. Acad. Sci. USA, 79:4118 (1982).
13. T. W. Mak, Personal Communication (1985).

DNA-PROTEIN CROSSLINKING OF PLATINUM COORDINATION COMPLEX IN
LIVING CELLS: IMPLICATION TO EVALUATE THE CYTOTOXIC EFFECTS OF
CHEMOTHERAPEUTIC AGENTS

L. S. Hnilica, R. Olinski, Z. M. Banjar, W. N. Schmidt
and R. C. Briggs

Departments of Biochemistry and Pathology, the A.B.
Hancock, Jr. Memorial Laboratory of the Vanderbilt
University Cancer Center and the Center in Molecular
Toxicology, Vanderbilt University School of Medicine
Nashville, Tennessee 37232, U.S.A.

INTRODUCTION

The first report by Rosenberg et al. (1) that, out of several platinum coordination complexes, the cis-diamminedichloroplatinum (II), (or cis-DDP) was most effective in inhibiting the growth of sarcoma 180 in mice has initiated intensive research into its mode of biological action, especially since its isomer, the trans-DDP, is essentially inactive (2,3). Because both isomers have been shown to bind DNA, it must be the stereospecificity of this binding which sets them apart as antitumor agents. Indeed, the elegant experiments of Lippard and his associates (4), who developed antibodies specific for intrastrand crosslinks of two adjacent guanine residues, showed that only cis-DDP was capabale of binding to the DNA in this fashion. The trans-isomer did not produce detectable crosslinks of this kind. These findings, supported by x-ray crystallography (5), together with the reports of others (6), suggest that the intrastrand DNA crosslinking by cis-DDP may be responsible for its antitumor activity.

In addition to forming adducts with DNA, both the cis- and trans-DDP produce crosslinks between the DNA and its associated proteins (7-9). Although in experiments with isolated nucleosomes the trans isomer has been shown to crosslink histones to DNA more efficiently than cis-DDP (8), studies of this phenomenon in living cells pointed out the cis-DDP as a more efficient nonhistone protein-DNA crosslinker (10). Since most investigators who addressed the DNA-Protein crosslinking by platinum coordination complexes used the alkaline elution method developed by Kohn et al. (11), the nature of the crosslinked proteins is not known. To address this problem, we have employed antisera to several nuclear protein fractions for the detection and initial identification of at least some of the proteins crosslinked in cells exposed to cis- or trans-DDP.

HeLa CELLS

The HeLa S_3 strain was used in our initial studies as well as in the cell cycle experiment. In a relatively simple experimental outline, the cells were exposed to various concentrations of cis- or trans-DDP for indicated time intervals and, after harvesting, solubilized in 2 % buffered SDS and centrifuged at 110,000 x g for 18 hrs. The DNA pellets were resuspended in 5 M urea, stirred for 2-3 hrs and centrifuged again as indicated. The final DNA pellets containing covalently bound proteins were sonicated and digested with DNAse I (10). The digests were electrophoresed in SDS-PAGE, the separated proteins transferred to nitrocellulose sheets, incubated with appropriate antisera and the antigen-antibody complexes were visualized by the peroxidase-antiperoxidase staining procedure (12,13).

Both the cis- and trans-DDP crosslinked chromosomal proteins to DNA (10) which could be detected either with antisera to 0.35 M NaCl extract of HeLa nuclei or to the 0.35 M residue. Both antisera recognized similar antigens although the antiserum to the 0.35 M NaCl residue recognized fewer antigens than that to the extract (Figs. 1 and 2). The extent of crosslinking, as determined by densitometric measurement of the PAP-stained proteins (14) was nearly linear, increasing with the rising cis- or trans-DDP concentration, reaching maximum and leveling between 3-5 mM. A similar increase in the DNA-crosslinked proteins was also seen when HeLa cells were incubated with 1 mM cis- or trans-DDP for increasing time intervals (Figs. 1 and 2, respectively). As can be seen in these figures, qualitative distribution of the crosslinked proteins did not change with time indicating the absence, in the DNA pellets, of protein-protein crosslinks. The same antigens became crosslinked regardless whether live cells or isolated nuclei were exposed to either of the two isomers (10). In general, the trans-DDP was less effective DNA-protein crosslinker in vivo than the cis-isomer (10).

The two antisera were also employed for analysis of proteins crosslinked to DNA in cells exposed to cis-DDP during the progression through their cell cycle. HeLa-S_3 cells, synchronized by double thymidine block (15) were incubated at 2 hr intervals with 1 mM cis-DDP for 1 hr as they progressed through S, G_2, M and then into G_1 and S phases of the subsequent cycle, solubilized in buffered 2 % SDS and analyzed as indicated. The DNA-protein crosslinking sequence for the 0.35 M NaCl extract antiserum is shown in Fig. 3A and B. Although many of the crosslinked antigens did not change qualitatively during the cell cycle, there were at least three prominent antigenic bands exhibiting cell cycle dependence of their crosslinking. E.g., antigens at approximate M_r 34 kD and 120 kD crosslinked only 8 hrs after the release of cells from G_1/S while another antigen, approximate M_r 44 kD became prominently crosslinked only during the first 8 hrs after the release (see arrows in Fig. 3A). Although very prominent in immunoblots, the approximate M_r 34 kD antigen was barely detectable in parallel gels stained with Coomassie Brilliant Blue. Controls, taken at the same time intervals as the experimental points but incubated without cis-DDP did not show any detectable DNA-protein crosslinks (Fig. 3B). Analysis of the antigens reactive with antiserum to the 0.35 M NaCl nuclear residue revealed that most of the antigens reacting in Fig. 3A (antiserum 0.35 M NaCl extract) could not be detected. However, two prominent antigens reacted. The approximate M_r 34 kD antigen started to crosslink at 8 hrs after the G_1/S release (as in Fig. 3A) and another one, approximate M_r 48 kD with a similar crosslinking schedule as the 34 kD protein.

Several conclusions can be drawn at this point. Both the cis- and trans-DDP are efficient DNA-protein crosslinkers, qualitatively affecting the same kinds of antigens. The crosslinking is selective, most likely reflecting the distances between the crosslinked proteins and DNA, since not all the proteins recognized by the antisera became crosslinked even

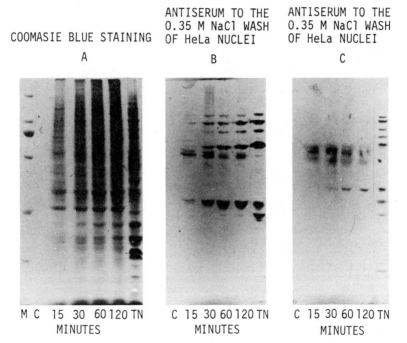

Fig. 1. Effects of the incubation time on DNA-protein crosslinking by cis-DDP. Equal numbers of HeLa cells were incubated with 1 mM cis-DDP for the indicated time intervals, solubilized in buffered 2 % SDS and centrifuged at 110,000 x g for 18 hrs. The resulting DNA pellet was analyzed for crosslinked proteins.

A: Coomassie Brilliant Blue staining of proteins equivalent to 60 µg of pelleted DNA. M = molecular weight standards (myosin 200 kd, β-galactosidase 116 kD, phosphorylase B 94 kD, bovine serum albumin 43 kD, carbonic anhydrase 32 kD). C = control DNA from untreated cells, TN = total proteins of HeLa nuclei (20 µg of DNA); the two prominent bands in this lane at approx. M_r 34 kD are H1 histones.

B and C: Duplicate gels transferred to nitrocellulose sheets (60 µg DNA) incubated with antiserum to 0.35 M NaCl nuclear extract (B) or nuclear residue (C) and developed with peroxidase-antiperoxidase staining.

at very high (5 mM) levels of cis-DDP over several hours of exposure. The selectivity of the crosslinking process is further supported by the concentration and time course experiments. Qualitatively, the same antigens became crosslinked over a wide concentration range (0.1 - 5.0 mM) of the crosslinker or over a period of up to 8 hrs. Little, if any protein-protein crosslinks co-sedimented with the DNA pellets. Such crosslinks would be easily detectable by their reaction with antibodies, increasing the heterogeneity of the immunoblots (16). The crosslinking selectivity is further supported by the cell cycle experiment. A number of weakly chromatin associated proteins (dissociable in 0.35 M NaCl)

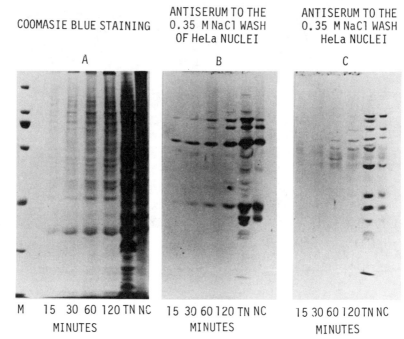

Fig. 2. Effects of the incubation time on DNA-protein crosslinking by trans-DDP. The experimental conditions were the same as in Fig. 1 except 1 mM trans-DDP was used for crosslinking and 100 μg DNA equivalent was applied to the experimental lanes. M = molecular weight standards as in Fig. 1. TN = total protein of HeLa nuclei (20 μg DNA). NC = DNA from isolated HeLa nuclei inucbated with 2 mM trans-DDP for 2 hrs (100 μg DNA).

must be in proximity to DNA since they became crosslinked by both the cis- or trans-DDP, i.e., short distance crosslinkers (5).

NOVIKOFF HEPATOMA

Although our experiments with HeLa cells show that both cis- and trans-DDP can selectively crosslink nuclear proteins to DNA in living cells, they did not provide us with information concerning the identity of the crosslinked proteins. Because of the availability, in our laboratory, of antisera to several defined fractions of Novikoff hepatoma chromatin-associated proteins, we used this system to ask what kind of proteins are involved in the crosslinking phenomenon. For better approximation of the in vivo conditions, rat bearing Novikoff hepatoma were injected with 7 mg/kg of cis-DDP six days after the transplant. Cells (3 ml) were removed from the animals by abdominal centhesis at 1, 8, 24, 48 and 72 hrs after the cis-DDP administration. The cells were then washed with buffered saline solution, solubilized in 4 % SDS and centrifuged at 110,000 x g for 18 hrs. The resulting pellets were resuspended in buffered 5 M urea and stirred for several hrs. Then the

HOURS AFTER RELEASE FROM THE G_1/S

1 3 5 7 8 10 12 14 16 18 20 22

HOURS AFTER RELEASE FROM THE G_1/S

1 3 5 7 8 10 12 14 16 18 20

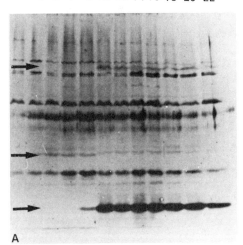
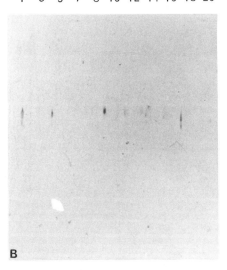

A   B

Fig. 3. Immunochemical detection of protein antigens crosslinked to the DNA during the cell cycle. HeLa-S_3 cells were released from G_1/S block, incubated with 1 mM cis-DDP for 1 hr at indicated time intervals, solubilized in buffered 2 % SDS, centrifuged and analyzed as described. Each lane contains proteins equivalent to 60 μg DNA in the ultracentrifugation pellet. Each experimental point is derived from the same number of HeLa cells.

A: Immunoblots of gels incubated with antiserum to 0.35 M NaCl extract of HeLa nuclei stained with peroxidase-antiperoxidase method. The approx. 34 kD, 44 kD and 120 kD antigens are indicated by arrows.

B: Controls of cells incubated in the absence of cis-DDP and treated as in (A). Each lane represents protein equivalent to 60 μg pelleted DNA.

solution was made 4 % in respect to SDS and centrifuged again as indicated. The final pellets, containing crosslinked proteins were processed as indicated for the HeLa cells experiments.

Three antisera were employed to detect crosslinked antigens: antiserum to dehistonized chromatin (17), to a nuclear matrix preparation (18) and to Novikoff hepatoma cytokeratin fraction (19). When administered in a therapeutically significant dose (7 mg/kg), cis-DDP crosslinked only few proteins which could be detected immunochemically. Indeed, as shown in Fig. 4, only 3 - 4 crosslinked protein bands were revealed with antiserum to dehistonized chromatin although a preparation of Novikoff hepatoma chromatin exhibited numerous protein bands when reacted with this antiserum (Fig. 4A). The crosslinking reached maximum at 24 hrs (Fig. 4D) and then declined 72 hrs after the administration of cis-DDP (Fig. 4F). Antiserum to nuclear matrix showed the crosslinking of 5 - 6 protein bands in the m.w. range between 40 - 70 kD. Again, the maximum crosslinking was observed at 24 hrs (Fig. 5E) after the

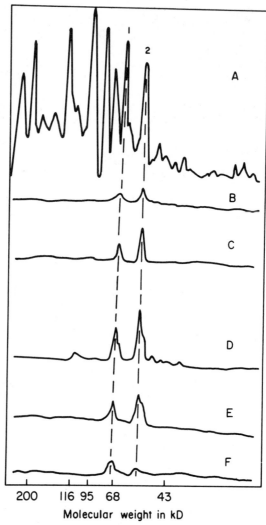

Fig. 4. Densitometric scans of DNA-associated proteins crosslinked in vivo by injecting Novikoff hepatoma-bearing rats with 7 mg/kg of cis-DDP intraperitoneally. The cells, harvested at indicated time intervals, were solubilized in 4 % buffered SDS and processed as described in the text. The nitrocellulose-bound proteins were reacted with antiserum to dehistonized chromatin and developed by peroxidase-antiperoxidase staining.

A: Preparation of total Novikoff hepatoma chromatin (10 µg DNA).

B - F: Proteins associated with the high speed 4 % SDS DNA pellets at 1, 8, 24, 48 and 72 hrs (respectively) after administration of cis-DDP. All lanes represent 80 µg DNA.

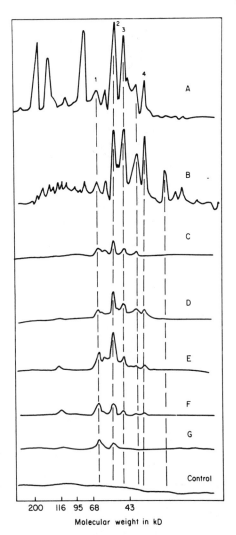

Fig. 5. Densitometric scans of DNA-associated proteins crosslinked in vivo by injecting Novikoff hepatoma bearing rats with 7 mg/kg of cis-DDP intraperitoneally. The cells were harvested and processed as indicated in the text and in the legend of Fig. 4. The transferred proteins were reacted with antiserum to Novikoff hepatoma nuclear matrix and developed by peroxidase-antiperoxidase staining.

A: Preparation of Novkoff hepatoma nuclear matrix (5 μg protein).

B: Preparation of Novikoff hepatoma chromatin (10 μg DNA).

C - G: Proteins associated with the high speed 4 % SDS DNA pellets at 1, 8, 24, 48 and 72 hrs (respectively) after the intraperitoneal injection of cis-DDP. All lanes contain 80 μg DNA. Control: 80 μg DNA equivalent of untreated Novikoff hepatoma cells.

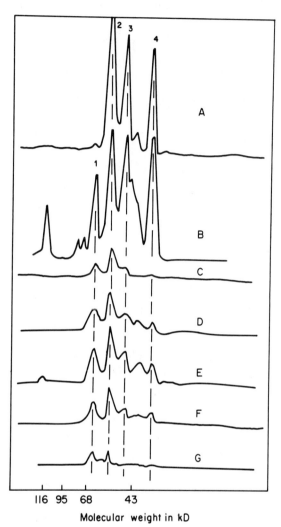

Fig. 6. Densitometric scans of DNA-associated proteins crosslinked in vivo by injecting Novikoff hepatoma bearing rats with 7 mg/kg of cis-DDP intraperitoneally. The cells were harvested and processed as indicated in the text and in the legend of Fig. 4. The transferred proteins were reacted with antiserum to the three principal Novikoff hepatoma cytokeratins (p39, p49 and p56) and developed by peroxidase-antiperoxidase staining.

A: Novikoff hepatoma cytoskeletal preparation (5 µg protein).

B: Preparation of Novikoff hepatoma chromatin (10 µg DNA).

C - G: Proteins associated with the 4 % SDS DNA pellets at 1, 8, 24, 48 and 72 hrs (respectively) after intraperitoneal injection of cis-DDP. All lanes contain 80 µg DNA.

administration of cis-DDP, declining thereafter (Fig. 5F and G). Untreated cells exhibited no detectable crosslinked proteins. Since three of the major crosslinked antigenic proteins (peaks 2, 3 and 4, approximate m.w. 56 kD, 50 kD and 40 kD, respectively) resembled the principal Novikoff hepatoma cytokeratins (19), an identical immunoblot to those in Fig. 4 and 5 was reacted with antiserum to Novikoff hepatoma cytokeratin preparation, containing the three principal cytokeratins (p56, p49 and p59, shown as peaks 2, 3 and 4 (respectively) in Fig. 6A). This antiserum reacted strongly with proteins in Novikoff hepatoma chromatin preparation (Fig. 6B) and identified the crosslinked protein peaks 2, 3 and 4 as cytokeratins. Again, maximum crosslinking occured at 24 hrs after cis-DDP administration (Fig. 6E) followed by a gradual removal of the crosslinked proteins (Fig. 6F and G). These results support our previous observations where a monoclonal antibody, specific for the p39 Novikoff hepatoma cytokeratin, was used to study the crosslinking of this protein to DNA (16).

The Novikoff hepatoma experiments support the conclusion reached with HeLa cells that cis-DDP is a potent and selective DNA-protein crosslinking agent. Of all the chromatin proteins, detectable by our antisera, the Novikoff hepatoma cytokeratins were among the most prominently crosslinked proteins. We do not know yet the biological significance of this finding.

CONCLUSIONS

Since its approval for human use by the U.S. Food and Drug Administration in December 1978, cis-DDP became one of the most widely used, clinically important anti-cancer drugs. Although the biological activity of cis-DDP seems to depend on its ability to form two Pt-N bonds between the N (7) atoms of two adjacent guanosine residues on the same strand (the biologically less active trans-DDP cannot form such crosslinks), the role of Pt-mediated DNA-protein crosslinking in cellular toxicity of this compound is not known and deserves attention. We have shown that at least in two biological systems, the HeLa-S_3 cells and transplantable Novikoff hepatoma in rats, the cis-DDP acts as efficient DNA-protein crosslinker. The chromosomal proteins became crosslinked in a highly selective pattern, changing with the progression of cells through their reproductive cycle. In Novikoff hepatoma, the principal cytokeratins p39, p49 and p56 were among the most prominently crosslinked proteins even at low (therapeutic) cis-DDP doses (7 mg/kg). Since intermediate filament proteins are notable for their cell specific distribution, changing with differentiation and carcinogenesis (20-22), this finding may be biologically important. Because of their in vivo crosslinking patterns, the intermediate filaments must penetrate the cell nucleus in close proximity to DNA (cis-DDP is a short distance crosslinker, approximately 4 Å), thus providing a continuum from desmosomal junctions on the cell membrane to its nucleus and DNA. These DNA-cytokeratin crosslinks were removed very slowly, being still detectable 72 hrs after the administration of cis-DDP to the Novikoff hepatoma-bearing rats.

Acknowledgments: The authors with to acknowledge the excellent editorial assistance of Ms. Doris Harris in preparation of this manuscript. Supported by grants from the National Cancer Institute (CA-26412 and CA-36459) and National Institute of Environmental Health Sciences (ES 00267). ZMB was supported by a grant from King Abdulaziz University, Jeddah, Saudi Arabia.

REFERENCES

1. B. Rosenberg, L. Van Camp, J. E. Trosko, and V. H. Mansour, Platinum compounds: a new class of potent antitumor agents, Nature, 222:385 (1969).
2. J. M. Hill, and R. J. Speer, Organo-platinum complexes as antitumor agents (review), Anticancer Res., 2:173 (1982).
3. A. C. M. Plooy, and P. H. M. Lohman, Platinum compounds with antitumor activity, Toxicology, 17:169 (1980).
4. S. J. Lippard, H. M. Ushay, C. M. Merkel, and M. C. Poirier, Use of antibodies to probe the stereochemistry of antitumor platinum drug binding do deoxyribonucleic acid, Biochemistry, 22:5165 (1983).
5. S. E. Sherman, D. Gibson, A. H. J. Wang, and S. J. Lippard, X-Ray structure of the major adduct of the anticancer drug Cisplatin with DNA: cis-$[Pt(NH_3)_2(\alpha pGpG)]$, Science, 230:412 (1985).
6. A. M. J. Fichtinger-Schepman, J. L. van der Veer, J.H.J. den Hartog, P. H. M. Lohman, and J. Reedijk, Adducts of the antitumor drug cis-diamminedichloroplatinum(II) with DNA: formation, identification and quantitation, Biochemistry, 24:707 (1985).
7. L. A. Zwelling, T. Anderson, and K. W. Kohn, DNA-protein and DNA interstrand cross-linking by cis- and trans-platinum(II) diamminedichloride in L1210 mouse leukemia cells and relation to cytotoxicity, Cancer Res., 39:365 (1979).
8. S. J. Lippard, and J. D. Hoeschele, Binding of cis- and trans-dichloro-diammineplatinum(II) to the nucleosome core, Proc. Nat. Acad. Sci. USA, 76:6091 (1979).
9. J. Filipski, K. W. Kohn, and W. M. Bonner, Differential crosslinking of histones and non-histones in nuclei by cis-Pt(II), FEBS Lett., 152:105 (1983).
10. Z. M. Banjar, L. S. Hnilica, R. C. Briggs, J. Stein, and G. Stein, Cis- and trans-diamminedichloroplatinum(II)-mediated cross-linking of chromosomal non-histone proteins to DNA in HeLa cells, Biochemistry, 23:1921 (1984).
11. K. W. Kohn, L. C. Erickson, R. A. G. Ewig, and C. A. Friedman, Fractionation of DNA from mammalian cells by alkaline elution, Biochemistry, 15:4629 (1976).
12. H. Towbin, T. Staehelin, and J. Gordon, Electorphoretic transfer of proteins from polyacrylamide gels to nitrocellulose sheets: procedure and some applications, Proc. Nat. Acad. Sci. USA, 76:1350 (1979).
13. W. F. Glass, R. C. Briggs, and L. S. Hnilica, Identification of tissue-specific nuclear antigens transferred to nitrocellulose from polyacrylamide gels, Science, 211:70 (1981).
14. F. P. Guengerich, P. Wang, and N. K. Davidson, Estimation of isozymes of microsomal cytochrome P-450 in rats, rabbits and humans using immunochemical staining coupled with sodium dodecyl sulfate-polyacrylamide gel electrophoresis, Biochemistry, 21:1698 (1982).
15. T. Borun, and G. S. Stein, The synthesis of acidic chromosomal proteins during the cell cycle of HeLa S-3 cells. II. The kinetics of residual protein synthesis and transport, J. Cell Biol., 52:308 (1972).
16. W. S. Ward, W. N. Schmidt, C. A. Schmidt, and L. S. Hnilica, Cross-linking of the Novikoff hepatoma cytokeratin filaments, Biochemistry, 24:4429 (1985).
17. M. Stryjecka-Zimmer, W. N. Schmidt, R. C. Briggs, and L. S. Hnilica, Immunological specificity of Novikoff hepatoma chromatin: isolation of three antigenic proteins, Int. J. Biochem., 14:591 (1982).
18. W. N. Schmidt, K. B. McKusick, C. A. Schmidt, L. H. Hoffman, and L. S. Hnilica, Nuclear matrix antigens in azo-dye induced primary

rat hepatomas, Cancer Res., 44:5291 (1984).
19. W. N. Schmidt, M. Stryjecka-Zimmer, W. F. Glass, R. C. Briggs, and L. S. Hnilica, Tissue specificity and distribution of Novikoff hepatoma antigenic proteins p39, p49 and p56, J. Biol. Chem., 256:8117 (1981).
20. G. Babbiani, T. Kapanci, P. Barrazone, and W. W. Franke: Immunochemical identification of intermediate filaments in human neoplastic cells: a diagnostic aid for the surgical pathologist, Am. J. Pathol., 104:206 (1981).
21. E. Lazarides, Intermediate filaments: a chemically heterogeneous, developmentally regulated class of proteins, Annu. Rev. Biochem., 51:219 (1982).
22. M. Osborn, M. Altmannsberger, E. Debus, and K. Weber; Conventional and monoclonal antibodies to intermediate filament proteins in human tumor diagnosis, in: "Cancer Cells: The transformed Phenotype", A. J. Levine, G. F. Vande Woude, W. C. Topp, and J. D. Watson, eds., Cold Spring Harbor Laboratory, Cold Spring Harbor, New York (1984).

ANTITUMOR ACTIVITY, PHARMACOLOGY AND

CLINICAL TRIALS OF ELLIPTINIUM (NSC 264-137)

Anette Kragh Larsen and Claude Paoletti

Unite de Biochimie et Enzymologie, INSERM U 140
CNRS LA 147, Institut Gustave-Roussy
94800 Villejuif, France

INTRODUCTION

Ellipticine and some of its derivatives are naturally occuring alkaloids found in plants of the Apocynaceae family. The antitumoral activities were described in 1967 by Dalton et al.(1) and later confirmed by Mathe et al. who reported the antitumor activity of 9-methoxy-ellipticine in patients with acute myeloblastic leukemia. The finding of drug-related side-effects such as hemolysis and nervous toxicity (3,4) prompted an intensive search for new derivatives at the Institut Gustave-Roussy in Villejuif, France.

These studies have led to the synthesis of new ellipticine derivatives such as 9-hydroxyellipticine (5,6) and 2-methyl-9-hydroxy-ellipticinium acetate (7) (Elliptinium, Celiptium^R). Both of these compounds exhibit a broad spectrum of antitumor activity against experimental tumors. However, in man 9-hydroxy-ellipticine displays no therapeutic effect, which probably is due to its rapid metabolism (8) in humans. In contrast, Elliptinium has shown favorable results, in particular in the treatment of patients with osteolytic metastasis of breast cancer (9). This commercially available drug has now completed phase II studies and is presently undergoing phase III clinial trials. In this paper, we present a general overview of the main properties of ellipticines, with special reference to elliptinium.

CYTOTOXICITY AND ANTITUMORAL PROPERTIES

Elliptinium is the first example of a quaternary ammonium derivative as an antitumor drug. Because of the positive charge, the cellular uptake of such molecules are affected by the membrane potential (10). This can lead to local drug accumulation as shown for certain cationic dyes which are concentrated within mitochondria and lysosomes (11,12). Both elliptinium and 2-N-methylellipticinium are rapidly taken up by tumor cells in vitro leading to an overconcentration of drug inside the cells. This also indicates the presence of intracellular structures with high drug affinity or numerous binding sites. The uptake seems to be species dependent; in chinese hamster cells elliptinium is overconcentrated about 300times after 4 hrs drug exposure (1 µg/ml) (13) while the corresponding value is about 100 in mouse sarcoma cells (AK Larsen, unpublished results). These finding warrent caution in extrapolation from in vitro studies to the clinical situation with respect to the intracellular drug concentration.

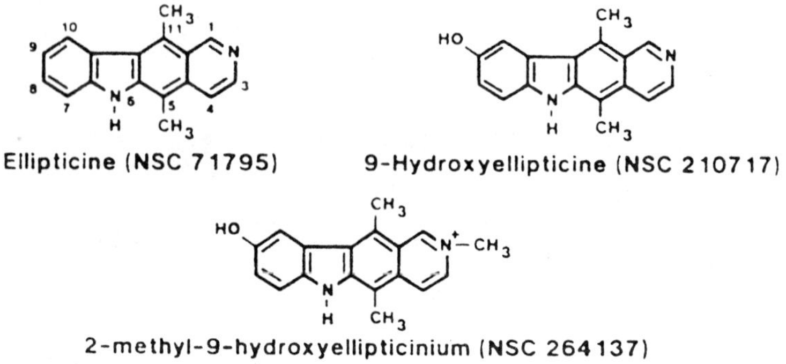

Ellipticine (NSC 71795)

9-Hydroxyellipticine (NSC 210717)

2-methyl-9-hydroxyellipticinium (NSC 264137)

Figure 1. Structure of ellipticine, 9-hydroxy-ellipticine and 2-methyl-9-hydroxyellipticinium (elliptinium).

In vitro studies of L1210 cells show that the toxic effect of these compounds are rapid and result in a marked rearrangement of the cellular architecture with the appearance of giant cells (14). DNA synthesis is preferentially inhibited while protein synthesis remains unaffected (15). The first toxic effect observed on the cells in vitro is a decrease in the rate of multiplication without appearance of cell lysis or immediate death. These cells lose their ability to clone in vitro on agar medium and to promote tumor when gratted to mice (14). Thus the antitumoral efficacy of ellipticine derivatives resides in a direct and preferential cytotoxic action upon malignant cells rather than in an indirect effect.

Elliptinium act on a broad range of experimental tumors. It is very active toward P388 and L1210 leukemias, ependymoblastoma, Gardner and Yoshida lymphosarcomas, active on B16 melanoma and squamous cell carcinoma and has borderline activity on Lewis lung carcinoma and no activity towards osteosarcoma (16). Daily subcutaneous injection of elliptinium (5 µg) in a high mammary tumor strain of mice was found to decrease the number of precancerous mammary hyperplastic alveolar nodules considerably without affecting the mammotropic hormone system (17).

When elliptinium was injected into mice inoculated with 10^5 L1210 cells per mouse it cured 2 of 31 mice at 5.6 x 10^{-8} M and 9/24 at 9.0 x 10^{-8} M. This activity is close to that of the most cytotoxic antitumoral drugs such as actinomycin (16).

MECHANISM OF ACTION

The nature of the ultimate intracellular target of ellipticine and its derivatives is still uncertain and might vary from one derivative to another. Among the various possibilities are [I] DNA, [II] ribosomal and transfer RNA, [III] topoisomerase II and [IV] other macromolecules. [I] All ellipticine derivatives with antitumor activity bind tightly to DNA in vitro with affinity coefficients ranging from 10^5 M^{-1} to 10^7 M^{-1} (5,14). Although a high affinity for DNA therefore seems to be required for expression of an antitumoral effect, this feature by itself is not enough since some ellipticines without antitumor effects such as e.g. 9-aminoellipticine bind equally tight to DNA. [II] Oxidative activation of elliptinium can lead to regioselective arylation of ribose in adenosine and guanosine (18), which might inhibit the formation of poly(ADP-ribose) as well as the methylation of certain tRNA's. [III] The involvement of topoisomerases, especially topoisomerase II is suggested by the following observations: a) treatment of mammalian cells with ellipticine and some of its derivatives results in the formation of protein-associated DNA-strand breaks (19). b) The incubation of ellipticine or elliptinium with DNA and purified mammalian DNA topoisomerase results in reversible protein-linked DNA strand breaks (20). Together, this suggests that the mechanism of DNA breakage induced by ellipticines is likely to be due to drug stabilization of a cleavable complex between DNA and topoisomerase II. [IV] Some ellipticine derivatives have high affinities for cytochrome P450 mixed function oxygenases and can be oxidized by these enzymes. However, the transformation is self-inhibited at high drug concentrations because the ellipticines are also inhibitors of these enzymes (21). Finally, it has been shown, that non-quaternized ellipticines exhibit high affinity in vitro for phospholipid monolayers to which they bind irreversible at low drug concentrations (21). A similar binding in vivo could lead to disturbance of the organization and the fluidity of the lipid matrix in membranes.

In order to understand the mechanism(s) of action better the effects of 9-OH ellipticine have been examined in sensitive and resistant chinese hamster cells. Drug uptake and retention studies did not reveal any differences between the sensitive parental cells and the resistant cells (13,22). Therefore, the resistance to ellipticine derivatives in these cells does not result from a decrease in drug accumulation but should rather be related to its mechanism of action. The results show that 9-OH ellipticine toxicity on sensitive and resistant cells involves at least two different mechanisms of action. a) In sensitive cells the lethal effects are independent of the effects on macromolecular synthesis whereas cell death occurs concomitantly with the inhibition of macromolecular synthesis in resistant cells (23). b) The cloning efficiency of sensitive cells is the same after 3 and 72 hours drug exposure suggesting that the interaction with the target is rapid and irreversible. In contrast, for the resistant cells there is a 8-fold difference in the cloning efficiency after 3 and 72 hours drug exposure. This suggests that the toxic effects are accumulative or, alternatively, that the damage can be repaired once the drug is removed (23). c) Finally in the sensitive cells, cycloheximide limits the 9-OH ellipticine toxicity when it is added before or during drug exposure. No such effect is observed in resistant cells (23). These findings can be interpreted as indicating that the drug toxicity in the sensitive cells relies on a protein which has to be present in the cells when the drug is added.

TOXICITY AND PHARMACOLOGY

The toxicity of ellipticine derivatives varies depending upon the nature of the derivative. Of nine related compounds tested in mice, the LD_8 after intraperitoneal injection fluctuated from 5 mg/kg to more than 500 mg/kg (16). The LD_0 for elliptinium in mice is 8.6 mg/kg (iv), 5.1 mg/kg (ip) and 250 mg/kg (po). No animals died after chronic administration of 0.7 mg/kg every 5 days for 120 days (16). In addition elliptinium has neither shown embryotoxicity nor teratogenicity in mice. Another important observation is the absence of haematological toxicity of this compound (16).

Intravenous injection of elliptinium to mice results in drug accumulation in the kidneys and salivary glands as shown by whole body autoradiography (24). In rats, the main route of elimination is the bile (56 % of the dose by 24 hours) while 24 % is recovered in the urine. Of the biliary material, 3 % is in the form of a glutathione adduct, 80 % as a O-glucuronide and 17 % as unchanged drug (25). An overview of the metabolic pathways is shown in figure 2.

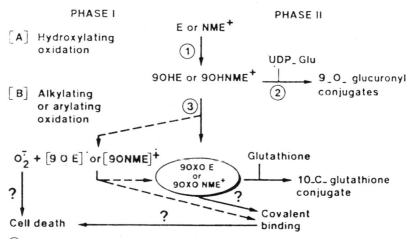

① Microsomal (nuclear) cyt P450 mixed oxygenases (liver, other organs?)
② Glucuronyltransferases (liver).
③ Peroxydases, some oxidases (most cells).

Figure 2. Metabolism of ellipticine and some of its derivatives.

The presence of glutathione conjugates can be considered as experimental evidence for the involvement of oxidative activation of Elliptinium. This is an interesting prospect since oxidative activation is a characteristic which is shared with several other antitumor drugs possessing a para-quinone or para-quinone-like structure such as mitomycin C, anthracyclines, lapachol and streptonigrin. This might contribute to the cytotoxic action of these substances. Another interesting point is that these drugs often exhibit a sufficient selectivity towards malignant cells. One difference between normal and cancer cells is that most malignant cells show low levels of superoxide dismutase (26), the key enzyme in protecting aerobic cells toward oxidation (27). This would make tumor cells more sensitive than normal cells towards the cytotoxic action of drugs generating oxidizing species.

CLINICAL TRIALS

The first clinical trial of ellipticine derivatives was undertaken by Mathe et al. (2) who observed several remissions of acute myeloblastic leukemia after treatment with 9-methoxyellipticine. Ellipticinium was first studied in phase I trials by Juret et al. (9) using a weekly administration over one hour at doses up to 100 mg/m^2. Thus ellipticinium is much less toxic in man than in mice since an equivalent dose to mice is lethal (16).

Elliptinium has demonstrated clinical activity on advanced breast cancer, hepatomas, non-Hodgkin's lymphomas and in renal cell carcinoma (28,30). Use of elliptinium in metastatic breast cancer patients refractory to all other treatment results in about 20 % objective remissions (29,31). The drug seems to have a selective activity on cutaneous or subcutaneous and lymph node metastasis, where the response rate was about 30 %, in contrast to 5-10 % for pleural and lung metastasis (31).

When elliptinium is administered once weekly the major toxicities are nausea and vomiting, dryness of the mouth and anorexia. No marrow or cardiac toxicity is observed. Some studies describe the presence of drug-induced antibodies that might result in hemolysis (28,31). The immunogenicity of the drug appears to be schedule-dependent: no anti-drug antibodies have been detected in patients treated for 3 consecutive days every 3 weeks at which dosage no significant drug accumulation occurs in the plasma (31).

The lack of alopecia, cardiac toxicity and myelotoxicity deserves emphasis. This suggests that elliptinium could be incorporated into combinations where myelosuppression represents the dose-limiting toxic effects.

REFERENCES

1. L. K. Dalton, S. Demerac, B. C. Elmes, I. W. Loder, J. M. Swan, and T. Teitei, Synthesis of the tumor inhibitory alkaloids ellipticine 9-methoxyellipticine and related pyrido 4,3-b-carbazols. Austr. J., 20:2715 (1967).
2. G. Mathe, M. Hayat, F. de Vassal, L. Schwartzenberg, M. Schneider, J. R. Schlumberger, C. Jasmin, and C. Rosenfeld, Methoxy-9-ellipticine lactate III. Clinical screening: its action in acute myeloblastic leukemia, Rev. Eur. Etud. Clin. Biol., 15:541 (1970).
3. E. H. Herman, D. P. Chadwick, and R. M. Mhatre, Comparison of the acute hemolytic and cardiovascular action of ellipticine (NSC 71795) and some ellipticine analogs, Cancer Chemother., 58:637 (1974).
4. R. H. Liss, and C. J. Kensler, Radioautographic methods for physiologic, disposition and toxicology studies, in: "Advances in modern toxicology; new concepts in safety evaluation". Hemisphere Publishing Corp., Washington D.C. (1976).
5. J. B. LePecq, N. Dat-Xuong, C. Gosse, and C. Paoletti, A new antitumoral agent: 9-hydroxy ellipticine. Possibility of a rational design of anticancerous drugs in the series of DNA intercalating drugs, Proc. Nat. Acad. Sci. USA, 71:5078 (1974).
6. J. B. LePecq, C. Gosse, N. Dat-Xuong, S. Cros, and C. Paoletti, Antitumoral activity of 9-hydroxy ellipticine (NSC 210717) on L1210 mice leukemia, Cancer Res., 36:3067 (1976).
7. C. Paoletti, J. B. LePecq, N. Dat-Xuong, P. Lesca, and P. Lecointe, New anticancer derivatives in the ellipticine series, Curr. Chemother., 1195 (1978).
8. N. Van-Bac, C. Moisand, A. Gouyette, G. Muzard, N. Dat-Xuong, J. B. LePecq, and C. Paoletti, Metabolism and disposition studies of 9-hydroxyellipticinium acetate in animals, Cancer Treat. Rep., 64:879 (1980).
9. P. Juret, A. Tanguy, J.Y. Le Talaer, J. S. Abbatucci, N. Dat-Xuong, J. B. LePecq, and C. Paoletti, Preliminary trials of 9-hydroxy-2-methyl ellipticinium (NSC 264137) in advanced human cancers, Eur. J. Cancer, 14:205 (1978).
10. J. Y. Charcosset, A. Jacquemin-Sablon, and J. B. LePecq, Effect of

membrane potential on the cellular uptake of 2-N-methyl-ellipticinium by L1210 cells, Biochem. Pharmacol., 33:2271 (1984).
11. L. V. Johnson, M. L. Walsh, B. J. Bockus, and L. B. Chen, Monitoring of relative mitochondrial membrane potential in living cells by fluorescence microscopy, J. Cell. Biol., 88:526 (1981).
12. P. Harikumar, and J. P. Reeves, The lysosomal proton pump is electrogenic, J. Biol. Chem., 258:10403 (1983).
13. J. Y. Charcosset, B. Salles, and A. Jacquemin-Sablon, Uptake and cytofluorescence localization of ellipticine derivatives in sensitive and resistant chinese hamster lung cells, Biochem. Pharmacol., 32:1037 (1983).
14. C. Paoletti, S. Cros, W. Dat-Xuong, P. Lecointe, and A. Moisand, Comparative cytotoxic and antitumoral effects of ellipticine derivatives on mouse L1210 leukemia, Chem. Biol. Interact., 25:45 (1979).
15. E. Garcia-Giralt, and A. Macieira-Coelho, Methoxy-9-ellipticine. Analysis in vitro of the mechanism of action, Eur. J. Clin. Biol. Res., 15:539 (1970).
16. C. Paoletti, J. B. LePecq, N. Dat-Xuong, P. Juret, H. Garnier, J.L. Amiel, and J. Rouesse, Antitumor activity, pharmacology and toxicity of ellipticines, ellipticinium and 9-hydroxy derivatives: preliminary clinical trials of 2-methyl-9-hydroxy ellipticinium (NSC 264-137). Recent Results, Cancer Res., 74:107 (1984).
17. H. Nasagawa, M. Homma, H. Namidi, and K. Niki, Inhibition by hydroxy-N-methyl Elliptinium of precancerous mammary hyperplastic alveolar module formation in mice, Eur. J. Cancer Clin. Oncol., 20:273 (1984).
18. J. Bernadou, B. Meunier, G. Meunier, C. Auclair, and C. Paoletti, Regioselective arylation of ribose in adenosine and guanosine with the antitumor drug 2-N-methyl-9-hydroxyellipticinium acetate Proc. Nat. Acad. Sci. USA, 81:1297 (1984).
19. L. A. Zwelling, S. Michaels, L. C. Ericson, L. C. Ungerleider, R.S. Nichols, and K. W. Kohn, Protein-associated DNA strand breaks in L1210 cells treated with the DNA intercalating agents, 4'-(9-acridinylamino) methanesulfon-m-anisidine (m-AMSA) and adriamycin, Biochemistry, 20:6553 (1981).
20. K. M. Tewey, G. L. Chen, E. M. Nelson, and L. F. Liu, Intercalative antitumor drugs interfere with the breakage-reunion reaction of mammalian DNA topoisomerase II, J. Biol. Chem., 259:9182 (1984).
21. C. Paoletti, C. Auclair, P. Lesca, J. F. Tocanne, C. Malvy, and M. Pinto, Ellipticine, 9-hydroxyellipticine, and 9-hydroxy-ellipticinium: some biochemical properties of possible pharmacologic significance, Cancer Treat. Rep., 65 (suppl. 3):107 (1981).
22. B. Salles, J. Y. Charcosset, and A. Jacquemin-Sablon, Isolation and properties of chinese hamster lung cells resistant to ellipticine derivatives, Cancer Treat. Rep., 66:327 (1982).
23. J. Y. Charcosset, J. P. Bendirdjian, M. F. Lantieri, and A. Jacquemin-Sablon, Effects of 9-OH-ellipticine on cell survival, macromolecular synthesis and cell cycle progression in sensitive and resistant cells, Cancer Res., 45:4229 (1985).
24. N. Van-Bac, C. Moisand, A. Gouyette, G. Muzard, N. Dat-Xuong, J.B. LePecq, and C. Paoletti, Metabolism and disposition studies of 9-hydroxyellipticine and 2-methyl-9-hydroxyellipticinium acetate in animals, Cancer Treat. Rep., 64:879 (1980).
25. M. Majetouh, B. Montsarrat, R. C. Rao, B. Meunier, and C. Paoletti, Identification of the glucuronide and glutathione conjugates of the antitumor drug 2N-methyl-9-hydroxyelliptinium acetate (Celiptium), Drug Metab. Disp., 12:111 (1984).
26. L. W. Oberley, and C. R. Buettner, Role of superoxide dismutase in

cancer, Cancer Res., 39:1141 (1980).
27. J. M. McCord, B. B. Keele, and I. Fridovich, An enzyme based theory of obligate anaerobiosis: the physiological function of superoxide dismutase, Proc. Nat. Acad. Sci. USA, 68:1024 (1971).
28. A. Clarysse, A. Brugarolas, P. Siegenthaler, R. Abele, F. Cavalli, R. De Jager, R. Grenard, M. Rozencweig, and H. Hansen, Phase II study of 9-hydroxy-2N-methylellipticinium acetate, Eur. J. Cancer Clin. Oncol., 20:243 (1984).
29. P. Juret, J. F. Heron, J. E. Couette, T. Delozier, and J. Y. Le Talaer, Hydroxy-9-methyl-2-ellipticinium for osseous metastases from breast cancer: a 5-year experience, Cancer Treat. Rep., 66:1909 (1982).
30. J. L. Amiel, J. Rouesse, J. P. Droz, P. Caille, J. P. Travagli, C. Theodore, T. Le Chevalier, J. P. Ducret, J. M. Bidard, H. Garnier, and C. Paoletti, Chimotherapie des cancers du rein, metastases par le N-methylhydroxyellipticinium (NHME), Nouv. Presse Med., 10:1504 (1981).
31. J. G. Rouesse, T. Le Chevalier, P. Caille, J. M. Mondesir, H. Sancho-Granier, F. May-Levin, M. Spielmann, R. de Jager, and J. L. Amiel, Phase II study of Elliptinium in advanced breast cancer, Cancer Treat. Rep., 69:707 (1985).

THERAPEUTIC EFFICACY OF OXAZAPHOSPHORINES BY IMMUNOMODULATION

R. Voegeli, J. Pohl, T. Reissmann, J. Stekar and P. Hilgard

Department of Experimental Cancer Research
Asta-Werke AG Degussa Pharma Gruppe
D-4800 Bielefeld 14, FRG

Cyclophosphamide and a few related compounds of the class of oxazaphosphorines are widely used in the chemotherapy of malignant diseases. Their antitumour efficacy is based on their cytotoxic activity against quickly proliferating cells. Beside this, these agents are strong immunosuppressants and a particularly high sensitivity of antibody-producing, proliferating B cells was postulated (1). In contrast, it could be shown that cyclophosphamide in non-toxic doses enhanced delayed-type hypersensitivity reactions, which indicated a selective effect of cyclophosphamide on T cells (2,3).

The augmenting effect of cyclophosphamide on the delayed-type hypersensitivity reaction is documented in humans, too (4). It could also be shown, that after administration of cyclophosphamide to cancer patients, the concanavalin A-inducible suppressor activity remained impaired for a much longer time than the mitotic response of peripheral blood mononuclear cells to phytohaemagglutinin (5). Bast et al. made an investigation into the cell surface phenotype of human peripheral mononuclear cells following various doses of cyclophosphamide (6). Low doses (100 - 600 mg/m²) temporarily decreased the level of circulating B-cells, T suppressor clones were depressed by slightly higher doses (200 - 600 mg/m²). Higher doses beyond 600 mg/m² affected all T cell populations equally.

It is difficult to conceive that the main cytotoxic activity of cyclophosphamide, namely DNA cross-linking, should be the only reason for the impairment of suppressor cell function. A hypothesis was put forward by L'Age-Stehr and Diamantstein that oxazaphosphorines modified cell surfaces, resulting in the disappearance of regulatory structures on suppressor cells or in the appearance of new antigenic sites on tumour cells with subsequent induction of autoreactive cytolytic cells (7). Grunicke et al. could indeed show cell membrane alterations after oxazaphosphorine treatment (8), but positive evidence for consequences of these alterations is yet lacking. In conclusion, there is some experimental evidence for the view, that there is a contribution of a modulation of the host's defence to the therapeutic efficacy of cyclophosphamide in cancer treatment.

Strongly supporting this view is the existence of experimental tumours which can be cured with doses of cyclophosphamide far below the usual effective dose. The one studied most thoroughly is probably the MOPC-315 plasmacytoma with which the group of Sheldon Dray at the University of Illinois is working. They found a remarkable phenomenon: Low doses of cyclophosphamide (15 mg/kg bodyweight) were curative, but only when the drug was administered at a late stage of tumor growth, when the tumour was palpable. Earlier treatment rarely resulted in total remission, in contrary, it prevented the cure when the drug was administered again in the usually effective stage (9). The curative effect of cyclophosphamide in low doses could be completely abrogated by anti-thymocyte serum (10), whereas the effect of high-dosed cyclophosphamide remained almost unchanged (11).

In our laboratory, we worked with the immature monocytic myeloic leukemia L5222 which is transplantable in the syngeneic BDIX rat. In vitro it is about as sensitive to cytotoxic drugs as the rodent leukemias L1210 or P388. When rats inoculated with one million leukemic cells i.p. were treated with cyclophosphamide or other oxazaphosphorines on day 5, a surprising result was obtained: The dose-response curve was biphasic (fig. 1). Cures were obtained with low doses (1 - 20 mg/kg bodyweight), whereas higher doses only resulted in prolonged life span. At very high doses (100 - 147 mg/kg bodyweight) again some cures could be obtained, but in this dose range toxic deaths began to occur and limited further increase of dosing. When spleens of treated animals were removed and cell suspensions were transfered to new hosts, we found that up to day five after treatment with low dose cyclophosphamide the tumour cells in the spleen remained viable and induced leukemia and killed the new host, whereas spleens of animals treated with the high dose were free of transplantable tumour cells within less than twenty-four hours. The pharmacokinetics of cyclophosphamide imply that the cell kill cannot be due to a direct effect of the drug in the low dose range, but rather had to be host-mediated, as the drug is totally excreted before day five. When longterm-survivors were reinoculated with leukemia cells, animals treated with low dosed drug were found to be resistant against the tumour for more than three months. Survivors of high-dose treatment, in contrast, showed no difference to the controls in survival time and rate after reinoculation. Apparently the L5222 line propagated in our laboratory induced a defence mechanism in the syngeneic host, which was sufficient to eradicate the malignant cells, provided a suppressor mechanism was eliminated or at least weakened by low cyclophosphamide doses.

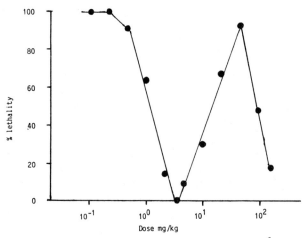

Figure 1. Lethality of animals transplanted with 10^6 L5222-cells i.p. after treatment with cyclophosphamide i.v. on day 5. Animals were observed for 60 days.

We now have a stable derivative of cyclophosphamide, mafosfamide, available, which undergoes non-enzymatic hydrolysis in aequous solution to active 4-hydroxy-cyclophosphamide. This molecule should not only allow the in vitro investigation of oxazaphosphorine effects on the immune system. As the immunomodulatory dose-response is probably bell-shaped, it will be of great importance to ensure predictable plasma levels of 4-hydroxy-cyclophosphamide. Cyclophosphamide has the disadvantage that it must be activated enzymatically in the liver. Therefore its plasma level is dependent on the enzyme activity which may vary between individuals and with time.

The dose-response of mafosfamide in the L5222 system shows a similar pattern as cyclophosphamide (fig. 2). Its chemical behaviour, preclinical pharmacology and toxicology are well documented (12). The general pharmacology of mafosfamide is similar to that of cyclophosphamide, its pharmacokinetics and pharmacodynamics, however, are distinctly different. With respect to immunomodulation, mafosfamide appears to be superior to its parent compound: After mafosfamide treatment the humoral antibody production in rats upon immunisation with sheep red blood cells showed a decrease of antibody titres only after intravenous administration of 100 mg/kg bodyweight, a dose which is already about 30 % of the LD_{50} value. On a molar basis, mafosfamide has nearly the same general toxicity as cyclophosphamide, but with regard to the specific toxicity such as myelosuppression, urotoxicity and, in particular, immunosuppression, mafosfamide is less than half as toxic (13).

The experimental evidence suggests that oxazaphosphorines could play a significant role in the immunotherapy of cancer patients. Theoretical considerations and preliminary experimental data indicate that oxazaphosphorines, which do not require biotransformation, are preferable compounds, because their plasma levels are fully predictable and their immunotherapeutic range is considerably broader than that of the classical drugs. Since the immunopharmacological doses of mafosfamide are likely to be well below the maximally tolerated dose in humans, the recently completed conventional phase I studies can serve as a toxicological background for future immunopharmacological trials (14).

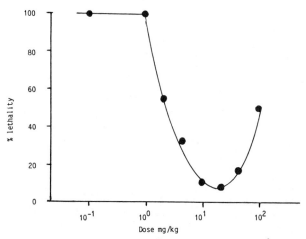

Figure 2. Lethality of animals transplanted with 10^6 L5222-cells i.p. after treatment with mafosfamide i.v. on day 5. Animals were observed for 60 days.

As with most immunomodulators, the dose-response curve for mafosfamide will probably be bell-shaped, and the dose range at which the maximum efficacy is to be expected has to be determined in special clinical phase I trials. The acute toxicities seen with low dose mafosfamide are minor and are unlikely to impair the quality of life of the patients. As late toxicity, however, the carcinogenic potential of alkylating agents has to be considered; particularly, if the drugs are to be used as adjuvant therapy to potentially curative surgery. Long-term-follow up of patients after therapy with relatively high doses of cyclophosphamide has however not yet revealed a higher risk for secondary tumours (15). Since the carcinogenicity of alkylating agents follows a linear dose-response relationship, it is unlikely that immunopharmacological doses of mafosfamide would constitute a significant risk, if treatment was limited to six or twelve months.

In conclusion, there is broad evidence that under adequate conditions oxazaphosphorines may well turn out to be highly specific and active biological response modifiers and mafosfamide seems to be the most suitable compound in this series.

REFERENCES

1. I. McConnell, A. Munro, and H. Waldman, "The Immune System: A Course on the Molecular and Cellular Basis of Immunity", The Alden Press, Oxford (1981).
2. P. W. Askenase, B. J. Hayden, and R. K. Gershon, Augmentation of delayed-type hypersensitivity by doses of cyclophosphamide which do not affect antibody response, J. Exp. Med., 141:697-702 (1975).
3. A. Mitsuoka, M. Baba, and S. Morikawa, Enhancement of delayed hypersensitivity by depletion of suppressor T cells with cyclophosphamide in mice, Nature, 262:77-78 (1976).
4. D. Bernd, M. J. Mastrangelo, P. F. Engstrom, A. Paul, and H. Maguire, Augmentation of the human immune response by cyclophosphamide, Cancer Res., 42:4862-4866 (1982).
5. D. Bernd, H. C. Maguire, Jr., and M. J. Mastrangelo, Impairment of concanavalin A-inducible suppressor activity following administration of cyclophosphamide to patients with advanced cancer, Cancer Res., 44:1275-1280 (1984).
6. R. C. Bast, Jr., E. L. Reinherz, C. Maver, P. Lavin, and S. F. Schlossman, Contrasting effects of cyclophosphamide and prednisolone on the phenotype of human peripheral blood leukocytes, Clin. Immunol. Immunopathol., 28:101-114 (1983).
7. J. L'Age-Stehr and T. Diamantstein, Induction of autoreactive T lymphocytes and their suppressor cells by cyclophosphamide, Nature, 271:663-665 (1978).
8. H. Grunicke, H. Putzer, F. Scheidl, and E. Wolff-Schreiner, The cell surface as a target for alkylating agents, in: "The control of tumour growth and its biological bases", W. Davis, C. Maltoni, and S. Tanneberger, eds., Akademie-Verlag, Berlin (1983).
9. J. C. D. Hengst, M. B. Mokyr, and S. Dray, Importance of timing in cyclophosphamide therapy of MOPC-315 tumor-bearing mice, Cancer Res., 40:2135-2141 (1980).
10. J. C. D. Hengst, M. B. Mokyr, and S. Dray, Cooperation between cyclophosphamide tumoricidal activity and host antitumor immunity in the cure of mice bearing large MOPC-315 tumors, Cancer Res., 41:2163-2166 (1981).
11. M. B. Mokyr, J. C. D. Hengst, and S. Dray, Role of antitumor immunity in cyclophosphamide-induced rejection of subcutaneous nonpalpable MOPC-315 tumors, Cancer Res., 42:974-979 (1982).

12. Proc. Satellite Meeting 10th Int. Symposium on the Biological Characterization of Human Tumors, Brighton, Invest. New Drugs, 2:129-259 (1984).
13. J. Pohl, P. Hilgard, W. Jahn, and H. J. Zechel, Experimental toxicology of ASTA Z 7557 (INN mafosfamide), Invest. New Drugs, 2:201-206 (1984).
14. Internal Report, Asta-Werke AG
15. K. Powell, A. Buzdar, T. Smith, and G. Blumenschein, Subsequent malignant neoplasma in stage II, III breast cancer patients treated with and without adjuvant combination chemotherapy, Proc. ASCO, 1:C-301 (1982).

THE ROLE OF CELLULAR GLUTATHIONE IN PROTECTING MAMMALIAN CELLS

FROM X RADIATION AND CHEMOTHERAPY AGENTS IN VITRO

Dennis C. Shrieve

Department of Radiation Oncology
University of California
San Francisco, CA 94143

INTRODUCTION

Clonogenic cells in solid tumors are heterogeneous in their response to cancer therapeutic modalities, i.e., ionizing radiation and chemotherapy. The heterogeneity may be due to cellular factors, arising in clonally-derived tumors from genetic or epigenetic drift, or to microenvironmental factors arising from intratumor gradients in, for example, oxygen tension (pO_2), glucose or other nutrients and pH. These factors may directly influence cellular response to treatment or may exert indirect influence by affecting, for example, cell proliferation.

Cytotoxic and mutagenic damage in cells is modifiable by the presence of low molecular weight thiols (1). Glutathione (GSH) (γ-glutamylcysteinylglycine) is the major endogenous cellular thiol, being present in millimolar concentrations in most animal and plant cells and in microorganisms. GSH may have as a major function protection of cells from environmental insult (e.g., oxidative stress) and may also affect cellular sensitivity to cancer treatment modalities.

We have been interested in studying the role of cellular thiols, in particular GSH, in modifying cellular sensitivity to cancer chemotherapy agents and X radiation, including hypoxic cell radiosensitizers. Our approach was initially to use buthionine sulfoxime (2) (BSO) to deplete cells of GSH and then to determine the effects on subsequent cytotoxic treatments in vitro under both hypoxic and aerated conditions.

This paper will discuss our results and those of others that indicate that cellular GSH plays an important role in modifying cellular sensitivity to cancer treatment modalities and suggest that depletion of cellular GSH may be an effective means of sensitizing resistant tumor cells to treatment.

RESULTS AND DISCUSSION

Effects of GSH Depletion by BSO on Cellular Growth and Viability

When EMT6/SF mouse tumor cells or Chinese hamster V79 cells were exposed to various concentrations of BSO (0.1 - 1000 µM) cellular GSH, expressed as nmol/mg Protein, decreased exponentially with a half-time dependent on the concentration of BSO used. A maximal rate of depletion, corresponding to a half-time of 1.9 h, was achieved by a concentration of 50 µM or greater. (Constitutive [GSH]: 25 - 30 nmol/mg protein).

Treatment of cells with BSO concentrations of up to 1 mM did not affect cellular growth or viability when exposure time was limited to less than 24 h. Similar results were obtained by us for EMT6/SF cells and V79 cells. When exposure times were extended beyond 24 h, growth inhibition and cytotoxicity were observed. This delayed effect of BSO treatment on cells in culture was first reported by Midander and Revesz (3) and clearly shows that GSH depletion by BSO can be cytotoxic. In our experiments GSH was depleted by treating cells with 50 µM BSO for 12 - 14 h, which depleted cellular GSH to less than 5 % of control but did not affect cellular growth or viability.

Effects of GSH Depletion on Cellular Radiosensitivity

It is generally agreed that the development of hypoxic foci in solid tumors poses a major obstacle to the control of local disease by conventional radiation therapy (4). The reason for this is the well known "radiobiological oxygen effect": Cells X irradiated in the absence of oxygen show a three-fold resistance compared to aerated cells, based on colony forming ability of irradiated cells. Attempts to overcome this radioresistance of hypoxic tumor cells have included development of oxygen-mimicking radiosensitizers, such as the 2-nitroimidazoles (5) and methods of depleting cells of endogenous radioprotectors, which theory predicts should preferentially sensitize hypoxic cells (6,7). Cellular GSH has often been regarded as the most important endogenous cellular radioprotector. Our studies (9,10) and those of others (see refs. 8 and 9) have investigated the effects of GSH depletion on the radiosensitivity of aerated and hypoxic cells and the effects on the radiosensitizing efficiency of oxygen and 2-nitroimidazole radiosensitizers.

Our studies (9,10) and those of most other investigators have shown that depletion of cellular GSH to < 5 % of control or to "nondetectable levels" sensitized only hypoxic and not aerated cells to X radiation. However, the radiosensitivity of GSH-depleted, hypoxic cells is found to still be approximately 2.5-fold less than that of aerated cells. These studies seem to indicate that cellular GSH is partially responsible for the radioresistance of hypoxic cells, but that 50 - 70 % of the magnitude of the oxygen effect is GSH-independent (7,8,10).

Hypoxic cell radiosensitizers (e.g., misonidazole) have entered clinical trials in a number of countries (Denmark, UK, USA, Canada, South Africa and Japan). Although a slight benefit has been observed in some tumor sites, it is generally agreed that the dose of sensitizer that can be given to patients without neurotoxicity is too low to allow for maximal radiosensitization of hypoxic tumor cells (11). One approach to improving this situation has been to deplete intracellular GSH, based on the rationale that intracellular thiols compete with radiosensitizers for repair or fixation of radiation-induced damage (6). This approach, investigated now by several groups in both in vitro and in vivo systems has produced mixed results (see ref. 8 and 9).

In many of these studies an increase in misonidazole efficiency - by as much as 30-fold on a log concentration vs. effect basis - has been reported. Our results with V79 cells (8), as well as those of some others (8,9), clearly showed that GSH depletion did not affect in any way the radiosensitizing efficiency of misonidazole (Table 1). The discrepancies in the results can be partially accounted for by differences in methods of analysis. Most authors finding increases in radiosensitizing efficiency following depletion of GSH failed to account for the radiosensitization of hypoxic cells by GSH depletion alone. When this is done the large increases in misonidazole efficiency reported are reduced to factors of less than 3, and in most studies are indistinguishable from 1 (see ref 9 for discussion).

Table 1. Enhancement Ratios for Anoxic Cells Irradiated in the Presence of MISO With or Without Prior GSH Depletion

[MISO]	SER_{MISO}*		SER_{TOT}*
	Anoxic V79	GSH depleted anoxic V79	
0	1.00	1.00	1.41 (SER_{BSO})*
0.2 mM	1.33	1.41	2.00
1.0 mM	1.73	1.71	2.42
5.0 mM	2.24	2.36	3.33

* SER: Sensitizer Enhancement Ratio, the ratio of X ray doses required to give the same level of survival compared to the relevant anoxic control (see ref. 8).

The precise role that cellular GSH plays in modifying hypoxic cell radiosensitivity and the radiosensitizing efficiency of electron affinic compounds remains unclear. Recent evidence suggesting that human tumor cells may have extremely high GSH contents (12) emphasizes the need to consider cellular GSH as a potentially important factor in determining the success of radiation therapy (or chemotherapy, see below).

Effects of GSH Depletion on Hypoxic and Aerated Cell Chemosensitivity

Our studies (10) have shown that EMT6/SF cells were more sensitive to the chemotherapy agents cis-dichlorodiammino Pt(II) (DDP), mitomycin C (MitC), L-phenylalanine mustard (L-PAM) and nitrogen mustard (HN_2) when treated under hypoxic rather than aerated conditions (Table 2). The magnitude of this effect ranged from 1.4 (HN_2) to 4.1 (MitC). Hypoxic cells were more resistant to actinomycin D (ActD) compared to aerated cells.

Aerated EMT6/SF cells were sensitized to DDP, MitC and HN_2 by depletion of GSH (Table 2) by factors ranging from 1.9 - 2.1. A much larger sensitization was seen in aerated cells treated with L-PAM (ER = 6.5). Hypoxic cells were also sensitized to these agents by depletion of GSH. ERs ranged from 1.4 (DDP) to 4.3 (L-PAM). GSH depletion did not further sensitize hypoxic cells to MitC. GSH depletion slightly protected aerated cells from killing by ActD.

Our results indicate that cellular GSH plays an important role in modifying response to cancer chemotherapeutic agents. The protective

Table 2. Effects of Hypoxia and GSH Depletion on the Sensitivity of EMT6/SF Cells to Chemotherapy Agents or X rays

ENHANCEMENT RATIOS (SF = 0.1)[a]

AGENT	HYPOXIA	BSO (O_2)	BSO (N_2)
DDP	2.9	1.9	1.4
MitC	4.1	2.1	1.0
L-PAM	3.1	6.5	4.3
HN_2	1.4	1.9	2.6
ActD	(10)[b]	(2)	(2)
X Rays	(3.1)	1.0	1.3

[a] Enhancement Ratios (ER) are ratios of doses required to give surviving fraction (SF) = 0.1 under various conditions:

Hypoxia-ERs reflect sensitivity of hypoxic cells relative to aerated cells.

BSO (O_2)-ERs reflect sensitization (or protection) of aerated cells by GSH depletion.

BSO (N_2)-ERs reflect sensitization (or protection) of cells by GSH depletion.

[b] ERs in parentheses indicate that the treatment condition protected cells from the specified agent.

effect of GSH against the cytotoxicity of L-PAM is a particularly large one. For this reason BSO is being considered as an adjuvant to L-PAM treatment in patients with ovarian carcinoma. The critical normal tissue toxicity following L-PAM in humans is to the bone marrow. Recent evidence that BSO may not enhance the toxicity of L-PAM to bone marrow in a mouse model (13) is encouraging; however, one must consider that, in a patient treated with BSO in sufficient concentrations to sensitize tumor cells, the sensitivity of many normal tissues to the chemotherapy agent may be increased. It seems that many studies of normal tissue effects of BSO depletion in mouse and large animal systems are called for before clinical application is attempted.

The mechanisms involved in radiosensitization and chemosensitization by depletion of cellular GSH are not known at this time; however, it seems clear that GSH and related enzyme systems (e.g., GSH reductase, GSH peroxidase and GSH S-transferase) are involved in modifying cellular response to X radiation and chemotherapy. These cellular systems may be responsible for cellular resistance to some agents and should be considered as potential targets for chemosensitization.

Supported by the American Cancer Society Grant No. PDT-225 and Gant No. CA38847 awarded by the USPHS.

REFERENCES

1. B. A. Arrick, and C. F. Nathan, Glutathione metabolism as a determinant of therapeutic efficacy: A review, Cancer Res., 44:4224 (1984).
2. O. W. Griffith, and A. Meister, Potent and specific inhibition of glutathione synthesis by buthionine sulfoximine (S-n-butyl homocysteine sulfoximide), J. Biol. Chem., 254:7558 (1979).
3. J. Midander, and L. Revesz, Toxic and growth inhibitory effects of cellular glutathione depletion by treatment with buthionine sulfoximine (BSO), Radiosens. Newslett., 254:7558 (1979).
4. R. P. Bush, R. D. T. Jenkin, W. E. C. Allt, F. A. Beale, H. Bean, A. J. Dembo, and J. F. Pringle, Definitive evidence for hypoxic cells influencing cure in cancer therapy, Br. J. Cancer, 37 (Suppl. III):302 (1978).
5. J. M. Brown, Clinical Trials of radiosensitizers: What should we expect? Int. J. Radiat. Oncol. Biol. Phys., 10:425 (1984).
6. C. J. Koch, and R. L. Howell, Combined radiation-protective and radiation-sensitizing agents II. Radiosensitivity of hypoxic or aerobic Chinese hamster fibroblasts in the presence of cysteamine and misonidazole: Implications for the 'oxygen effect', Radiat. Res., 87:265 (1981).
7. C. J. Koch, C. C. Stobbe, and E. A. Bump, The effect on the K_m for radiosensitization at 0 °C of thiol depletion by diethylmaleate pretreatment: Quantitative differences found using radiation sensitizing agent misonidazole or oxygen, Radiat. Res., 98:141 (1984).
8. D. C. Shrieve, J. Denekamp, and A. I. Minchinton, Effects of glutathione depletion by buthionine sulfoximine on radiosensitization by oxygen and misonidazole in vitro, Radiat. Res., 102:283 (1985).
9. D. C. Shrieve, and J. Denekamp, Glutathione and misonidazole: independent or interactive mechanisms of action? Br. J. Radiol., 58:383 (1985)
10. D. C. Shrieve, and J. W. Harris, Effects of glutathione depletion by buthionine sulfoximine on the sensitivity of EMT6/SF cells to chemotherapy agents or X radiation, Int. J. Radiat. Oncol. Biol. Phys., in press (1986).
11. S. Dische, Misonidazole in the clinic at Mt. Vernon, Cancer Clin. Trials, 3:175 (1980).
12. T. L. Phillips, J. B. Mitchell, W. de Graff, A. Russo, and E. Glatstein, Variation in the sensitizing efficiency for SR2508 in human cells dependent on glutathione content, Int. J. Radiat. Oncol. Biol. Phys., in press (1986).
13. A. Russo, Z. Tochner, T. L. Phillips, J. Carmichael, W. de Graff, N. Friedman, J. Fischer, and J. B. Mitchell, In vivo modulation of glutathione by buthionine sulfoximine: Effect in marrow response to melphalan, Int. J. Radiat. Oncol. Biol. Phys., in press (1986).

THE ROLE OF PAPILLOMA VIRUSES IN HUMAN CANCER

Harald zur Hausen and Ethel-Michele de Villiers

Deutsches Krebsforschungszentrum
Im Neuenheimer Feld 280
D-6900 Heidelberg, Fed. Rep. Germany

ABSTRACT

Twelve types of papillomaviruses have been isolated thus far from papillomatous and Bowenoid lesions of the human genital tract. Four types, HPV 6, 11, 16 and 18 are most frequently found. HPV 6 and 11 cause the typical genital warts (condylomata acuminata) and mild dysplastic lesions of the cervix characterized by a high degree of koilocytotic atypia. HPV 16 and 18 are preferentially found in Bowenoid papulosis and Bowen's disease at external genital sites. Moderate and severe cervical dysplasias with little or no koilocytosis appear to be common manifestations of these infections at cervical sites. HPV 16 is found in approximately 50 % of all cervical, penile and vulvar cancers, HPV 18 in close to 20 %. The majority of the remaining tumours reveals evidence for infections with additional types of papillomaviruses. Several cell lines have been identified containing either HPV 18 or HPV 16 genomes. The state of viral DNA in Bowenoid precursor lesions differs from that of cervical carcinomas. The former contain episomal DNA whereas integration seems to be a regular event in carcinomas. Integration regularly affects the E1-E2 open reading frames of HPV 16 or 18 DNA. Fusion transcripts from the integrated HPV DNA (E6-E7 region) and adjacent host cell DNA have been documented. The available data support a causative role of specific HPV infections in the etiology of human genital cancer.

INTRODUCTION

During the past 10 years the plurality of human papillomavirus types has been well established (see review zur Hausen and Schneider, 1985). At present 40 distinct types of human papillomaviruses (HPV) have been analyzed. It is likely that this number will increase in the future.

Papillomaviruses reveal a very characteristic infectious cycle: they appear to require the availability of cells still capable to divide for succesful infection. The uptake of viral DNA by such cells, usually exposed to the surface in microlesions of epidermis or mucosa or at specific sites (e.g. the transition zone of the cervix) results in its episomal persistence. The persisting DNA stimulates enhanced proliferation. Its own independent replication seems to be blocked by host cell factors. During subsequent steps in differentiation and keratinization this block obviously is released resulting in viral DNA replication, synthesis of structural proteins and the maturation of viral particles. Thus, the keratinized superficial layer of a wart containing infectious particles indicates an underlying proliferating layer of cells harbouring only viral DNA which stimulates cell proliferation.

The dependence of particle maturation on specific stages on cell differentiation is probably the main reason for the present unability to propagate papillomaviruses in tissue culture. The analysis of their biological functions is further restricted by a remarkable host specificity with a pronounced preference of individual types for specific types of tissue. Thus, human papillomaviruses do not produce recognizable changes in animal hosts.

The DNA of several papillomavirus types has been sequenced. In contrast to polyoma-type viruses transcripts are read from one strand only revealing the same polarity. Several open reading frames (ORF) exist, two of them (L1 and L2) appear to code for structural proteins, a varying number of additional ORF's (E1 to E8) seem to contain information responsible for early functions (see review Pfister, 1984). E6 and E5 of bovine papillomaviruses apparently code for transforming functions (Schiller et al., 1984; Yang et al., 1985), E2, E4 reveal a trans-activating activity for the E6 (Spalholz et al., 1985), E7 region and E1 and possibly also E7 somehow regulate the episomal state of persisting papillomavirus DNA (Lusky and Botchan, 1984). Very little is known on the proteins coded for by early ORF's. Their expression in bacterial vector systems is presently being explored.

Genital Papillomavirus Infections

Twelve types of papillomaviruses have been isolated from the human genital tract (zur Hausen, 1986). The majority of these types has only rarely been found in genital tumors. The most prevalent types clearly are HPV6, 11, 16 and to a lesser extent 18.

HPV 6 and 11 are closely related, 82 % of their nucleotides are identical (Dartmann et al., in print). These viruses are found in typical genital warts (condylomata acuminata). In these tumors HPV 6 is found in about 60 %, HPV 11 in about 30 %. Invasively growing, non-metastasizing giant condylomata have been described, frequently labelled as Buschke-Löwenstein tumors. HPV 6 DNA was found in almost all of these tumors so far analyzed, the only exceptional one contained HPV 11 (Boshart et al., unpublished data).

The histology of genital warts is characterized by the exophytic papillary growth and the typical appearance of koilocytotic cells. The latter represent the sites of viral DNA replication, protein synthesis and particle maturation, thus being an expression of cytopathogenic changes induced by events leading to the synthesis of infectious virus.

Although condylomatous changes are also noted upon infections of the vaginal wall, most notably HPV 11 infections of cervical sites reveal a different pattern (see review zur Hausen and Schneider, in print). Kolposcopically they are observed as flat dysplastic lesions which histologically show features of mild dysplasias (cervical intraepithelial neoplasia, CIN-1) usually with extensive koilocytosis.

HPV 6 and 11 infections are extremely rare at non-genital epidermal sites. They occur, however, at low frequency at oral sites or within the respiratory tract (de Villiers et al., 1985). The most frequent non-genital predilection site is infection of the vocal cords, resulting in laryngeal papillomatosis (Gissmann et al., 1982; Mounts et al., 1982). This represents a serious clinical condition, sometimes spreading into the bronchial tract. Laryngeal papillomatosis occurs at higher frequency as a perinatal infection, most likely during delivery due to maternal genital warts. HPV 11 is found in this condition more often than HPV 6.

HPV 6 or 11 - containing condylomatous proliferations are occasionally found in the buccal mucosa, the lips, or located directly on the tongue. The histology shows less koilocytotic changes, probably reflecting a substantially reduced virus production at these sites.

Recently a technique has been developed which permits to study directly the causative role of papillomavirus infection for the induction of dysplastic and papillomatous proliferations (Kreider et al., 1985). Inoculation of normal cervical tissue beneath the renal capsule of nude mice results in the development of cysts outlined at their inner surface by cervical epithelium. These cysts persist for several months. Infection of the cells with HPV 11 prior to inoculation results in dysplastic proliferations revealing extensive koilocytosis and within the koilocytes virus-specific antigens.

In HPV 16 and 18 infections, the macroscopic and microscopic pattern of the induced lesions differs markedly from that described for HPV 6 and 11. Most available data result from HPV 16-containing proliferations. This virus type is by far the most frequent HPV found in Bowenoid papulosis or genital Bowen's disease which represent rather discrete and inconspicuous-looking white or reddish plaques found at vulvar, penile and perianal sites (Ikenberg et al., 1984). Their histology, however, reveals marked atypia and usually all characteristics of a carcinoma in situ. Spontaneous regression of these tumors does occur, although the majority of these lesions seems to persist for long periods of times, in many instances for years and possibly for decades.

HPV 16 infections of the cervix are associated with marked atypia, changes characteristic for CIN II or CIN III and carcinoma in situ (Crum et al., 1985). Usually little or no koilocytosis is noted in these proliferations, although in some biopsies an extensive koilocytosis may originate from simultaneous infections with additional HPV types.

Simultaneous infections with either HPV 6/11 or HPV 16 and 18 appear to be relatively frequent. A recent survey of smears analyzed from more than 3000 women indicated the presence of both groups of viruses in about 5 % of this population, respectively (de Villiers et al., unpubl. data). More than 2 % of these women revealed evidence for infections by more than one of these agents. These figures have not been obtained from a randomized group of patients, thus they may represent overestimates for the real prevalence of these infections.

Papillomaviruses in Anogenital Cancer

The original isolations of HPV 16 and HPV 18 DNA were both obtained from cervical cancer samples (Dürst et al., 1983; Boshart et al., 1984). Thus, it was an interesting question to analyze other tumor samples for the presence of HPV DNA. Up to now a large number (200) of cervical cancer biopsies and a limited number of penile and vulvar cancers have been analyzed (Dürst et al., 1983; Boshart et al., 1984; Scholl et al., 1985). HPV 16 DNA is found in close to 50 % of biopsies tested from various parts of the world. Individual results differ in the range between 30 and 80 %. HPV 18 DNA is less frequently found. In our own studies the percentage of positive biopsies (including penile and vulvar cancer) comes close to 20 %. In additional tumors other types of papillomaviruses have been detected. In a few samples HPV 11 DNA (Gissmann et al., 1983), in others more recently HPV 31 (Lorincz and Temple, 1985), HPV 33 (Orth et al., pers. communication) or HPV 36 (Lorincz and Temple, pers. communication) have been found. Approximately 80 % of all of these biopsies contain specific types of HPV DNA. In the majority of the remaining tumors hybridization under conditions of low stringency discloses the presence of HPV-related, yet not identified sequences, most likely due to the presence of new types of papillomaviruses. Thus, we encounter the regulator presence of DNA from specific HPV types in cervical, vulvar and penile cancer.

Perianal and anal cancer has also been found to frequently contain HPV 16 DNA (Beckmann et al., 1985).

Metastatic tissue derived from cervical cancer usually contains HPV DNA in similar quantities as the primary tumor tissue.

A number of cell lines have been derived from cervical cancer, among others the HeLa cells. Southern blot analysis of these cells revealed the presence of HPV DNA in the majority of lines thus far tested (Boshart et al., 1984; Schwarz et al., 1985; Yee et al., 1985; Pater and Pater, 1985). Interestingly, the otherwise rare HPV 18 DNA is present in the majority of lines analyzed up to now including the HeLa cells. The copy number of HPV genomes in these cells differs considerably from about 1 genome copy in C4-1 cells (Schwarz et al., 1985) to up to 600 or more copies in the Caski line.

The availability of HPV-positive cell lines permitted a convenient testing for the state of persisting HPV DNA. At least the vast majority of HPV DNA in these lines, but also HPV DNA in fresh biopsy samples from cervical cancer contains integrated sequences. The integrated viral DNA is frequently amplified usually involving the flanking host cell DNA. Some primary tumors contain in addition to the integrated sequences episomal viral DNA.

It is interesting to note that HPV 16 positive precursor lesions (Cervical dysplasias and Bowenoid papulosis) seem to contain only episomal DNA (Dürst et al.,1985). It appears therefore that malignant conversion is associated with a shift in the state of persisting viral DNA.

The persistence and chromosomal localization of viral DNA can be visualized by in situ hybridizations, most easily in the Caski line with the large numnber of viral genome copies. Several integration sites are evident from these studies, revealing high copy numbers for individual sites (A. Mincheva, L. Gissmann and H. zur Hausen, unpubl. studies).

The integrational pattern reveals some specificity (Schwarz et al., 1985). In the majority of primary tumors, metastases and in all positive cell lines tested so far, at least some of the viral DNA molecules integrate within the E1-E2 open reading frames. This obviously disrupts an intragenomic regulation which has recently been reported by Spalholz and her associates (1985) for bovine papillomavirus genomes.

The integrated viral DNA is transcribed, in cell lines exclusively involving the E6-E7 open reading frames (Schwarz et al., 1985). In addition fusion transcripts between E6-E7 and adjacent host cell sequences are formed. Their biological significance is at present unknown. Transcriptions from the same regions appear to be also a regular feature of primary tumors, although in some of them the transcriptional pattern appears to be more complicated.

Papillomavirus DNA in Non-Genital Cancer

Specific types of papillomaviruses, most notably HPV 5 have been detected in squamous cell carcinomas of patients with a rare condition, epidermodysplasia verruciformis (Orth et al., 1980).

Our own group became interested in the presence of papillomavirus DNA in carcinomas of the human respiratory tract and the oral mucosa. Since papillomavirus infections seem to interact synergistically with chemical or physical carcinogens in carcinoma development (zur Hausen, 1977, 1982), human tumors clearly linked etiologically to chemical factors (smoking) like laryngeal and lung cancers were the primary target for this investigation. The analysis of more than 100 individual cancer biopsies led indeed to the carcinomas which are listed in Table 1. It is interesting to note that 5 of these tumors contained HPV 16 sequences, demonstrating that this type of infection also occurs at extragenital sites leading again to a remarkable association with malignant growth.

Some rather preliminary data point to the existence of additional, yet not fully characterized HPV types in other laryngeal and lung carcinomas. The availability of some HPV-containing carcinomas derived from this region provides an encouraging baseline for the concept that carcinomas of the oral mucosa and the respiratory tract may originate from an interaction of cell persistently infected by specific HPV types and environmental chemical factors. This, in addition, would open a new pathway for strategies in preventing very common types of human cancers.

Tab. 1. HPV DNA in Cancer of the Oral Mucosa and the Respiratory Tract

HPV-Type	Site of Cancer	Authors
HPV-11	Buccal mucosa	Löning et al., 1985
HPV-16	Buccal mucosa	Löning et al., 1985
HPV-2	Tongue	de Villiers et al., 1985
HPV-16	Tongue	de Villiers et al., 1985
HPV-16	Tongue	de Villiers et al., 1985
HPV-16	Larynx	Scheurlen et al., 1985
HPV-30	Larynx	Kahn et al., 1985
HPV-16	Lung	Stremlau et al., 1986

REFERENCES

1. A.M. Beckmann, J.R. Daling, and J.K. McDougall, Human papillomavirus DNA in anogenital carcinomas, J. Cell Biochem. Suppl., 9c: 68(1985)
2. M.Boshart, L. Gissmann, H. Ikenberg, A. Kleinheinz, W. Scheurlen, and H. zur Hausen, A new type of papillomavirus DNA, its presence in genital cancer biopsies and in cell lines derived from cervical cancer, EMBO J., 3: 1151-1157 (1984)
3. C.P. Crum, M. Mitao, R.U. Levine, and S. Silverstein, Cervical papillomaviruses segregate within morphologically distinct pre-cancerous lesions, J. Virol., 54:675-681 (1985)
4. K.Dartmann, E. Schwarz, L. Gissmann, and H. zur Hausen, The nucleotide sequence and genome organization of human papillomavirus type 11. Submitted for publication.
5. E.M. de Villiers, C. Neumann, J.-Y. Le, H. Weidauer, and H. zur Hausen, Infection of the oral mucosa with defined types of human papilloma-viruses, Med. Microbiol. Immunol., in print
6. E.M. de Villiers, H. Weidauer, H. Otto, and H. zur Hausen, Papilloma-virus DNA in human tongue carcinomas, Int. J. Cancer, 36: 575-579 (1985)
7. M.Dürst, L. Gissmann, H. Ikenberg, and H. zur Hausen, A papillomavirus DNA from a cervical carcinoma and its prevalence in cancer biopsy samples from different geographic regions, Proc. Nat. Acad. Sci.USA 80: 3812-3815 (1983)
8. M.Dürst, A. Kleinheinz, H. Hotz, and L. Gissmann, The physical state of human papillomavirus type 16 DNA in benign and malignant genital tumors, J. Gen. Virol., 66:1515-1522 (1985)
9. L. Gissmann, V. Diehl, H.-J. Schultz-Coulon, and H. zur Hausen, Molecular cloning and characterization of human papillomavirus DNA derived from a laryngeal papilloma, J. Virol., 44: 395-400 (1982)
10. L.Gissmann, H. Wolnik, H. Ikenberg, U. Koldovsky, H.G. Schnürch, and H. zur Hausen, Human papillomavirus type 6 and 11 sequences in genital and laryngeal papillomas and in some cervical cancers, Proc. Nat. Acad. Sci. USA, 80:560-563 (1983)
11. H.Ikenberg, L. Gissmann, G. Gross, E.-I. Grussendorf, and H. zur Hausen, Human papillomavirus type 16-related DNA in genital Bowen's disease and in Bowenoid papulosis, Int. J. Cancer, 32:563-565(1983)
12. T.Kahn, E. Schwarz, and H. zur Hausen, Molecular cloning and characterization of the DNA of a new human papillomavirus (HPV 30) from a laryngeal carcinoma, Int. J. Cancer, in print
13. J.W. Kreider, M.K. Howett, S.A. Wolfe, G.L. Bartlett, R.J. Zaino, T.V. Sedlacek, and R. Mortel, Morphological transformation in vivo of human uterine cervix papillomavirus from condylomata acuminata, Nature, 317:639-641 (1985)
14. T.Löning, H. Ikenberg, J. Becker, L. Gissmann, I. Hoepfer, and H. zur Hausen, Analysis of oral papillomas, leukoplakias and invasive carcinomas for human papillomavirus related DNA, J. Invest. Dermatol., 84:417-420 (1985)
15. A.T. Lorincz, W.D. Lancaster, and G.F. Temple, Detection and characterization of a new type of human papillomavirus, J. Cell Biochem., Suppl. 9c: 75 (1985)
16. M.Lusky, and M.R. Botchan, Characterization of the bovine papilloma-virus plasmid maintenance sequences, Cell, 36:391-401 (1984)
17. P.Mounts, K.V. Shah, and H. Kashima, Viral etiology of juvenile and adult onset squamous papilloma of the larynx, Proc. Nat. Acad. Sci. USA, 79:5425-5429 (1982)
18. G.Orth, M. Favre, F. Breitburd, O. Croissant, S. Jablonska, M. Obalek, M. Jarzabek-Chorzelska, and G. Rzesa, Epidermodysplasia verruci-formis: A model for the role of papillomaviruses in human cancer, in: "Viruses in Naturally Occuring Cancers" (M. Essex, G. Todaro

and H. zur Hausen, eds.) Cold Spring Harbor Labor. Press, Cold Spring Harbor, New York, pp 259-282 (1980)
19. M.M. Pater, and A. Pater, Human papillomavirus types 16 and 18 sequences in carcinoma cell lines of the cervix, Virology, 145: 313-318 (1985)
20. H.Pfister, Biology and biochemistry of papillomaviruses, Rev.Physiol. Biochem. Pharmacol., 99:111-181 (1984)
21. W.Scheurlen, A. Stremlau, L. Gissmann, D. Höhn, H.-P. Zehner, and H. zur Hausen, Rearranged HPV 16 molecules in an anal carcinoma and in a laryngeal carcinoma, Submitted for publication
22. J.T. Schiller, W.C. Vass, and D.R. Lowry, Identification of a second transforming region in bovine papillomavirus DNA, Proc. Nat. Acad. Sci. USA, 82:7880-7884 (1984)
23. S.M. Scholl, E.M. Pillers, R.E. Robinson, and P.J. Farrell, Prevalence of human papillomavirus type 16 DNA in cervical carcinoma samples in East Anglia, Int. J. Cancer, 35:215-218 (1985)
24. E.Schwarz, U.K. Freese, L. Gissmann, W. Mayer, B. Roggenbuck, A. Stremlau, and H. zur Hausen, Structure and transcription of human papillomavirus sequences in cervical carcinoma cells, Nature, 314:111-114 (1985)
25. B.A. Spalholz, Y.-C. Yang, and P.M. Howley, Transactivation of a bovine papilloma virus trranscriptional regulatory element by the E2 gene product, Cell, 42:183-191 (1985)
26. A.Stremlau, L. Gissmann, H. Ikenberg, M. Stark, P. Bannasch, and H. zur Hausen, Human papillomavirus type 16-related DNA in an anaplastic carcinoma of the lung, Cancer, 55:737-740 (1985)
27. Y.C. Yang, H. Okayama, and P.M. Howley, Bovine papillomavirus contains multiple transforming genes, Proc. Nat. Acad. Sci. USA, 82:1030-1034 (1985)
28. C.Yee, I. Krishnan-Howlett, C.C. Baker, R. Schlegel, and P.M. Howley, Presence and expression of human papillomavirus sequences in human cervical carcinoma cell lines, Am. J. Pathol., 119:361-366(1985)
29. H.zur Hausen, Human genital cancer: synergism between a two virus infections or between a virus infection and initiating events, Lancet, 2:1370-1372 (1982)
30. H.zur Hausen, and A. Schneider, The role of papillomaviruses in human anogenital cancer, in: "The Papillomaviruses" (P.M. Howley and N.P. Salzman, eds.) in print

ANTI-VIRAL VACCINE CONTROL OF EB VIRUS-ASSOCIATED CANCERS

M. A. Epstein

Nuffield Department of Clinical Medicine
University of Oxford, John Radcliffe Hospital
Headington, Oxford OX3 9DU, U.K.

INTRODUCTION

Over the past 21 years Epstein-Barr (EB) virus (1) has been implicated by a huge body of evidence (2,3) in the causation of two important human cancers, endemic Burkit's lymphoma (BL) (4) which provides the paradigm for virus-induced tumours in man (5), and undifferentiated nasopharyngeal carcinoma (NPC) (6) which has a high incidence amongst a significant section of the world population. More recently, the virus has also been found to be associated with the lymphomas which arise with undue frequency in immunosuppressed organ graft patients (7-9), and in individuals suffering from AIDS (10).

It has long been clear that both in BL and NPC the virus requires certain essential co-factors to bring about malignant change, and at least in the case of BL, possible mechanisms for this are beginning to emerge. It has been suggested that in areas where BL is endemic, EB virus acts by transforming target B lymphocytes and that these undergo unrestrained cell divisions such that there is an increased likelihood of one or other of three specific chromosomal translocations coming about (11). These translocations are thought to ensure that the c-myc oncogene is moved from its normal site on chromosome 8 to the immediate vicinity of one of the Ig genes active in the lymphoid cell destined to give rise to the tumour (12,13) where it would come under the influence of the Ig gene promoter with subsequent selection of a myc oncogene-driven clone of malignant BL cells. However, the human B-lym oncogene has also been implicated in BL (14) and how this might interact with the myc gene, and the exact significance of each has yet to be determined. Indeed, serious doubts regarding the relevance of either in the induction of endemic BL have recently been raised (15), and it is certain that in animal experiments EB virus can directly, rapidly, and potently activate the chain of events leading to the development of multiple malignant lymphomas (16). Such a direct role is not entirely surprising in the light of findings on transformation by EB virus <u>in vitro</u>; this transformation has often been categorized as being merely a form of "immortalization" (17), yet careful studies have shown that in addition to this phenomenon some cells are indeed changed by the virus in such a way as to possess from the outset many of the attributes of malignancy (18).

Irrespective of how the foregoing controversy will ultimately resolve, the question of the main co-factor in endemic BL has recently become clearer. Already in 1969 (19) compelling reasons had been advanced for believing that hyperendemic malaria fulfilled this function and was responsible for determining the geographical distribution of the tumour, and now studies have been reported which explain new aspects of the interaction between malaria and EB virus. Like all other herpesviruses, EB virus gives rise to a life-long infection to which the healthy carrier develops specific immunological responses. These are excellent at damping down the virus but never enough to eliminate it from the as yet unidentified immunologically privileged site in which it persists. Thus, a delicate balance evolves; on the host side this depends both on virus-neutralizing antibodies which restrict the spread of the agent from one target cell to another, and on a full cellular immunological response capable of destroying infected cells because they display virus-determined antigens. After the immediate non-specific responses of primary infection, specific cytotoxic, HLA-restricted, T lymphocytes play a central role in maintaining the vital surveillance of the virus carrier state (20). The very great importance of this surveillance is shown by the direct consequences when it fails: genetic defects lead to death after primary infection because of uncontrolled proliferation of EB virus transformed cells and widespread virus replication, whilst therapeutic immunosuppression in graft recipients is responsible for the undue frequency with which malignant lymphomas emerge.

In the case of acute malaria caused by <u>Plasmodium falciparum</u> it has been clearly demonstrated that control by EB virus-specific T lymphocytes is dramatically impaired (21), and this impairment resembles that of immunosuppressed renal allograft recipients (22). Furthermore, not only are T cell numbers decreased and B cells, the target for EB virus, increased, but the percentage of T helper (T4) cells is reduced in relation to T suppressor (T8) cells in much the same way as in AIDS, another condition in which EB virus lymphomas have been found (10). However the damage to EB virus-specific T lymphocyte control arises in malaria, it is not at all difficult to envisage that in the hyperendemic disease with its repeated attacks throughout the year and year after year, such damage will favour the unrestrained growth of EB virus-carrying B cells with the emergence of malignant lymphomas in a substantial number of individuals (23).

Much less is known about comparable mechanisms in the causation of undifferentiated NPC, principally because the absence of tumour cell lines has hampered laboratory studies. Nevertheless, it is generally accepted that EB virus plays a key role in causation (24), that there is a strong racial predisposition to the tumour (6), and that an environmental co-factor also operates, perhaps a tumour-promoting substance present in traditional herbal remedies (25). But even in the absence of detailed insights, large scale serological surveys for EB virus antibody patterns characteristic of NPC have been used for the mass screening of populations at risk and have been successful in the very early detection of tumours (26).

RATIONALE FOR A VACCINE AGAINST EB VIRUS

There can be no doubt that the accumulation of more and more detailed information on the biological behaviour of EB virus and on its role in human malignancy is both scientifically important and intellectually satisfying. But because of the strong evidence implicating the virus in some way in the causation of BL and NPC, it has seemed for many years that such activities would be more worthwhile if they included investigations which would lead to the development of methods for intervention against the infection which might, in turn, reduce the incidence of the associated cancers; it was in this context that proposals were first put forward for a vaccine against EB virus (27). Although endemic BL is not of very great numerical significance and in high incidence areas there are many far more pressing public health and medical problems, undifferentiated NPC is of importance in World Cancer terms since it is the most common tumour of men and the second most common of women amongst Southern Chinese (6) and Eskimos (28), and there are areas with moderately high incidence levels right across North Africa (29), in East Africa (30), and through most of South East Asia (6).

The numerical argument would carry little weight in the absence of persuasive precedents for the control of virus-induced animal cancers by antiviral vaccination. In this connection the naturally-occuring herpesvirus-induced lymphomas of Marek's disease in chickens (31) have been virtually eliminated from commercial flocks by inoculation with apathogenic virus (32,33) and thus provided the first example of antiviral vaccination affecting the frequency of a cancer. Later work with the malignant lymphoma which can be induced experimentally by inoculation of Herpesvirus saimiri in South American subhuman primates (34) has shown that animals given killed virus vaccine were protected against challenge infection and did not get tumours (35). Furthermore, in the Marek's disease system, antigen-containing membranes from cells infected with Marek's disease herpesvirus markedly reduced lymphoma incidence when used as an experimental vaccine (36) and even soluble viral antigens extracted from such cells protected in the same way (37). Similar approaches with EB virus have long appeared worthy of investigation.

A SUBUNIT VACCINE BASED ON EB VIRUS MEMBRANE ANTIGEN

It has been known for many years that the virus-neutralizing antibodies developed by EB virus infected individuals are directed against the virus-determined cell surface membrane antigen (MA) (38-41), and this, by analogy with the Marek's disease situation mentioned above (36,37), prompted the suggestion that MA should form the basis for an antiviral vaccine (27).

Investigations into the molecular structure of MA have identified two high molecular weight glycoprotein components of 340 and 270 kD (gp340 and gp270) (42-46) and the equivalence of human antibodies to MA and antibodies which neutralize EB virus has been formally explained by the finding of the same glycoproteins in both the viral envelope and the cell membrane MA (45). Not surprisingly, therefore, monoclonal antibodies which react with both MA components neutralize EB virus (47, 48) and purified gp340 can elicit virus-neutralizing antibodies in experimental animals (49,50).

REQUIREMENTS FOR A VACCINE BASED ON EB VIRUS MEMBRANE ANTIGEN

A suitable Experimental Animal Model

To demonstrate protection by a vaccine it is essential to have available in the laboratory an animal capable of being infected by EB virus and responding with lesions. Only two kinds of animal are known to give lesions after experimental infection with EB virus, the owl monkey (Aotus) (51-53) and the cottontop tamarin (Saguinus oedipus

oedipus) (54-56). However, it has not proved possible to repeat the work with the owl monkey and the explanation was provided when it was shown that this "species" included at least 10 different karyotypes (57-59); animals from several of these karyotypes cannot interbreed and show considerable variation in susceptibility to certain infections. The cottontop tamarin is therefore the species of choice for experimental studies with EB virus in vivo, and although it was placed on the endangered species list some years ago, the necessary husbandry and nutritional conditions have been worked out for breeding (60-64) and flourishing colonies have been established. Notwithstanding, these tamarins are extremely rare and costly, and only small numbers can be used for each experiment; the constraints are similar to those operating in work with hepatitis B virus where biological tests require the use of chimpanzees. And because of the constraints it is necessary to test out all methodologies with banal laboratory animals which will make antibodies to EB virus for example even through they cannot be infected, before applying them to tamarins. Last, because of the small numbers of tamarins available, any dose of virus to be given as a challenge after vaccination must be capable of causing lesions in 100 % of unprotected animals.

EB virus-induced Malignant Lymphomas in Cottontop Tamarins

EB virus suspensions were prepared with special care (16) and were titrated at least 3 times in foetal cord blood lymphocytes (65) to take account of inherent variations in sensitivity shown by cells from different cord bloods. Preliminary experiments indicated that only those virus preparations with more than 10^5 lymphocyte transforming doses per ml regularly caused lesions, and the titrations were used to identify such material.

Young adult cottontop tamarins from a succesful breeding colony were given $10^{5.3}$ lymphocyte-transforming units of EB virus in 7 ml suspending fluid by i.p. (2/3) and i.m. (1/3) injections and lesions arose in all the animals in 2 to 3 weeks; this rapid 100 % induction of disease occured with different batches of virus and in various experiments (Table 1). The i.m. injection was given in the thigh to allow clinical progress of the lesions to be followed easily in the subcutaneous inguinal nodes and the rather large volume of inoculum was chosen to ensure wide dispersal of virus at the injection sites. The lesions progressed in just over half the tamarins and the animals were killed for ethical reasons when terminal disease developed; the lesions in the remaining animals regressed over 8 to 14 weeks, much as has been reported in the past (54-56).

Table 1. EB virus-induced lymphomas in unprotected cottontop tamarins.

Experiment No.	Lymphocyte-transforming units of virus injected	Route of injection	Animals with gross multiple lesions
1.	$10^{5.3}$	iv;ip;im	2/2
2.	$10^{5.3}$	ip;im	2/2
3.	$10^{5.3}$	ip;im	4/4
4.	$10^{5.3}$	ip;im	2/2
5.	$10^{5.3}$	ip;im	4/4
6.	$10^{5.3}$	ip;im	2/2
7.	$10^{5.3}$	ip;im	2/2

Macroscopically, the lesions presented as white tumour masses (not infiltrates) in the spleen, kidneys, liver, gut wall, adrenals, and thoracic inlet, together with whitish lymph nodes enlarged up to 20 mm diameter (from the normal 1 - 2 mm) in the inguinal, iliac, axillary, mesenteric, paravertebral, submandibular, mediastinal, and abdominal regions. Each animal had tumours at most of these sites and before the onset of regression there was no clinically detectable difference between those which progressed and those which did not. Histologically, the diseased nodes and organs were filled with large-cell malignant lymphomas of the large non-cleaved (follicular centre-cell) and immunoblastic types (66) and hybridization studes for EB virus DAN (67), and Southern blotting and hybridization for immunoglobulin (Ig) gene rearrangements (68,69) showed that EB virus was present in the tumours, that the apparent number of virus genomes per cell varied for individual tumours from 2 to 25, and that each tumour was composed of only one or a few B lymphocyte clones as distinct from polyclonal lymphoproliferation. Equally interestingly, the different tumours from a given animal arose from different cell clones. Cell lines were established in vitro from all the tumours investigated and were shown by Southern blotting and Ig gene probing to have arisen from the tumour cells. The lines consisted of cells which expressed the EB virus nuclear antigen (EBNA) and the great majority shed infectious virus into the culture medium.

The tumours must thus be regarded as EB virus-induced, and in view of their histology and mono- or oligo-clonality, they must also be accepted as malignant. In this connection it should be noted that cytogenetic investigations failed to detect any consistent chromosomal abnormality in the tumour cell-derived lines such as would be compatible with the translocations deemed to be important for the activation of onc genes in the induction of BL (11-13). Although subtle inapparent chromosome changes of some other type cannot be excluded, the speed with which the tumours followed the inoculation of EB virus makes it plain that the latter played a crucial and central role in causation.

A full account of this work has been given elsewhere (16) together with a discussion of its importance in relation to the oncogenicity of EB virus and the relevance of the tamarin tumours as a model for the lymphomas of human graft recipients.

An Assay for EB Virus Membrane Antigen gp340

For the elaboration of an efficient purification procedure for gp340 a sensitive assay is essential to quantify the product and permit modification designed to maximize yields. A quantitative radioimmunoassay (RIA) for gp340 was developed based on very small amounts of extremely pure radio-iodinated glycoprotein. The antigenicity of this ^{125}I-gp340 was established by the finding that up to 75 % was precipitated specifically by naturally occuring human antisera to MA, and the radio-labelled material was thereafter used in a conventional competition RIA to quantify unlabelled gp340 in test samples. One unit of antigen was arbitrarily defined as the amount which caused a 50 % reduction in the binding of ^{125}I-gp340 to a standard reference serum (70).

Preparation Methods for EB Virus Membrane Antigen gp340

As regards sources of MA gp340 and gp270, most EB virus-producing lymphoid cell lines synthesize roughly equal amounts of the two molecules, but in this respect the B95-8 lymphoblastoid line (71) is anomalous in expressing almost exclusively gp340, thus providing an important advantage for molecular weight-based purificiation procedures.

With the indispensable help of the RIA a preparative sodium dodecyl sulphate poylacrylamide gel electrophoresis (SDS-PAGE) method was worked out for the isolation of pg340 from B95-8 cell membranes. A most important step involved renaturation to ensure that the final product was in an antigenic form. This was achieved by excising the gp340 from the gels and eluting in the presence of urea (72). It is possible that the dilution of eluted pg340 in urea solutions leads to sufficient removal of SDS to allow refolding to occur when the urea is subsequently removed by dialysis against buffer containing non-ionic detergent, and certainly this method resulted in a fifty-fold increase in the recovery of antigenic activity as compared to other methods and provided a homogeneous product as judged by analytical SDS-PAGE. A full account of this work has already been reported (73).

gp340 has also been purified using a monoclonal immunoabsorbant (74). The antigen was collected from detergent extracts of B95-8 cell membranes by application during upward flow to a mouse monoclonal antibody-Sepharose control column with irrelevant specificity to remove material which bound nonspecifically, followed by application to a specific anti-MA gp340 monoclonal antibody-Sepharose column. Bound material was eluted, neutralized, concentrated, and fractionated by gel filtration through a Sephacryl S-300 column. The final gp340 product was antigenically active, 95 % pure, and available in mg amounts (74).

Enhancement of Immunogenicity of EB Virus Membrane Antigen gp340

gp340 made by the molecular weight based method proved only weakly immunogenic in mice and rabbits after repeated injection and the use of Freund's adjuvant. To eliminate the need for these two disadvantageous procedures, gp340 was instead incorporated in liposomes (75,76) and comparative immunogenicity studies determined the best routes and methods of administration. Liposomes containing gp340 gave good titres of EB virus-neutralizing antibodies in mice, rabbits, and finaly tamarins, after rather few inoculation, and all the sera were specific in that they reacted only with MA gp340 and failed to recognize any other molecules from the surface or the interior of B95-8 cells (49,50).

A Sensitive Test for Antibodies to EB Virus Membrane Antigen gp340

In order to exploit immunogenicity studies to the full, a highly sensitive test to measure antibody responses was required. Accordingly, a rapid enzyme-linked immunosorbant assay (ELISA) was developed based on gp340 purified by the immunoaffinity monoclonal antibody chromatography method. 96-well microtitre plates were coated with the gp340 followed by treatment with blocking solution; thereafter, serial dilutions of test sera were added to individual wells and, after washing, the wells were exposed to an enzyme detection system. This consisted of rabbit anti-human IgG peroxidase conjugate, the binding of the conjugate being assessed after addition of substrate (orthopheylene diamine) for a fixed time, by measuring the optical density of the reaction product in an ELISA reader at 492 nm. This ELISA (77) has proved one thousand-fold more sensitive than conventional immunofluorescence assays and has made it possible accurately to follow the sequential production of antibodies to gp340 during the immunization of animals.

VACCINATION OF COTTONTOP TAMARINS

The direction of the present vaccine programme was determined by the assumption that strategies which worked with the herpesvirus of Marek's disease and the lymphomas it causes in chickens would be applicable to EB virus (72), a seemingly carcinogenic herpesvirus of man. As already explained, EB virus MA components were selected for use as a subunit vaccine both because of the concordance between human antibodies to MA and virus-neutralizing antibodies, and because plasma membranes from cells infected with Marek's disease herpesvirus were efficacious as an experimental vaccine in chickens (36). Accordingly, the first and simplest step in the EB virus system was to determine whether plasma membranes from EB virus-infected cells could be used successfully in a comparable manner.

Protection With MA-positive Cell Membranes

For the experiments with cell membranes, B95-8 cells showing maximum expression of MA were disrupted and the plasma membranes were collected by sucrose density gradient centrifugation (78). The resulting material was injected i.p. 8 times into 2 tamarins at fortnightly intervals, each inoculum containing on average 2500 units of MA gp340 as assessed by RIA (70). Serological responses were measured by indirect immunofluorescence, by the highly sensitive ELISA (77), and by virus neutralization tests which assess the ability of a given serum to prevent transformation by the virus of foetal cord blood lymphocytes (65,41); all the neutralizations were repeated (up to 3 times) and included control antisera of known titre, to take account of inherent variations in sensitivity shown by cells from different cord bloods. Specific antibodies were rapidly induced, eventually reached high titre, and were then powerfully virus-neutralizing; since the neutralization tests were all repeated more than once, neutralizing titres were recorded as being greater than the minimum obtained in each set of tests. The tamarins were considered to have been satisfactorily immunized when their levels of neutralizing antibody equaled or exceeded the highest titres found in sera from naturally infected human subjects.

The immunized animals were challenged with $10^{5.3}$ lymphocyte-transforming units of virus (65), the standard 100 % tumour-inducing dose (Table 1); to further ensure exclusion of equivocal results, 2 additional normal control animals were given the challenge material at the same time. The immunized tamarins remained entirely free of clinically detectable lesions whereas the normal control animals developed the expected gross, multiple tumours in 2 - 3 weeks (Table 2).

Table 2. Protection of cottontop tamarins against EB virus-induced lymphomas by vaccination with B95-8 cell membranes.

Tamarin	Antibody to B95-8 cell surface (immunofluorescence)	ELISA antibody titre	Virus neutralising antibody	Lesions after $10^{5.3}$ units of virus i.p.;i.m.
Vaccinated	+++	1:800	strong*	none
Vaccinated	+++	1:2000	strong*	none
Control	-	0	none	gross multiple
Control	-	0	none	gross multiple

* 1 ml serum neutralised > 100 000 lymphocyte transforming units of virus (from lowest value in repeated tests)

Protection with SDS-PAGE Purified MA gp340

Since it had been shown that purified gp340 prepared by the molecular weight-based method which conserved immunogenicity (73) induced virus-neutralizing antibodies in various species when injected after incorporation in liposome (49,50), the protective effects of such antibodies were investigated in tamarins, as well as the levels of antibody required for protection. In a pilot experiment 4 animals were selected in whom differeing amounts of neutralizing antibody were present after i.p. immunization with liposomal gp340, and all were challenged with the 100 % lymphomagenic dose of virus (Table 1). Table 3 (A) shows that only the one animal whose serum was strongly neutralizing was completely protected, whereas the other animals with much lower, or undetectable, levels of neutralizing antibody were not protected at all.

Table 3. Protection of cottontop tamarins against EB virus-induced lymphoma by vaccination with MA gp340 purified from B95-8 cell membranes by the molecular weight-based method.

Tamarin	Antibody to MA (immunofluorescence)	ELISA antibody titre	Virus neutralising antibody	Lesions after $10^{5.3}$ units of virus i.p.; i.m.
A. Pilot Experiment				
Vaccinated	+++	1:1000	strong*	none
Vaccinated	+	1:100	none	gross multiple
Vaccinated	+	< 1:10	slight	gross multiple
Vaccinated	+	**	none	gross multiple
B. Confirmatory Experiment				
Vaccinated	+++	1:800	strong+	2 small, transient
Vaccinated	+++	1:1000	strong++	single, transient
Control	-	0	none	gross multiple
Control	-	0	none	gross multiple

*   1 ml serum neutralised > 10 000 lymphocyte transforming units of virus
** serum sample haemolysed - not testable
+   1 ml serum neutralised 100 000 lymphocyte transforming units of virus
++   1 ml serum neutralised > 100 000 lymphocyte transforming units of virus

(* + ++ from lowest value in repeated tests)

To confirm this result 2 further tamarins were given gp340 in liposomes 17 times i.p. at fortnightly intervals, each dose containing on average 2250 RIA units (70). Antibodies were monitored as before and when these had reached high levels (Table 3 (B)) the vaccinated animals were challenged with the 100 % pathogenic dose of EB virus (Table 1), along with 2 normal control animals. The unprotected tamarins developed gross, multiple tumours in 2 - 3 weeks whilst those which had been vaccinated remained free of tumours (Table 3 (B)); one protected animal exhibited minor transient swelling of a single inguinal lymph node draining the site of i.m. injection in the thigh, and the other developed similar inguinal node enlargement to 5 mm which rapidly regressed, accompanied by minor transient enlargement of a single mesenteric node. These results have recently been published in full (79).

Failure of Protection with Monoclonal Antibody Purified MA gp340

The monoclonal antibody immunoaffinity chromatography method for purifying MA gp340 produced such excellent material that 5 inoculations were sufficient to induce high titre antibodies in each of 4 tamarins. The neutralizing capacity of the sera from these 4 animals was the same as that of the tamarins successfully protected by vaccination with SDS-PAGE purified material (cf. Tables 3 and 4), yet when challenge virus was given in the usual dose, all 4 animals developed gross, multiple tumours in 2 - 3 weeks (Table 4) (80).

Table 4. Failure of vaccination with affinity chromatography-purified MA gp340 to protect tamarins against EB virus-induced lymphomas.

Tamarin	Antibody to MA (immunofluorescence)	ELISA antibody titre	Virus neutralising antibody	Lesions after $10^{5.3}$ units of virus i.p.;i.m.
Vaccinated	+++	1:1000	strong^x	gross multiple
Vaccinated	+++	1:1000	strong^x	gross multiple
Vaccinated	+++	1:1000	strong^x	gross multiple
Vaccinated	+++	1:1600	strong^x	gross multiple
Control	-	0	none	gross multiple
Control	-	0	none	gross multiple

x 1 ml serum neutralized > 100 000 lymphocyte transforming units of virus (from lowest value in repeated tests).

DISCUSSION

The foregoing experiments demonstrate that gp340 prepared by the molecular weight based SDS-PAGE method and incorporated in liposomes can be used as an efficient prototype subunit vaccine to protect experimental animals against tumour induction by a massive challenge dose of EB virus. The protected tamarins have remained well for more than 6 months.

The fact that similar protection was not afforded by gp340 purified by the immunoaffinity chromatography methods raises important points. It was known from the outset that the monoclonal antibody preparation method bound, and therefore collected, only about 50 % of the MA epitopes (74), so that failure of the inoculated product to elicit full protection against EB virus was not wholly unexpected. What does surprise is the finding that the powerful neutralizing antibodies in the animals immunized with this material although not protective, were indistinguishable in the neutralization tests from those which were. Thus, neutralizing antibody _per_ _se_ cannot be taken as a certain measure of immunity, and the SDS-PAGE purification method must obviously be isolating a component important specifically for protection which is not selected by the monoclonal antibody used. Alternatively, exposure of pg340 to the denaturing conditions during the SDS-PAGE procedure may result in the exposure of normally cryptic epitopes, even after renaturation has been brought about. Investigation of differences in the repertoire of immunological responses of tamarins depending on which of the two types of immunogen they received could well reveal the nature of the molecule(s) crucial for protection.

Both endemic BL and undifferentiated NPC are diseases of developing countries and it has long been known that primary EB virus infection occurs at an early age in the social and hygiene conditions of the third world (81). Vaccinations are thus required for the very young and since undifferentiated NPC is a preponderantly a disease of middle and later life (6), vaccine protection would have to be maintained for a great many years. However, these logistic requirements are no different from those relating to hepatitis B vaccination for the prevention of primary liver cancer in high risk populations for which plans on a 30 year prospective basis well advanced.

MA components prepared in the way discussed here have never been thought suitable for anything beyond the present verificative prototype vaccine designed to demonstrate the capacity of these subunits, when injected, to stimulate immunological responses giving protection against the virus. It is thus important to know something of the general structure of gp340 and of the contribution, if any, of the sugar moiety to antigenicity. For this, gp340 was analysed after treatment with a battery of glycosidases and V8 protease, with and without preliminary exposure during synthesis to tunicamycin. These investigations have shown that carbohydrate represents more than 50 % of the total mass of gp340, that it is both O- and N-linked, that V8 protease fragments are antigenic, and that specific antibody appears to bind the protein, not the sugar (82,83). The seemingly major importance of the protein in the immunogenicity of gp340 means that for use in man new and sophisticated procedures can be exploited, and recent progress in biotechnology is likely to facilitate matters greatly.

The region of EB virus DNA carrying the gene coding for MA has already been identified (84) and the nucleotide sequence of this gene is also known (85). The expression of the gene in pro- or eukaryotic cells is currently being investigated and there is also an excellent chance that it may prove possible to incorporate it into the genome of vaccinia virus and thus permit its direct expression during vaccination (86-88). It is also possible that synthetic gp340 peptides could be used as immunogens perhaps in conjunction with powerful new adjuvants (89). The recent experiments discussed above on the vaccine protection of tamarins establish a strong case for determined efforts in these directions to ensure the speedy production of MA gp340 in a form suitable for use as a vaccine in man.

FUTURE PROSPECTS

The EB virus system is unusually well placed for the preliminary testing of a vaccine based on gp340 in the human context. It is well known that in Western countries a considerable number of children escape silent primary infection with the virus, the proportion of each age group increasing with increasing standards of living; young adults can readily be screened to detect those who have remained uninfected and who are therefore at risk for delayed primary infection which is accompanied in the post-childhood age groups by the clinical manifestations of infectious mononucleosis in 50 % of cases (90). Screening could be applied to groups of new students entering Universities to determine who is uninfected followed by a double-blind vaccine trial amongst informed, consenting volunteers in the "at risk" category. The effectiveness of immunization in preventing infection and reducing the expected incidence of infectious mononucleosis would rapidly become apparent.

Thereafter, the effect of vaccination and consequential prevention of disease should be assessed in a high incidence region for endemic BL. This tumour has a peak incidence at the age of about 7 years (4) and the influence of vaccination on this would therefore be evident within a decade. If this were successful there would then be inescapable reasons for tackling the more difficult problem of intervention against undifferentiated NPC, a disease of later life (6) requiring the maintenance of immunity over many years.

REFERENCES

1. M. A. Epstein, B. G. Achong, and Y. M. Barr, Virus particles in cultured lymphoblasts from Burkitt's lymphoma, Lancet, 1:702 (1964).
2. M. A. Epstein, and B. G. Achong, eds., "The Epstein-Barr virus", Springer, Berlin, Heidelberg, and New York (1979).
3. M. A. Epstein, and B. G. Achong, eds., "The Epstein-Barr virus: recent advances", Heinemann, London - in press (1985).
4. D. Burkitt, A lymphoma syndrome in tropical Africa, in: "International review of experimental pathology", 2, G. W. Richter, and M. A. Epstein, eds., Academic Press, New York and London (1963).
5. G. M. Lenoir, G. T. O'Connor, and C. L. M. Olweny, eds., "Burkitt's lymphoma: a human cancer model", IARC, Lyon (1985).
6. K. Shanmugaratnam, Studies on the etiology of nasopharyngeal carcinoma, in: "International review of experimental pathology", 10, G. W. Richter, and M. A. Epstein, eds., Academic Press, New York and London (1971).
7. I. Penn, Malignancies associated with immunosuppressive or cytotoxic therapy, Surgery, 83:492 (1978).
8. L. J. Kinlen, A. G. R. Shiel, J. Peto, and R. Doll, Collaborative United Kingdom-Australasian study of cancer in patients treated with immunosuppressive drugs, Brit. Med. J., 2:1461 (1979).
9. J. Weintraub, and R. A. Warnke, Lymphoma in cardiac allotransplant recipients: clinical and histological features and immunological phenotype, Transplantation, 33:347 (1982).
10. J. L. Ziegler, W. L. Drew, R. C. Miner, L. Mintz, E. Rosenbaum, J. Gershon, E. T. Lennette, J. Greenspan, E. Shillitoe, J. Beckstead, C. Casavant, and K. Yamamoto, Outbreak of Burkitt-like lymphoma in homosexual men, Lancet, 11:631 (1982).
11. G. M. Lenoir, J. L. Preud'homme, A. Bernheim, and R. Berger, Correlation between immunoglobulin light chain expression and variant translocation in Burkitt's lymphoma, Nature, 298:474 (1982).
12. R. Dalla-Favera, M. Bregni, J. Erikson, D. Patterson, R. C. Gallo, and C. M. Croce, Human c-myc onc gene is located on the region of chromosome 8 that is translocated in Burkitt lymphoma cells,

Proc. Nat. Acad. Sci. USA, 79:7824 (1982).

13. R. Traub, I. Kirsch, C. Morton, G. Lenoir, D. Swan, S. Tronick, S. Aaronson, and P. Leder, Translocation of the c-myc gene into the immunoglobulin heavy chain locus in human Burkitt's lymphoma and murine plasmacytoma cells, Proc. Nat. Acad. Sci. USA, 79:7837 (1982).

14. A. Diamond, G. M. Cooper, J. Ritz, and M.-A. Lane, Identification and molecular cloning of the human Blym transforming gene activated in Burkitt's lymphomas, Nature, 305:112 (1983).

15. P. H. Duesberg, Activated proto-onc genes: sufficient or necessary for cancer? Science, 228:669 (1985).

16. M. L. Cleary, M. A. Epstein, S. Finerty, R. F. Dorfman, G. W. Bornkamm, J. K. Kirkwood, A. J. Morgan, and D. Sklar, Individual tumours of multifocal EB virus-induced malignant lymphomas in tamarins arise from different B-cell clones, Science, 228:722 (1985).

17. G. Miller, Biology of the Epstein-Barr virus, in: "Viral oncology", G. Klein, ed., Raven Press, New York (1980).

18. M. Zerbini, and I. Ernberg, Can Epstein-Barr virus infect and transform all the B-lymphocytes in human cord blood? J. Gen. Virol., 64:539 (1983).

19. D. P. Burkitt, Etiology of Burkitt's lymphoma - an alternative hypothesis to a vectored virus, J. Nat. Cancer Inst., 42:19 (1969).

20. A. B. Rickinson, D. J. Moss, L. E. Wallace, M. Rowe, I. S. Misko, M. A. Epstein, and J. H. Pope, Long term T-cell mediated immunity to Epstein-Barr virus, Cancer Res., 41:4216 (1981).

21. H. C. Whittle, J. Brown, K. marsh, B. M. Greenwood, P. Seidelin, H. Tighe, and L. Wedderburn, T cell control of B cells infected with EB virus is lost during P. falciparum malaria, Nature, 312:449 (1984).

22. J. S. H. Gaston, A. B. Rickinson, and M. A. Epstein, Epstein-Barr-virus-specific T-cell memory in renal-allograft recipients under long-term immunosuppression, Lancet, 1:923 (1982).

23. M. A. Epstein, Burkitt's lymphoma - clues to the role of malaria, Nature, 312:398 (1984).

24. G. Klein, The relationship of the virus to nasopharyngeal carcinoma, in: "The Epstein-Barr virus", M. A. Epstein, and B. G. Achong, eds., Springer, Berlin, Heidelberg and New York, (1979).

25. Y. Ito, M. Kawanishi, T. Harayama, and S. Takabayashi, Combined effects of the extracts from Croton tiglium, Euphorbia lathyris, or Euphorbia tinucalli and n-butyrate on Epstein-Barr virus expression in human lymphoblastoid P_3HR-1 and Raji cells, Cancer Letters, 12:173 (1981).

26. Y. Leng, L. G. Zhang, H. Y. Li, M. G. Jan, Q. Zhang, Y. S. Wang, and G. R. Su, Serological mass survey for early detection of nasopharyngeal carcinoma in Wuzhou City, China, Int. J. Cancer, 29:139 (1982).

27. M. A. Epstein, Epstein-Barr virus - is it time to develop a vaccine program? J. Nat. Cancer Inst., 56:697 (1976).

28. A. Lanier, T. Bender, M. Talbot, S. Wilmet, C. Tschopp, W. Henle, G. Henle, D. Ritter, and P. Terasaki, Nasopharyngeal carcinoma in Alaskan Eskimos, Indians and Aleuts: a review of cases and study of Epstein-Barr virus, HLA and environmental risk factors, Cancer, 46:2100 (1980).

29. M. Cammoun, G. V. Hoerner, and N. Mourali, Tumors of the nasopharynx in Tunisia: an anatomic and clinical study based on 143 cases, Cancer, 33:184 (1974).

30. P. Clifford, A review: on the epidemiology of nasopharyngeal carcinoma, Int. J. Cancer, 5:287 (1970).

31. L. N. Payne, J. A. Frazier, and P. C. Powell, Pathogenesis of Mareks

disease, in: "International review of experimental pathology", 16, G. W. Richter, and M. A. Epstein, eds., Academic Press, New York and London (1976).

32. A. E. Churchill, L. N. Payne, and R. C. Chubb, Immunization against Marek's disease using a live attenuated virus, Nature, 221:744 (1969).

33. W. Okazaki, H. G. Purchase, and B. R. Burmester, Protection against Marek's disease by vaccination with a herpesvirus of turkeys, Avian Dis., 14:413 (1970).

34. L. V. Melendez, R. D. Hunt, M. D. Daniel, F. G. Garcia, and C. E. O. Fraser, Herpesvirus saimiri. II. An experimental induced primate disease resembling reticulum cell sarcoma, Lab. Animal Care, 19:378 (1969).

35. R. Laufs, and H. Steinke, Vaccination of non-human primates against malignant lymphoma, Nature, 253:71 (1975).

36. O. R. Kaaden, and B. Dietzschold, Alterations of the immunological specificity of plasma membranes of cells infected with Marek's disease and turkey herpesviruses, J. Gen. Virol., 25:1 (1974).

37. F. Lesnick, and L. J. N. Ross, Immunization against Marek's disease using Marek's disease virus-specific antigens free from infectious virus, Int. J. Cancer, 16:153 (1975).

38. G. Pearson, F. Dewey, G. Klein, G. Henle, and W. Henle, Relation between neutralization of Epstein-Barr virus and antibodies to cell membrane antigens, J. Nat. Cancer Inst., 45:989 (1970).

39. L. Gergely, G. Klein, and I. Ernberg, Appearance of Epstein-Barr virus associated antigens in infected Raji cells, Virology, 45:10 (1971).

40. G. Pearson, G. Henle, and W. Henle, Production of antigens associated with Epstein-Barr virus in experimentally infected lymphoblastoid cell lines, J. Nat. Cancer Inst., 46:1243 (1971).

41. A. de Schryver, G. Klein, J. Hewetson, J. Rocchi, W. Henle, G. Henle, D. J. Moss, and J. H. Pope, Comparison of EBV neutralization tests based on abortive infection or transformation of lymphoid cells and their relation to membrane reactive antibodies (anti MA), Int. J. Cancer, 13:353 (1974).

42. L. F. Qualtiere, and G. Pearson, Epstein-Barr virus-induced membrane antigens from EBV-superinfected Raji cells, Int. J. Cancer, 23:808 (1979).

43. B. C. Strnad, R. H. Neubauer, H. Rabin, and R. A. Mazur, Correlation between Epstein-Barr virus membrane antigen and three large cell surface glycoproteins, J. Virol., 32:885 (1979).

44. D. A. Thorley-Lawson, and C. M. Edson, The polypeptides of the Epstein-Barr virus membrane antigen complex, J. Virol., 32:458 (1979).

45. L. F. Qualtiere, and G. R. Pearson, Radioimmune precipitation study comparing the Epstein-Barr virus membrane antigens expressed on P_3HR-1 virus-superinfected Raji cells to those expressed in a B95-8 virus-transformed producer culture activated with tumor-promoting agent (TPA), Virology, 102:360 (1980).

46. J. R. North, A. J. Morgan, and M. A. Epstein, Observations on the EB virus envelope and virus-determined membrane antigen (MA) polypeptides, Int. J. Cancer, 26:231 (1980).

47. G. J. Hoffman, S. G. Lazarowitz, and S. D. Hayward, Monoclonal antibody against a 250,000-dalton glycoprotein of Epstein-Barr virus identifies a membrane antigen and a neutralizing antigen, Proc. Nat. Acad. Sci. USA, 77:2979 (1980).

48. D. A. Thorley-Lawson, and K. Geilinger, Monoclonal antibodies against the major glycoprotein (gp350/220) of Epstein-Barr virus neutralize infectivity, Proc. Nat. Acad. Sci. USA, 77:5307 (1980)

49. J. R. North, A. J. Morgan, J. L. Thompson, and M. A. Epstein, Purified EB virus gp340 induces potent virus-neutralizing antibodies

when incorporated in liposomes, Proc. Nat. Acad Sci. USA, 79:7504 (1982).

50. A. J. Morgan, M. A. Epstein, and J. R. North, Comparative immunogenicity studies on Epstein-Barr (EB) virus membrane antigen (MA) with novel adjuvants in mice, rabbits and cottontop tamarine J. Med. Virol., 13:281 (1984).

51. M. A. Epstein, R. D. Hunt, and H. Rabin, Pilot experiments with EB virus in owl monkeys (Aotus trivirgatus) I. Reticuloproliferative disease in an inoculated animal, Int. J. Cancer, 12:309 (1973).

52. M. A. Epstein, H. Rabin, G. Ball, and L. V. Melendez, Pilot experiments with EB virus in owl monkeys (Aotus trivirgatus) II. EB virus in a cell line from an animal with reticuloproliferative disease, Int. J. Cancer, 12:319 (1973).

53. M. A. Epstein, H. zur Hausen, G. Ball, and H. Rabin, Pilot experiments with EB virus in owl monkeys (Aotus trivirgatus) III. Serological and biochemical findings in an animal with reticuloproliferative disease, Int. J. Cancer, 15:17 (1975).

54. T. Shope, D. Dechairo, and G. Miller, Malignant lymphoma in cottontop marmosets after inoculation with Epstein-Barr virus, Proc. Nat. Acad. Sci. USA, 70:2487 (1973).

55. G. Miller, T. Shope, D. Coope, L. Waters, J. Pagano, G. W. Bornkamm, and W. Henle, Lymphoma in cotton-top marmosets after inoculation with Epstein-Barr virus: tumor incidence, histologic spectrum, antibody responses, demonstration of viral DNA, and characterization of viruses, J. Exp. Med., 145:948 (1977).

56. G. Miller, Experimental carcinogenicity by the virus in vivo, in: "The Epstein-Barr virus", M. A. Epstein, and B. G. Achong, eds., Springer, Berlin, Heidelberg and New York (1979).

57. N. S. F. Ma, Chromosome evolution in the owl monkey, Aotus, Amer. J. Phys. Anthropol., 54:293 (1981).

58. N. S. F. Ma, T. C. Jones, A. C. Miller, L. M. Morgan, and E. A. Adams, Chromosome polymorphism and banding patterns in the owl monkey (Aotus), Lab. Animal Sci., 26:1022 (1976).

59. N. S. F. Ma, R. N. Rossan, S. T. Kelley, J. S. Harper, M. T. Bedard, and T. C. Jones, Banding patterns of the chromosomes of two new karyotypes of the owl monkey, Aotus, captured in Panama, J. Med. Primatol., 7:146 (1978).

60. H. M. Brand, Husbandry and breeding of a newly established colony of cotton-topped tamarins (Saguinus oedipus). Lab. Animals, 15:7 (1981).

61. J. K. Kirkwood, M. A. Epstein, and A. J. Terlecki, Factors influencing population growth of a colony of cotton-top tamarins, Lab. Animals, 17:35 (1983).

62. J. K. Kirkwood, Effects of diet on health, weight and litter size captive cotton-top tamarins Saguinus oedipus oedipus, Primates, 24:515 (1983).

63. J. K. Kirkwood, and S. J. Underwood, Energy requirements of captive cotton-top tamarins (Saguinus oedipus oedipus), Folia Primatolog. 42:180 (1984).

64. J. K. Kirkwood, M. A. Epstein, A. J. Terlecki, and S. J. Underwood, Rearing a second generation of cotton-top tamarins Saguinus oedipus oedipus, Lab. Animals, in press (1985).

65. D. J. Moss, and J. H. Pope, Assay of the infectivity of Epstein-Barr virus by transformation of human leucoytes in vitro, J. Gen. Virol., 17:233 (1972).

66. R. F. Dorfman, J. S. Burke, and S. Berard, A working formulation of non-Hodgkins lymphomas: background, recommendations, histological criteria, and relationship to other classifications, in: "Malignant lymphomas", S. Rosenberg, and H. Kaplan, eds., Academic Press, New York and London (1982).

67. G. W. Bornkamm, C. Desgranges, and L. Gissmann, Nucleic acid hybridization for the detection of viral genomes, Curr. Topics Microbiol. Immunol., 104:287 (1983).
68. A. Arnold, J. Cossman, A. Bakhshi, E. S. Jaffe, T. A. Waldmann, and S. J. Korsmeyer, Immunoglobulin-gene rearrangements as unique clonal markers in human lymphoid neoplasms, New Engl. J. Med., 309:1593 (1983).
69. M. L. Cleary, J. Chao, R. Warnke, and J. Sklar, Immunoglobulin gene rearrangement as a diagnostic criterion of B cell lymphoma, Proc. Nat. Acad. Sci. USA, 81:593 (1984).
70. J. R. North, A. J. Morgan, J. L. Thompson, and M. A. Epstein, Quantification of an EB virus-associated membrane antigen (MA) component, J. Virol. Methods, 5:55 (1982).
71. G. Miller, T. Shope, H. Lisco, D. Stitt, and M. Lipman, Epstein-Barr virus: transformation, cytopathic changes, and viral antigens in squirrel monkey and marmoset leukocytes, Proc. Nat. Acad. Sci. USA, 69:383 (1972).
72. B. Bowen, J. Steinberg, U. K. Laemmli, and H. Weintraub, The detection of DNA binding proteins by protein blotting, Nucleic Acids Res., 8:1 (1980).
73. A. J. Morgan, J. R. North, and M. A. Epstein, Purification and properties of the gp340 component of Epstein-Barr (EB) virus membrane antigen (MA) in an immunogenic form, J. Gen. Virol., 64:445 (1983).
74. B. J. Randle, A. J. Morgan, S. A. Stripp, and M. A. Epstein, Large-scale purification of Epstein-Barr virus membrane antigen gp340 using a monoclonal immunoabsorbent, J. Immunological Methods, 77:25 (1985).
75. B. Morein, A. Helenius, K. Simons, R. Pettersson, L. Kääriäinen, and V. Schirrmacher, Effective subunit vaccines against an enveloped animal virus, Nature, 276:715 (1978).
76. E. K. Manesis, C. H. Cameron, and G. Gregoriadis, Hepatitis B surface antigen-containing liposomes enhance humoral and cell-mediated immunity to the antigen, FEBS Lett., 102:107 (1979).
77. B. J. Randle, and M. A. Epstein, A highly sensitive enzyme-linked immunosorbent assay to quantitate antibodies to Epstein-Barr virus membrane antigen gp340, J. Virological Methods, 9:201 (1984).
78. M. J. Crumpton, and D. Snary, Preparation and properties of lymphocyte plasma membrane, Contemp. Topics Mol. Immunol., 3:27 (1974).
79. M. A. Epstein, A. J. Morgan, S. Finerty, B. J. Randle, and J. K. Kirkwood, Protection of cotton-top tamarins against EB virus-induced malignant lymphoma by a prototype subunit vaccine, Nature, in press (1985).
80. M. A. Epstein, B. J. Randle, S. Finerty, and J. K. Kirkwood, Not all potently neutralizing, vaccine-induced antibodies to EB virus ensure protection of susceptible experimental animals, submitted to press (1985).
81. W. Henle, and G. Henle, Seroepidemiology of the virus, in: "The Epstein-Barr virus", M. A. Epstein, and B. G. Achong, eds., Springer, Heidelberg and New York (1979).
82. C. M. Edson, and D. A. Thorley-Lawson, Synthesis and processing of the three major envelope glycoproteins of Epstein-Barr virus, J. Virol., 46:547 (1983).
83. A. J. Morgan, A. R. Smith, R. N. Barker, and M. A. Epstein, A structural investigation of the Epstein-Barr (EB) virus membrane antigen glycoprotein, gp340, J. Gen. Virol., 65:397 (1984).
84. M. Hummel, D. A. Thorley-Lawson, and E. Kieff, An Epstein-Barr virus DNA fragment encodes messages for the two major envelope glycoproteins (gp350/300 and gp220/200), J. Virol., 49:413 (1984)
85. M. Biggin, P. J. Farrell, and B. G. Barrell, Transcription and

DNA sequence of the BamHI L fragment of B95-8 Epstein-Barr virus, EMBO J., 3:1083 (1984).
86. G. L. Smith, M. Mackett, and B. Moss, Infectious vaccinia virus recombinants that express hepatitis B virus surface antigen, Nature, 302:490 (1983).
87. B. Moss, G. L. Smith, J. L. Gerin, and R. H. Purcell, Live recombinant vaccinia virus protects chimpanzees against hepatitis B, Nature, 311:67 (1984).
88. M. Mackett, T. Yilma, J. K. Rose, and B. Moss, Vaccinia virus recombinants: expression of VSV genes and protective immunization of mice and cattle, Science, 227:433 (1985).
89. B. Morein, B. Sundquist, S. Höglund, K. Dalsgaard, and A. Osterhaus, Iscom, a novel structure for antigenic presentation of membrane proteins from enveloped viruses, Nature, 308:457 (1984).
90. J. C. Niederman, A. S. Evans, L. Subrahmanyan, and R. W. McCollum, Prevalence, incidence and persistence of EB virus antibody in young adults, New Engl. J. Med., 282:361 (1970).

BOVINE LEUKOSIS VIRUS AS A MODEL FOR HUMAN RETROVIRUSES

Claudine Bruck, Richard Kettmann, Daniel Portetelle, Dominique Couez and Arsene Burny

Dept. of Molecular Biology, University of Brussels
67, rue des Chevaux, 1640, Rhode-Saint-Genese, Belgium

INTRODUCTION

The most common neoplasm of the bovine species is, by far, lymphoid leukosis. Enzootic bovine leukosis (EBL) has been recognized as a neoplasm of infectious origin for half a century. The agent was identified as a retrovirus of exogenous origin (1). This virus, named bovine leukemia virus, was found to be unrelated to any known retrovirus family, until the discovery of the human T cell lymphotropic viruses (HTLV) in 1980 (2). BLV, HTLV-I and HTLV-II share a number of biochemical, biological and immunological features which suggest that the three leukemia viruses belong to a new family of retroviruses. According to the same criteria, a more distant relationship to HTLV-III was found. Recently, primate viruses related to the HTLV viruses (STLV-I and III) were identified (3-5). HTLV-III was shown to belong to the lentivirus family by its extensive sequence homology to the ovine VISNA virus (6). BLV, the HTLV and STLV leukemia viruses, HTLV-III, STLV-III and other lentiviruses were shown to form a unique group of retroviruses characterized by the presence of a tat gene (tat = trans-acting transcriptional activation) in their genome (7,8). The tat gene product is believed to play a major role in the induction of the transformed phenotype by BLV and HTLV-I and II (7).

In this report, we shall focus on the mechanisms of BLV-induced leukaemogenesis and on the molecular characterization of the BLV envelope glycoprotein gp51.

MECHANISMS OF BLV-INDUCED LEUKEMOGENESIS

Infection occurs through close contact with BLV-infected animals. Body secretions and blood are responsible for transmission of the disease within a herd, as evidenced by the correlation of disease transmission with intensive husbandry and nursing practices, reuse of needles and the occurence of specific biting insects in tropical climates (9). Experimental transmission has shown that the inoculation of infected cells is much more efficient in transmitting BLV infection than cell-free virus preparation. Enzootic bovine leukemia is a chronic disease evolving over an extended period of time. BLV persists indefinitely in the host, inducing a variety of symptoms. The different patterns of disease observed after BLV infection are relevant to our understanding of BLV-induced leukaemogenesis.

A. Asymptomatic BLV-infected Cattle

The only constant feature observed after BLV infection is a strong and persistent antibody response to BLV antigens. Antibodies are directed mainly against the envelope glycoprotein gp51 and the major core protein p24 (10). In spite of their high titer, BLV antibodies do not seem to be protective against the onset of the leukemic phase: although subject to major fluctuations, the antibody titer rises constantly during the progression of the disease and reaches maximal levels at the animal's death in the tumor phase. The site and mechanism of the BLV persistence at the origin of this antigenic stimulation remains a mystery since infected animals show no overt sign of viremia. The majority of BLV-infected animals remains healthy, showing no economic down performances. The only signs of BLV infection are the production of BLV antibodies and the occurrence of a minority of B lymphocytes harbouring proviral DNA sequences integrated in the cellular DNA.

B. BLV-Infected Cattle with Persistent Lymphocytosis (PL)

Depending on ill-defined parameters, including genetic and environmental factors, between 30 and 70 % of BLV carriers develop a form of the disease known as persistent lymphocytosis (PL). PL is charcterized by an increase in circulating B-lymphocytes as evidenced by the presence of surface immunoglobulins and Fc receptors on the circulating cells. Depending on the animal, this B-lymphocyte population represents 40 to 80 % of the total lymphocyte population (vs. 15 - 20 % in healthy individuals) (11). In animals with PL, the lymphocyte count rises to an elevated level, but remains stable for several years. The occurrence of PL does not seem to influence the progression of the disease towards the tumor phase: animals can develop lymphoid tumors without preliminary PL and many animals with PL will never produce lymphoid tumors. In spite of their excessive in vivo proliferation, PL lymphocytes show no sign of malignant transformation: they appear as small resting lymphocytes and do not show transformation-specific growth properties in soft agar. Interestingly, the subpopulation of BLV-infected lymphocytes is polyclonal with respect to BLV integration sites (12) and exhibits no gross chromosomal abnormalities. The transcription of viral genes in vivo is inhibited through the action of a repressor protein (13). Viral production can be initiated by in vitro culture of infected lymphocytes (14).

C. BLV-Infected Cattle in the Tumor Phase

Only a few percent (0.1 to 10 %, depending on non-defined genetic factors) of BLV-infected animals, whether or not in PL, develop lymphoid tumors (14). Several important characteristics of BLV-induced tumors are as follows:

[1] According to functional and cytological criteria, the vast majority of EBL tumors are of the B-cell type (15).

[2] In all cases of enzootic bovine leukemia, proviral DNA is present in the cells forming the tumor mass. This strongly suggests that the presence of BLV is necessary to induce tumorization.

[3] Most tumors are a rather homogeneous lymphoid cell population, harbouring one to four copies of proviral DNA. Tumors appear to be monoclonal with respect to proviral integration sites (12). The integration sites, however, are not conserved from one tumor to the next and different tumors harbour the provirus in different chromosomes (16).

[4] Whether unique or multiple, the proviral copy(ies) may be complete or deleted. Deletions affecting the 3' end have not been encountered so far.

[5] BLV-induced tumors display extensive chromosomal abnormalities. Aneuploidy occurs frequently, but no consistent abnormality has been described so far (17-19).

[6] BLV proviruses are present in tumours in a repressed state. In contrast to PL lymphocytes, BLV transcription could not be induced to detectable levels after in vitro culture of several tumors and establishment of pure tumor cell lines (20).

BLV, like HTLV, contains a tat sequence (trans-acting transcriptional activation) at its 3' end. Expression of BLV genes, probably through expression of the tat activity, triggers the transcription of genes controlled by the BLV LTR enhancer-promoter system, in a trans-acting way (21). It is not known whether the tat protein is directly or indirectly involved in this regulation.

The discovery of a gene totally exogenous to the host genome, but with the ability to interfere with the viral promoter-enhancer, has led to the suggestion that tat might also be able to activate other cellular genes. A new model for leukaemogenesis was developed, in which the erroneous expression of cellular genes (proto-oncogenes or other genes involved in cell cycle regulation) is considered as the first step towards malignancy (7). The IL-2 receptor gene is a good candidate for a cellular target gene of the tat protein: the expression of the IL-2 receptor protein is systematically increased in human T cells following in vitro HTLV-I infection, concomitantly with the acquisition of long term in vitro growth properties (22). However, IL-2 receptor expression alone is unlikely to direct the progression of these cells towards a malignant phenotype, as suggested by the long latency period befor tumor appearance and the monoclonal nature of these tumors. In vitro infection of bovine lymphocytes by BLV could not be achieved so far, and it is thus difficult to assess the direct influence of BLV infection on the growth of its target cells. However, the stimulating effect of BLV infection is illustrated by the increase in the B lymphocyte count during persistent lymphocytosis.

In vivo BLV tumors have given several clues as to the mechanism of leukaemogenesis: i) BLV proviral integration is necessary, but not sufficient to direct tumor development, ii) the site of BLV integration in the cellular genome is not relevant for leukaemogenesis, iii) the expression of BLV genes, including the tat gene, is not necessary to maintain the transformed phenotype, iv) BLV-induced tumors are characterized by extensive abnormalities in chromosome count (aneuploidy).

Several mechanism can be envisaged to be responsible for the abnormal chromosome pattern: If aneuploidy is the only aberrant feature, the polykaryocyte-inducing activity of BLV proteins (expressed during the early stages of BLV infection) could be the triggering factor. The discovery of the tat activity, however, points to another interesting possibility: erroneous expression of proteins involved in DNA replication (e.g.: unbalanced expression of histone genes) can impair the correct chromosome replication and division process and has been shown to lead to chromosomal abnormalities in yeast (23,24). The appearance of chromosome abnormalities, in turn, has been linked with progression of the tumor phenotype towards metastasis in *in vivo* tumors (25). We propose to link these observations with the transcription regulation effect of the tat gene product: The tat protein could itself interfere with the expression rate of proteins involved in DNA replication or other proteins triggering chromosome defects. "Chromosome anarchy" resulting from this perturbation could give rise to the statistically rare occurrence of cells in which genes involved in growth control are sufficiently deregulated or altered to produce cells growing as tumors *in vivo*.

This model, although lacking experimental confirmation, responds to the criteria imposed by the analysis of BLV-induced tumors: the presence of the virus is indispensable to initiate the tumor process, but its expression is not required once the first events have been achieved; the statistically rare occurrence of a complex series of events leads to monoclonal tumors occurring after a long latency; the resulting tumor contains extensive chromosomal abnormalities.

Chromosome aneuploidy has been previously described in human chronic lymphoid leukemias (26) and specific chromosomal rearrangements have been shown to be associated with tumor development in a variety of human leukemias: examples are the Philadelphia chromosome translocation in chronic myelocytic lymphoma (CML) (27), a translocation involving the myc proto-oncogene and the immunoglobulin genes in Burkitt lymphoma (28), the bcl1 and bcl2-specific chromosome breaks in B cell lymphomas (29). The extensive abnormalities described in BLV tumors display no obvious degree of consistency and, although aneuploidy is frequent, translocations have not been described. These date require reassessment since bovine karyotypes are difficult to interpret and chromosome banding experiments were not performed. Moreover, a relatively complex pattern of abnormalities is expected as a result of the nonspecific nature of the supposed agent of chromosome damage. Both aneuploidy (gene dosage effect), and translocations could result in altered expression of one or several genes regulating cell growth.

A more detailed karyotype analysis of a great number of BLV- and HTLV-induced tumors is necessary to assess the validity of this hypothesis in these two viral systems. The HTLV system moreover offers the availability of an *in vitro* infection system which allows to follow the early events of leukemia induction.

MOLECULAR CHARACTERIZATION OF THE BLV ENVELOPE GLYCOPROTEIN

Although BLV infected lymphocytes do not express detectable levels of viral antigens *in vivo* and thus escape immune detection, passive administration of colostral antibodies to young sheep is protective against simultaneous BLV infection (30). This suggests that efficient vaccination against primary BLV infection is possible. The viral envelope glycoproteins gp30 and gp51 are the best candidates for this purpose, since they are responsible for receptor binding and hence for the induction of virus-neutralizing antibodies.

In order to characterize the epitope composition of the external glycoprotein gp51, monoclonal antibodies to this protein were obtained. Competition antibody binding assays have shown that these antibodies identify eight independent antigenic sites on the gp51 molecule (sites A through H) (31). These epitopes are also apparent on the 72,000 dalton gp51-gp30 precursor molecule. Three of these antigenic sites (sites F, G and H) were found to be recognized by virus-neutralizing antibodies. Limited proteolytic digestion of gp51 with urokinase and immunoprecipitation studies with the different monoclonal antibodies have shown that antigenic sites F, G and H, together with site E, are part of a 15,000 hypoglycosylated fragment of gp51 (32). The remaining sites A, B, C and D are on a heavily glycosylated fragment of 30,000 daltons. The analysis of the amino acid sequence derived from the nucleotide sequence of the env gene (33) allowed to attribute sites A, B, C and D to the carboxyterminal half of gp51 (this segment contains the vast majority of the potential glycosylation sites), whereas epitopes E, F, G and H were localized at the aminoterminus (this segment contains many tyrosine residues, explaining why the 15,000 daltons urokinase cleavage product is heavily labeled by tyrosine-specific iodination). The virus neutralizing sites F, G and H are highly conformational: the binding activity of the virus-neutralizing antibodies is lost after disruption of the disulfide bonds within gp51. Moreover, the nonglycosylated protein moiety of the gPr72env precursor synthesized after tunicamycin treatment of BLV-producing cells is not immunoprecipitable by the virus-neutralizing monoclonal antibodies, although the 5 other antigenic sites remain intact (34). It is interesting to note that epitopes F, G and H are on the poorly glycosylated subfragment of gp51. These results emphasize the importance of the tertiary structure and carbohydrate moiety for the reactivity with our virus-neutralizing monoclonal antibodies. They do, however, not prove that the integrity of the tertiary structure of gp51 is required for the elicitation of virus-neutralizing antibodies, since only a limited number of epitopes were studied.

Anti-gp51 monoclonal antibodies were used to analyse antigenic variation among BLV isolates of different geographical origins (34). 4 different virus isolates, among them the standard FLK-BLV isolate against which the antibodies were raised, were tested for their reactivity with the monoclonal reagents. These different isolates showed extensive variation of their reactivity with the virus-neutralizing antibodies: one isolate, originating from an American cow with persistent lymphocytosis (VdM), had lost epitope F (phenotype F-G+H+), two other isolates from a Belgian (MM285) and a French case were missing epitopes G and H (F+G-H-). None of the isolates analysed in our study had lost all three epitopes. No variation was detected within one herd. Interestingly, in spite of these differences, sera from BLV infected animals of many different origins contain cross-neutralizing antibodies.

The sequence of cloned env genes of three different BLV isolates was obtained and compared to the published sequence of the reference BLV isolate (FLK-BLV). Two of these isolates were MM285 (F+G-H-) and VdM (F-G+H+). The third clone was derived from tumor lymphocytes of a Belgian cow and its phenotype is unknown. Each of the three isolates differed from the FLK-BLV gp51 sequence by several point mutations. However, it was possible to tentatively attribute the loss of epitope F in VdM to a point mutation at residue 62, replacing a glutamine residue by a highly basic lysine residue (35). Experiments attempting to confirm epitope localization with use of synthetic peptides are underway.

REFERENCES

1. R. Kettmann, D. Protetelle, M. Mammerickx, Y. Cleuter, D. Dekegel, M. Galoux, J. Ghysdael, A. Burny, and H. Chantrenne, Bovine Leukemia Virus: an exogenous RNA Oncogenic Virus, Proc. Nat. Acad.

Sci. USA, 73:1014-1018 (1976).
2. B. J. Poiesz, F. W. Ruscetti, A. F. Gazdar, P. A. Bunn, J. D. Minna, and R. C. Gallo, Detection and isolation of type C retrovirus particles from fresh and cultured lymphocytes of a patient with cutaneous T cell lymphoma, Proc. Nat. Acad. Sci. USA, 77:7415-7419 (1980).
3. P. J. Kanki, R. Kurth, W. Becker, G. Dressman, M. F. McLane, and M. Essex, Antibodies to Simian T-lymphotropic retroviruses type III in African Green Monkeys and recognition of STLV-III viral proteins by AIDS and related sera, Lancet, I:1330-1332 (1985).
4. P. J. Kanki, M. F. McLane, N. W. King, N. L. Letvin, R. D. Hunt, P. Sehgal, M. D. Daniel, R. C. Desrosiers, and M. Essex, Serologic identification and characterization of a macaque T-lymphotropic retrovirus closely related to HTLV-III, Science, 228:1199-1201 (1985).
5. N. L. Letvin, M. D. Daniel, P. K. Sehgal, R. C. Desrosiers, R. D. Hunt, L. M. Waldron, J. J. Mac Key, D. K. Schmidt, L.V. Chalifoux, and N. W. King, Induction of AIDS-Like Disease in Macaque Monkeys with T-Cell Tropic Retrovirus STLV-III, Science, 230:71-73 (1985)
6. D. Soniqo, M. Alizon, K. Staskus, D. Klatzmann, S. Cole, O. Danos, E. Retzel, P. Tidlais, A. Haase, and S. Wain-Hobson, Nucleotide Sequence of the Visna Lentivirus: Relationship to the AIDS virus, Cell, 42:369-382 (1985).
7. J. G. Sodroski, C. A. Rosen, and W. A. Haseltine, Trans-acting transcriptional activation of the long terminal repeat of human T-lymphotropic viruses in infected cells, Science, 225:381-385 (1985).
8. F. Wong-Staal, and R. C. Gallo, Human T-lymphotropic retroviruses, Nature, 317:395-403 (1985).
9. M. J. Van der Maaten, and J. M. Miller, Susceptibility of cattle to BLV infection by various routes of exposure, in: "Advances in Comparative Leukemia Research", P. Bentvenltzen, ed., Elsevier, Amsterdam (1978).
10. D. Protetelle, M. Mammerickx, F. Bex, A. Burny, Y. Cleuter, D. Dekegel, J. Ghysdael, R. Kettmann, and H. Chantrenne, Purification of BLV pg70 and p24. Detection by radioimmunoassay of antibodies directed against these antigens, in: "Bovine Leukosis: Various Methods of Molecular Virology", A. Burny, ed., CEC, Luxembourg, p. 131-152 (1977).
11. J. F. Ferrer, R. R. Marshak, D. A. Abt, and S. J. Kenyon, Persistent lymphocytosis in cattle: its cause, nature and relation to lympho-sarcoma, Ann. Rech. Vet., 9:851-857 (1978).
12. R. Kettmann, G. Marbaix, Y. Cleuter, D. Portetelle, M. Mammerickx, and A. Burny, Genomic integration of BLV provirus and lack of viral expression in the target cells of cattle with different responses to BLV infection, Leukemia Res., 4:509-519 (1980).
13. P. Gupta, S. V. S. Kashmiri, and J. F. Ferrer, Transcriptional control of the bovine leukemia virus genome: role and characterization of a non-immunoglobulin plasma protein from BLV-infected cattle, J. Virol., 50:267-270 (1984).
14. A. Burny, C. Bruck, H. Chantrenne, Y. Cleuter, D. Dekegel, J. Ghysdael, R. Kettmann, M. Leclercq, J. Leunen, M. Mammerickx, and D. Portetelle, Bovine Leukemia Virus: Molecular Biology and Epidemiology, in: "Viral Oncology", G. Klein, ed., Raven Press, New York (1980).
15. A. L. Parodi, M. Mialot, F. Crespeau, D. Levy, H. Salmon, G. Nogues, and R. Girard-Marchand, Attempt for a new cytological and cyto-immunological classification of bovine malignant lymphoma (BML) (Lymphosarcoma), Current Topics in Vet. Med. Anim. Sci., 15: 561-572 (1982).
16. D. Gregoire, D. Couez, J. Deschamps, S. Heuertz, M. C. Hors-Cayla,

J. Szpirer, C. Szpirer, A. Burny, G. Huez, and R. Kettmann, Different bovine leukemia virus-induced tumors harbor the provirus in different chromosomes, J. Virol., 50:275-279 (1984).
17. W. C. D. Hare, and R. A. McFeeley, Chromosome abnormalities in lymphatic leukemia in cattle, Nature, 209:108-110 (1966).
18. W. C. D. Hare, R. A. McFeeley, D. A. Abt, and J. R. Feierman, Chromosomal studies in bovine lymphosarcoma, J. Nat. Cancer Inst., 33:105-118 (1964).
19. W. C. D. Hare, T. J. Yang, and R. A. McFeeley, A survey of chromosome findings in 47 cases of bovine lymphosarcoma (leukemia), J. Nat. Cancer Inst., 38:383-392 (1967).
20. R. Kettmann, Y. Cleuter, D. Gregoire, and A. Burny, Role of the 3' Long Open Reading Frame Region of Bovine Leukemia Virus in the Maintenance of Cell Transformation, J. Virol., 54:899-901 (1985).
21. C. A. Rosen, J.G. Sosroki, R. Kettmann, A. Burny, and W.A. Haseltine, Transactivation of the bovine leukemia virus long terminal repeat in BLV-infected cells, Science, 227:320-322 (1985).
22. B. J. Poiesz, F. W. Ruscetti, J. W. Mier, A. M. Woods, and R. C. Gallo, T-cell lines established from human T lymphocytic neoplasias by direct response to T-cell growth factors, Proc. Nat. Acad. Sci. USA, 77:6815-6819 (1980).
23. L. H. Hartwell, and D. Smith, Altered fidelity of mitotic chromosome transmission in cell cycle mutants of Saccharomyces cerevisiae, Genetics, 110:381-395 (1985).
24. L. H. Hartwell, Communication at the "Oncogene Symposium", Boston, May 13, 1985.
25. F. Gilbert, Chromosome abnormality, gene amplification and tumor progression, Progr. Clin. Biol. Res., 175:151-159 (1985).
26. J. N. Hurley, ShuMan Fu, H. G. Kunkel, R. S. K. Chaganti, and J. German, Chromosome abnormalities of leukemic B lymphocytes in chronic lymphocytic leukemia, Nature, 283:76-78 (1980).
27. N. Heisterkamp, J. R. Stephenson, J. Groffen, P. F. Hansen, A. de Klein, C. R. Bartram, and G. Grosveld, Localization of the c-abl oncogene adjacent to a translocation break point in chronic myelocytic leukemia, Nature, 306:239-242 (1983).
28. R. Taub, I. Kirsch, C. Morton, G. Lenoir, D. Swan, S. Tronick, S. Aaronson, and P. Leder, Translocation of the c-myc gene into the immunoglobulin heavy chain locus of human Burkitt lymphoma and murine plasmacytoma cells, Proc. Nat. Acad. Sci. USA, 79: 7837-7841 (1982).
29a. Y.Tsujimoto, J. Yunis, L. Onorato-Showe, J. Erikson, P. C. Nowell, and C. Croce, Molecular Cloning of the chromosomal break-point of B cell lymphomas and leukemias with the t (11;14) chromosome translocation, Science, 224:1403-1406 (1984).
29b. Y.Tsujimoto, L. R. Finger, J. Yunis, P. C. Nowell, and C. Croce, Cloning of the chromosome break-point of neoplastic B cells with the (14;18) chromosome translocation. Science, 226:1097-1099 (1984).
30. M. Mammerickx, D. Portetelle, A. Burny, and J. Leunen, Zentralblt. Veterinärmed., B27:291-303 (1980).
31. C. Bruck, S. Mathot, D. Portetelle, C. Berte, J. D. Franssen, P. Herion, and A. Burny, Monoclonal antibodies define eight independent antigenic regions on the bovine leukemia virus (BLV) envelope glycoprotein gp51, Virology, 122:342-352 (1982).
32. C. Bruck, D. Portetelle, A. Burny, and J. Zavada, Topographical analysis of monoclonal antibodies of BLV gp51 epitopes involved in viral functions, Virology, 122:353-362 (1982).
33. N. R. Rice, R. M. Stephens, D. Couez, J. Deschamps, R. Kettmann, A. Burny, and R. V. Gilden, The nucleotide sequence of the env gene and post-env region of BLV, Virology, 138:82-93 (1984).

34. C. Bruck, N. Rensonnet, D. Portetelle, Y. Cleuter, M. Mammerickx, A. Burny, R. Mamoun, B. Guillemain, M. Van der Maaten, and J. Ghysdael, Biologically active epitopes of BLV gp51: their dependence on protein glycosylation and genetic variability, <u>Virology</u>, 136:20-32 (1984).
35. D. Couez, Manuscript in preparation.

HUMAN T-LYMPHOTROPIC RETROVIRUSES AND THEIR ROLE IN HUMAN DISEASES

Robert C. Gallo and Flossie Wong-Staal

NIH, National Institute of Cancer
Laboratory of Tumor Cell Biology
Bethesda, Maryland 20205

INTRODUCTION

Information on a group of human retroviruses called Human T-lymphotropic Virus (HTLV) has rapidly amassed in the last few years (1). These are horizontally transmitted viruses which infect the subset of lymphocytes expressing helper function and the OKT4 marker. HTLV-I and HTLV-III are the etiologic agents of, respectively, adult T-cell leukemia (ATL) (2) and acquired immunodeficiency syndrome (AIDS) (3). Discovery of these viruses and the identification of their pathogenic roles represent the culmination of efforts in the preceeding decade directed at detection of retroviruses in humans, particularly in neoplasms of hematopoietic origin. In 1976, this laboratory identified a factor in the culture medium of mitogen-stimulated human lymphocytes which supported the long term growth of human T-cells (4). This factor, called T-cell growth factor or interleukin-2 (IL-2), was used to grow normal and neoplastic T-cells. Since normal T-cells required activation by mitogen or antigen to respond to IL-2 and T-cells from ATL patients responded directly to IL-2 without prior activation, it was possible to selectively grow neoplastic T-cells in culture and study them for the presence of retroviruses. This led first to identification and isolation of HTLV-I from ATL patients. Subsequently, using a similar protocol, HTLV-II (5) was obtained from a patient with T-cell Hairy Cell leukemia. We later showed that HTLV-I and HTLV-II infected cells constitutively expressed high levels of IL-2 receptors. When the disease AIDS was first recognized, its many parallels with HTLV-I related disorders led us to believe that a retrovirus of the same general family would be the etiological agent. The parallels included the modes of transmission, origin in the African continent, impairment of the immune system, and a likely defect in the same target cell, namely the OKT4+ helper T lymphocyte. This idea was proposed by us at the time that many other theories were in vogue, including cyclosporin-like fungal products, mutant viruses and non-specific antigen activiation. In 1983, Barre-Sinoussi reported the identification of a retrovirus from a lymph node of a patient with lymphadenopathy by culturing the cells in the presence of IL-2. Failure to obtain a permanent producing cell line to mass produce the virus had impeded further characterizing the virus and linking it to the disease. In the meantime, our laboratory had similar observations and similar difficulties in analysis of retroviruses in samples from patients with AIDS or AIDS-related complex (ARC). We knew that retroviruses not identified as HTLV-I or -II were frequently detected, but only transiently. To compound the problem, authentic HTLV-I and -II isolates were also obtained occasionally from these patients. The confusion was resolved when M. Popovic in our laboratory was able to transmit the virus to a permanently growing T-cell line which then became high virus producers (6). This allowed extensive seroepidemiological studies, repeated isolates to be characterized, and molecular cloning of the virus genome to be done (8). These studies definitively showed that a novel virus, which we called HTLV-III, was the etiological agent of AIDS.

THE TARGET CELL OF HTLV AND IMPAIRMENT OF IMMUNE FUNCTION BY VIRUS INFECTION

The most striking feature common to all three HTLVs is their remarkable lymphotropism, particularly for the OKT4+ lymphocyte, although the cellular receptor recognized by HTLV-I and -II is different from that of HTLV-III. An abnormal property of T-cells infected with HTLV-I is that they constitutively express IL-2 receptors which cannot be down regulated. The constitutive expression of IL-receptor is most likely induced by a transacting viral gene product (see Wong-Stall, this volume) and may be a key event in the immortalization of T-cells by HTLV-I and -II. Infection with HTLV-I and -II not only results in immortalization, but also in suppression or abrogation of immune processes in vivo and in vitro. Patients in infectious disease wards in endemic regions of Japan have a greater level of seropositivity for HTLV-I than the general population, and opportunistic infections (including Pneumocystis carinii) are often noted in ATL patients. Infection of immunocompetent T-cell lines in vitro resulted in the loss of various T-cell functions, including the loss of cytotoxicity, the ability to recognize specific antigens and macrophage dependent antigen recognition, as well as the gain of an indiscriminate response to all class II HLA allotypes (9). These immune suppressive effects may explain the observations that ATL-derived T4+ T-cells lacked helper function and that HTLV-I infected T-cells induced polyclonal B-cell activitation. The loss of specific antigen recognition does not appear to involve abnormal expression of the T-cell receptor α- or β-chains. The immunosuppressive effects on T-cells and macrophage function in vitro could be in part mediated by the trans-membrane envelope protein by analogy to the profound immuno-suppressive effects of the transmembrane protein of murine and feline leukemia viruses.

There is no obvious defect in IL-2 or IL-2 receptor expression in HTLV-III infected cells. The basis for immunosuppression in AIDS or ARC patients may be two fold: first, the virus specifically kills the T4+ helper T-lymphocyte so that the infected individuals are characteristically depleted in this cell population. One of the early symptoms of ARC and AIDS is the inversion of T4/T8 ratio in the peripheral blood so that the T8 cells predominate. Second, the remaining T4 lymphocytes may be functionally defective as well. We have shown that the cytopathic activity of the virus is intrinsic to its genome (10). Since HTLV-III is a complex virus with at least six genes compared to the usual three for retroviruses, it will take some time to sort out the relative contributions of viral and cellular factors toward the cyto-pathic and immunosuppressive properties of this virus (11).

CLINICAL IMPACT OF HTLV

HTLV-I was first recognized to be endemic in the southwestern parts of Japan and etiologically associated with an aggressive form of leukemia called ATL by Takatsuki and his colleagues. Subsequently, other endemic areas were found: the Carribean islands, southeastern United States, parts of South America, pockets in southern Italy, and many regions of Africa (12). The virus is also prevalent in European emigrant populations from these areas, such as the Surinam population in Holland, and the West Indian population in England. In the endemic areas, HTLV-I associated leukemias could account for as high as 70 % of all lymphoid leukemias of adulthood. Recent studies in Japan also showed that HTLV-I was transmitted through blood transfusions. Although HTLV-II is relatively rare, increasing evidence of HTLV-II infection is noted among intravenous drug abusers in England and in patients with various forms of chronic lymphoid leukemias. The potential thus exists that these potent leukemogens can be introduced into previously uninfected populations.

The clinical impact of HTLV-III infection is even greater than originally expected. Its role as the etiological agent of AIDS is now firmly established by extensive seroepidemiology and repeated virus isolation studies. Furthermore, at least two additional aspects have been recognized from these studies. First, the virus has a broad clinical spectrum, ranging from asymptomatic infection to profound immune suppression and development of multiple opportunistic infections to secondary malignancies such as Kaposi's sarcoma and B-cell lymphoma. The demonstrations of virus infection in brain and lung tissues also provides the probable etiological link of HTLV-III to neurological disorders and pneumonia seen in AIDS and ARC patients (13). Many clinicians feel the need to redefine and unify these diseases as HTLV-III associated disorders of which AIDS is only one sector. The second aspect concerns the mode of transmission of HTLV-III. Epidemiological data suggest that the major routes of transmission are sexual contact, congenital infection and blood transfusion. However, virus iosolates have now been obtained from other sources of body fluids in addition to blood and semen: saliva, cerebral spinal fluid, plasma and tears. These findings at least suggest that the virus could be transmitted through additional means, albeit infrequently, and there is anecdotal evidence in support of this. The incidence of HTLV-III infection is increasing in both homosexual and heterosexual populations in this country and in Europe (14). In Africa and Haiti, most infected people are heterosexual. Thus, we believe that the preponderance of virus among homosexuals in U.S. and Europe could simply be due to the extent of promiscuity and the fact that the virus first entered this population in these countries. Preliminary estimates are that 5 - 10 % of HTLV-III seropositivity individuals develop AIDS each year, and a higher rate develop other HTLV-III associated disorders. Considering the number of infected people has already been conservatively estimated at one million, public the health impact of this virus is extensive. There are three major fronts in our efforts to contain virus infection and pathogenesis. First, diagnostic reagents have been developed for blood bank assays to remove contaminated blood products. Test kits utilizing whole disrupted virus can be obtained commercially, but it is conceivable that these will be replaced by second generation diagnostic kits using proteins expressed by recombinant DNA in the near future (see Wong-Staal, this volume). Second, potential vaccine reagents are being developed in multiple systems and tested in non-human primate systems. Within one to two years, we would at least know the feasibility of obtaining a broadly cross-reactive and effective vaccine. Third, therapeutic reagents are being tested for effects on virus replication in vitro and for suppression of disease in vivo. Most of these drugs are specific inhibitors of reverse transcriptase, and several of these look promising in their efficacy in blocking virus replication.

CONCLUSION

The human T-lymphotropic retroviruses are a unique group of viruses which have received increasing attention from virologists, immunologists, epidemiologists, clinicians and molecular biologists alike. They are the first and to-date only human retroviruses isolated and are associated with two of the most aggressive and deadly diseases known to man. Their remarkable effects on T-cells (immortalization and cell death) and their novel mechanism of transcriptional regulation (TAT, see Wong-Staal, this volume) also provide novel models for the study of T-cell function and gene regulation. The most pressing priority of course is to meet the medical challenges posed by these viruses. Onces these are successfully dealt with, then perhaps the study of these viruses will have additional value as tools for elucidating the basic mechanisms of some important cellular functions.

REFERENCES

1. F. Wong-Staal, and R. C. Gallo, Human T-lymphotropic retroviruses, Nature, 317:395 (1985).
2. B. J. Poiesz, F. W. Ruscetti, A. F. Gazdar, P. A. Bunn, J. D. Minna, and R. C. Gallo, Detection and isolation of type-C retrovirus particles from fresh and cultured lymphocytes of a patient with cutaneous T-cell lymphoma, Proc. Nat. Acad. Sci. USA, 77:7417 (1980).
3. S. Broder, and R. C. Gallo, A pathogenic retrovirus (HTLV-III) linked to AIDS, N. Engl. J. Med., 311:1292 (1984).
4. M. Robert-Guroff, M. G. Sarngadharan, and R. C. Gallo, T-cell growth factor, in: "Growth and Maturation Factors", G. Guroff (ed.), Vol. 2, John Wiley and Sons, Inc., New York, p. 267 (1984).
5. V. S. Kalyanaraman, M. G. Sarngadharan, M. Robert-Guroff, I. Miyoshi, D. Blayney, D. Golde, and R. C. Gallo, A new subtype of human T-cell leukemia virus (HTLV-II) associated with a T-cell variant of Hairy Cell Leukemia, Science, 218:571 (1982).
6. M. Popovic, M. G. Sarngadharan, E. Read, and R. C. Gallo, Detection, isolation, and continuous production of cytopathic retroviruses (HTLV-III) from patients with AIDS and pre-AIDS, Science, 224:497 (1984).
7. R. C. Gallo, S. Z. Salahuddin, M. Popovic, G. M. Shearer, M. Kaplan, B. F. Haynes, T. J. Palker, R. Redfield, J. Oleske, B. Safai, G. White, P. Foster, and P. D. Markham, Frequent detection and isolation of cytopathic retroviruses (HTLV-III) from patients with AIDS and at risk for AIDS, Science, 224:500 (1984).
8. G. M. Shaw, B. H. Hahn, S. K. Arya, J. E. Groopman, R. C. Gallo, and F. Wong-Staal, Molecular characterization of human T-cell leukemia (lymphotropic) virus type III in the acquired immune deficiency syndrome, Science, 226:1165 (1984).
9. D. Volkman, M. Popovic, R. C. Gallo, and A. Fauci, Human T-cell leukemia/lymphoma virus-infected antigen-specific T-cell clones: indiscriminant helper function and lymphokine production, J. Immunol., 134:4237 (1985).
10. A. G. Fisher, E. Collalti, L. Ratner, R. C. Gallo, and F. Wong-Staal, A molecular clone of HTLV-III with biological activity, Nature, 316:262 (1985).
11. L. Ratner, W. Haseltine, R. Patarca, K. J. Livak, B. Starcich, S. F. Josephs, E. R. Doran, J. A. Rafalski, E. A. Whitehorn, K. Baumeister, L. Ivanoff, S. R. Petteway Jr., M. L. Pearson, J. A. Lautenberger, T. S. Papas, J. Ghrayeb, N. T. Chang, R. C. Gallo, and F. Wong-Staal, Complete nucleotide sequence of the AIDS virus, HTLV-III, Nature, 313:277 (1985).

12. W. A. Blattner, V. S. Kalyanaraman, M. Robert-Guroff, T. A. Kister, D. A. G. Galton, P. S. Sarin, M. H. Crawford, D. Catovsky, M. Greaves, and R. C. Gallo, The human type-C retrovirus, HTLV, in Blacks from the Caribbean region, and relationship to adult T-cell leukemia/lymphoma, Int. J. Cancer, 30:257 (1982).
13. G. M. Shaw, M. E. Harper, B. H. Hahn, L. G. Epstein, D. C. Gajdusek, R. W. Price, B. A. Navia, C. K. Petito, C. J. O'Hara, J. E. Groopman, E.-S. Cho, J. M. Oleske, F. Wong-Staal, and R. C. Gallo, HTLV-III infection in brains of children and adults with AIDS encephalopathy, Science, 227:177 (1985).
14. J. W. Curran et al., The epidemiology of AIDS: Current Status and Future Projects, Science, 229:1352 (1985).

MOLECULAR BIOLOGY OF THE HUMAN T-LYMPHOTROPIC RETROVIRUSES

Flossie Wong-Staal

Laboratory of Tumor Cell Biology, National Cancer Institute
National Institutes of Health
Bethesda, Md. 20205

INTRODUCTION

There are only three human retroviruses identified to date. Remarkably, they share a number of common properties, most notably the tropism for the OKT4+ helper T lymphocyte. For this reason, they have been named human T-lymphotropic viruses or HTLV (see Gallo and Wong-Staal, this volume). They are also associated with two naturally occurring diseases: HTLV type I with an aggressive form of leukemia called adult T-cell leukemia (ATL), and HTLV type III with the acquired immunodeficiency syndrome (AIDS). This chapter will summarize the molecular biology of this group of viruses, including both our current concept of the molecular mechanism of pathogenesis of these viruses as well as application of some of these studies for development of diagnostic and preventive measures.

The HTLVs belong to a subgroup of retroviruses with an unique property: the presence of a viral gene (tat) which encodes a protein that activates transcription (see Wong-Staal and Gallo, 1985). Members of this group can be further divided into two subgroups (Fig. 1): one represents the transforming viruses or viruses that induce excessive cellular proliferation. This includes HTLV-I, the agent associated with adult T-cell leukemia; STLV-I, a Simian virus closely related to HTLV-I found in many old world monkey species, which is also associated with a T-cell lymphoma in macaques; HTLV-II was first isolated from a T-cell hairy cell leukemia patient, but is probably the only virus in this group that has not been shown to be etiologically associated with a naturally occuring disease; bovine leukemia virus (BLV) induces enzootic bovine leukosis, and is distantly related to HTLV-I. The other subgroup represents the cytopathic retroviruses and includes HTLV-III, the AIDS agent; STLV-III, a related virus found in macaques and African green monkeys which is also associated with an AIDS-like disease in macaques (see Kanki et al., 1985) and the more distantly related ungulate lentiviruses (e.g., visna, equine infectious anemia virus, and caprine arthritis and encephalitis virus). A number of parallels exist between HTLV-III and the ungulate lentivirus, particularly visna virus, including persistence of unintegrated viral DNA (Shaw et al., 1984) dual lympho- and neurotropism (Shaw et al., 1985) similar morphology and distant nucleotide sequence homology (Gonda et al., 1985) and a similar genomic organization (Sonigo et al., 1985). Based on these observations, we and others have concluded that HTLV-III is a human lentivirus. The presence of a transacting viral protein (TAT) for transcriptional activation was first observed for HTLV-I and HTLV-II, and now extended to HTLV-III, STLV-I, BLV, and the ungulate lentiviruses. Although not yet tested, we believe STLV-III will also have such a gene function. It is speculated that the tat protein not only regulates transcription of viral genes, but also plays a role in the biological activities of this group of viruses.

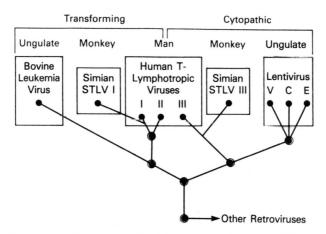

Fig. 1. Family tree of transactivating retroviruses. STLV-I and STLV-III: Simian T-lymphotropic virus types I and III. V = visna virus; E = equine infectious anemia virus; C = caprine arthritis and encephalitus virus.

Structures of the HTLV Genomes

The genomes of one or more isolates of each subgroup of HTLV have been cloned and their complete nucleotide sequence determined. Fig. 2 presents their genomic organizations as compared to two other retroviruses, namely, Moloney murine leukemia virus and Rous Sarcoma virus. HTLV-I and HTLV-II are essential identical in their organization and exhibit extensive homology throughout their genomes. HTLV-III has a distinct organization. However, all three HTLVs share some common features. First, the gag genes are small and code for only three proteins, lacking a small phosphoprotein which usually interpolates between the amino-terminal protein and the major capsid protein (p24 for the HTLVs). Second, all HTLVs contain in addition to the three viral

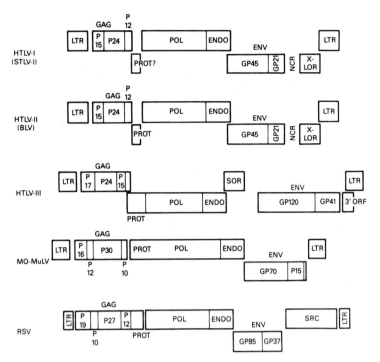

Fig. 2. Genetic structures of retroviruses. HTLV, human T-lymphotropic virus; BLV, bovine leukemia virus; MoMuLV, Moloney murine leukemia virus; RSV, Rous sarcoma virus; ENDO, endonuclease; PROT, pX; SOR, short open reading frame; 3' ORF, 3' open reading frame.

replicative genes gag, pol and env, one or more genes that are not cell derived. HTLV-I and HTLV-II contain a region between the env gene and 3' LTR which has a single highly conserved open reading frame designated x-lor (Seike et al., 1983; Haseltine et al. 1984). Based on nucleotide sequence analysis, HTLV-III has at least two additional genes. One (termed sor) is immediately located after pol and capable of encoding a protein of 203 amino acids; another (3' orf) is located after the env gene and overlaps with the 3' LTR (Ratner et al., 1985a). It has a maximal coding capacity of 216 amino acids. Protein products derived from both sor and 3' orf have been shown to react with some AIDS patient's sera (unpublished date with G. Franchini et al., N. Kan et al.), suggesting that these are indeed functional genes.

Demonstration that x-lor of HTLV-I Encodes the tat Function and Implication for the Mechanism of Transformation

A fragment of the HTLV-I genome containing only the x-lor gene was inserted into an expression vector utilizing the metallothionine promoter, initiator and splice donor. Mouse fibroblasts transfected with the plasmid synthesize a transcriptional activator specific for the HTLV-I LTR (Felber et al., 1985) and a 42 kD protein that is immuno-precipitable with an ATL patient's sera positive for the x-lor protein. Therefore, the x-lor gene could be equated with the tat gene by this experiment. Furthermore, the tat gene product was shown to be localized in the nuclei of the transfected cells. There is some preliminary evidence that like some of the DNA virus transforming proteins, HTLV-I tat not only regulates transcription of viral genes, but also of certain cellular genes. A model for HTLV transformation then could hypothesize that the tat gene turns on growth promoting gene(s) specific for T-cell proliferation.

Localization of the tat Function on the HTLV-III Genome

To localize the tat gene of HTLV-III, we carried out functional mapping of cDNA clones derived from mRNA of an infected cell line (H9/HTLV-expression vectors) and cotransfected into lymphoid cells (H9 or Jurkat) with a plasmid containing HTLV-III LTR sequences 5' to the bacterial chloramphenicol acetyl transferase (CAT) gene. Many cDNA clones were screened for activation of the CAT gene in this cotransfection assay and the results led to a single conclusion. For illustrative purposes, we shall focus on three of these cDNA clones (Fig. 3). Two clones, pCV-1 and pCV-3, each with about 1.8 kB inserts, correspond to mRNAs the synthesis of which involved two splicing events. This is a remarkable

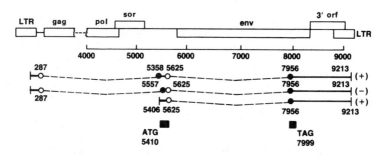

Fig. 3. Physical maps of HTLV-III cDNA clones and predicted amino acid sequence of tat open reading frame. The open and closed circles are for donor and acceptor splice sites, respectively. Overlined amino acid sequence is derived from a truncated DNA clone that is functional in transactivation (Seigel et al., in press).

coincidence with the tat-mRNA of HTLV-I, II and BLV (see Wong-Staal and Gallo, 1985 for review). The nucleotide sequence of pCV-3 has a slightly shorter second exon due to utilization of a splice acceptor site at nucleotide 5557 instead of 5358 for pCV-1. As a result, pCV-1 contains, in addition to 3' orf, a short open reading frame capable of encoding 86 amino acids. The pCV-3 clone lacks the initiator ATG, and part of the coding sequence of this open reading frame and probably corresponds to the 3' orf mRNA. Transactivation function was seen with pCV-1 and not pCV-3. Another clone, pPL12, corresponding to a partial transcript of the pCV-1 mRNA, contains this short open reading frame intact, and, it too, induces transactivation. From these analysis, we identified the tat gene to be a small tripartite gene with the major coding region located in the second exon. The predicted amino acid sequence of this gene (Fig. 3) showed a highly basic protein with an arginine-lysine rich stretch (8/9 residues) which may be important for nuclear transport or DNA binding of this protein. Deletion of the coding sequences of the carboxyl end up to and including this arginine-lysine rich region did not remove tha transactivating function (Sodroski et al., 1985; Seigel et al., in press). The complete tat protein of HTLV-III has now been expressed in bacteria and purified (Aldovini et al., unpublished) so that direct functional tests such as DNA binding properties and capacity for transcriptional activation in reconstituted systems are now feasible.

A Molecular Clone of HTLV-III with Cytopathic Activity

As a first step to dissect the genetic determinants of the cytopathic activity of HLTV-III, we have obtained a cloned virus genome which directs the synthesis of infectious and cytopathic virions (Fisher et al., 1985). Normal cord blood T cells transfected with this clone (HXB-2) contained viral DNA, expressed p15 and p24, extracellular reverse transcriptase, and complete virions with the characteristic cyclindrical cores as detected by electron microscopy. The percentage of virus infected cells increased steadily up to 18 days post transfection, suggesting that infectious virus particles were made. However, after 20 days, there is a drop in cell viability, rapidly leading to depletion of infected cells. At 30 days, the culture consisted of only uninfected cells. Furthermore, while more than half the cells were OKT4+ at Day 0, greater than 90 % of cells are of the OKT8+ phenotype at day 30, indicating selective killing of the OKT4+ cells. Therefore, the HTLV-III genome alone contains all the essential information for inducing the primary defect in AIDS, i.e., depletion of the T4+ helper cells. It is now possible to systematically manipulate this DNA genome to localize the genes responsible for these biological activities.

Expression of HTLV-III Antigens as Diagnostic Reagents

Once the etiological agent of AIDS has been identified, there is an urgent need to develop a blood bank assay to remove the hazards of blood transfusion. Although such assays with whole disrupted virus are available, increased sensitivitiy, economy and safety in implementation of these assays can be achieved through recombinant DNA technology. In collaboration with N. Chang and co-workers, we have expressed various portions of the HTLV-III genome in open reading frame vectors (Chang et al., 1985), and showed that peptides derived from gag, pol, env and 3' orf regions are immunoreactive and can be used potentially as diagnostic reagents. A bacterially expressed peptide derived from the transmembrane portion of the env gene has been purified and adapted to a solid phase immunoassay system (Chang et al., in press). The sensitivity of this assay far surpassed that using whole disrupted virus, and has not yielded any false positives after screening over 500 normal samples. General application of this assay should be widely available soon.

Heterogeneity of HTLV-III Isolates and Prospects for a Vaccine

Genomic variation of different HTLV-III isolates was first recognized by restriction endonuclease analysis of different HTLV-III proviruses (Shaw et al., 1984). We have now analyzed over 20 consecutive HTLV-III isolates and found a different restriction enzyme pattern for each of them (Wong-Staal et al., 1985). At a nucleotide sequence level, it was also shown that genomic variability of different degrees occurs among different isolates and that the env gene is the most highly divergent (Ratner et al., 1985b). The relatedness between HTLV-III and visna virus, which is known to undergo rapid genomic variation in its envelope gene as a consequence of immune selection (Clements et al., 1980) further points toward genomic diversity as an important biological property of HTLV-III.

We have determined the nucleotide sequences of three HTLV-III isolates, two of which (BH 10 and BH 8/5) are more closely related (Ratner et al., 1985a) differing in 1 - 2 % in nucleotide sequence, while the third, obtained from a Haitian man (RF), showed more substantial difference (8 - 10 %) from BH 10 and BH 8/5 (Starchich et al., in press). Taking these sequences as well as nucleotide sequences of two additional isolates (LAV and ARV) published from other laboratories (Wain-Hobson et al., 1985; and Sanchez-Pescador et al., 1985) we have scrutinized the nature of the divergence in the env gene. As shown in Fig. 4 within

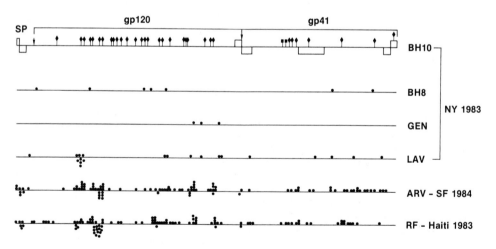

Fig. 4. Divergence of the env gene of HTLV-III, BH 10, BH 8, GEN are clones obtained from the H9/HTLV-IIIB cell line which had been infected with pooled virus from several patients. LAV was obtained from a French patient. ♦ glycosylation sites; ⊔ hydrophilic stretches; ⊓ hydrophobic stretches; ▌ cysteine residues; • nonconservative amino acid changes.

the envelope gene, the signal peptide and extracellular portion are regions that are most divergent. Furthermore, while changes in the transmembrane portion are all due to point mutations, changes in the exterior envelope sequences frequently result from insertions and deletions and appear as clustered mutations interspersed with segments that are highly conserved. Since epitopes that elicit neutralizing antibodies generally reside in the exterior glycoprotein portion, it is highly significant that there are conserved segments in this protein. These could represent antigenically important sites to be included in genetically engineered vaccine reagents.

Conclusions

The three human retroviruses that have been identified to-date share many common properties. Two particularly noteworthy common features are the tropism for the helper T-lymphocyte and the presence of a gene (tat) that regulates transcription of viral, and possibly cellular genes. HTLV-III contains in addition two noncell-derived genes of as yet undetermined functions. Furthermore, these viruses have distinct biological properties in vivo and in vitro. HTLV-I and HTLV-II are highly related, transform T cells efficiently in vitro and are associated with leukemia in vivo. In particular, HTLV-I has been unequivocally established as the etiological agent of ATL. HTLV-III on the other hand has a distinct virus genome which is more related to the ungulate lentiviruses. It is also highly cytopathic for normal T4 lymphocytes in vitro, and is the causative agent of AIDS and related disorders. The tat gene of HTLV-I has been shown to be critical for the transforming activity of this virus (Aldovini et al., in press). Elucidation of the role of the three HTLV-III specific genes (tat, sor and 3' orf) in the cytopathic effect of this virus is in progress.

Rapid progress has been made in generating diagnostic reagents using recombinant DNA technology. Second generation test kits should be available shortly, which may significantly reduce the hazards of blood transfusions. Optimism for an effective vaccine is somewhat tempered by the finding of heterogeneity among HTLV-III isolates, particularly in the env gene. However, we have also identified highly conserved regions in this gene, which probably represent functionally important domains (for example, that wich is responsible for binding to the cellular receptor). It is our hope that at least some epitopes that will elicit a neutralizing antibody response in infected individuals will reside in these conserved regions.

ACKNOWLEDGEMENTS

I am grateful to all my colleagues at the Laboratory of Tumor Cell Biology for primary contributions to this work, helpful discussions and moral support, in particular Drs. R. C. Gallo, G. M. Shaw, B. H. Hahn, S. K. Arya, G. Franchini, A. Aldovini, B. Starcich and S. Josephs.

REFERENCES

Aldovini, A., DeRossi, A., Feinberg, M., Wong-Staal, F., and Franchini, G., 1985, Molecular analysis of HTLV-I deletion mutant provirus: evidence for a double spliced x-lor mRNA, Proc. Nat. Acad. Sci. USA, in press.

Arya, S. K., Chan, G., Josephs, S. F., and Wong-Staal, F., 1985, Transactivator gene of human T-cell leukemia (lymphotropic) virus type III (HTLV-III), Science, 229:69.

Chang, N. T., Chandra, P. K., Barne, A. D., McKinney, S., Rhodes, D. P., Tam, S. H., Shearman, C., Huang, J., Chang, T. W., Gallo, R. C., and Wong-Staal, F., 1985, Expression of E. coli of open reading frame gene segments of type III human T-cell lymphotropic virus, Science, 228:93.

Chang, T. W. Kato, I., McKinney, S., Chanda, P., Barone, A. D., Wong-Staal, F., Gallo, R. C., and Chang, N. T., 1985, Detection of human T-cell lymphotropic virus (HTLV-II) infection with a sensitive and specific immunoassay employing a recombinant E. coli derived viral antigenic peptide, Biotech., in press.

Clements, J. E., Pedersen, F. S. Narayan, O., and Haseltine, W. A., 1980, Genomic changes associated with antigenic variation of visna virus during persistent infection, Proc. Nat. Acad. Sci. USA, 77:4454.

Felber, B., Paskalis, H. Kleinman-Ewing, C., Wong-Staal, F., and Pavlakis, G., 1985, The pX protein of HTLV-I is a transcriptional activator of its LTR, Science, 229:675.

Fisher, A. M., Collalti, E., Ratner, L., Gallo, R. C., and Wong-Staal, F., 1985, A molecular clone of HTLV-III with biological activity, Nature (London), 316:262.

Gonda, M. A., Wong-Staal, F., Gallo, R. C., Clements, J. E., Narayan, O., and Gilden, R. V., 1985, Sequence homology and morphologic similarities of HTLV-III and visna virus, a pathogenic lentivirus, Science, 227:173.

Hahn, B. H. Shaw, G. M., Arya, S. K. Popovic, M., Gallo, R. C., and Wong-Staal, F., 1984, Molecular cloning and characterization of the HTLV-III virus associated AIDS, Nature (London), 312:166.

Hahn, B. H. Gonda, M. A., Shaw, G. M., Popovic, M., Hoxie, J., Gallo, R. C., and Wong-Staal, F., 1985, Genomic diversity of the AIDS virus HTLV-III: Different viruses exhibit greatest divergence in their envelope genes, Proc. Nat. Acad. Sci. USA, 82:4813.

Haseltine, W. A., Sodroski, J. Patarca, R., Briggs, D., Perkins, D., and Wong-Staal, F., 1984, Structure of 3' terminal region of type II human T lymphotropic virus: evidence for new coding region, Science, 225:419.

Kanki, P. J., McLane, M. F., King, N. W., Letvin, M. L., Hunt, R. D., Sehgal, P., Daniel, M. D., Desrosier, R. C., and Essex, M., 1985, Serologic identification and characterization of a macaque T-lymphotropic retrovirus closely related to HTLV-III, Science, 228:1199.

Ratner, L., Haseltine, W., Patarca, R., Livak, K. Starcich, B., Josephs, S., Doran, E. R., Rafalski, J. A., Whitehorn, E. A., Baumeister, K., Ivanhoff, L., Petteway, Jr., S. R., Pearson, M. L., Lautenberger, J. A., Papas, T. S., Ghrayeb, J., Chang, N. T., Gallo, R. C., and Wong-Staal, F., 1985a, Complete nucleotide sequence of the AIDS virus, HTLV-III, Nature (London), 313:277.

Ratner, L. Gallo, R. C., and Wong-Staal, F., 1985b, HTLV-III, LAV and ARV are variants of the same AIDS virus, Nature (London), 313:636.

Sanchez-Pescador, R., Power, M. D., Barr, P. J., Steinmer, K. S., Stempien, M. M., Brown-Shimer, S. L., Gee, W. W., Renard, A., Randolph, A., Levy, J. A., Dina, D., and Luciw, P. A., 1985, Nucleotide sequence and expression of an AIDS-associated retrovirus (ARV-2), Science, 227:484.

Seigel, L., Ratner, L., Josephs, S. F., Derse, D., Feinberg, M., Reyes, G., O'Brien, S. J., and Wong-Staal, F., Transactivation induced by human T-lymphotropic virus type III (HTLV-III) maps to a viral sequence encoding 58 amino acids and lacks tissue specificity, Virology, in press.

Seiki, M. Hattori, S., Hirayama, and Yoshida, M., 1983, Human adult T-cell leukemia virus: complete nucleotide sequence of the provirus genome integrated in leukemia cell DNA, Proc. Nat. Acad. Sci. USA, 80:3618.

Shaw, G. M., Hahn, B. H., Arya, S. K., Groopman, J. E., Gallo, R. C., and

Wong-Staal, F., 1984, Molecular characterization of human T-cell leukemia (lymphoma) virus type III in the acquired immune deficiency syndrome, Science, 226:1165.

Shaw, G. M., Harper, M. E., Hahn, B. H., Epstein, L. G., Gajdusek, D. C., Price, R. W., Navia, B. A., Petito, C. K., O'Hara, C. J., Cho, E.-S., Oleske, J. M., Wong-Staal, F., and Gallo, R. C., 1985, HTLV-III infection in brains of children and adults with AIDS encephalopathy, Science, 227:177.

Sodroski, J., Patarca, R., Rosen, C., Wong-Staal, F., and Haseltine, W. A., 1985, Location of the trans-activating region of the genome of HTLV-III/LAV, Science, 229:74.

Sonigo, P. Alizon, M., Staskus, K., Klatzmann, D., Cole, S., Danos, O., Retzel, E., Tiollais, P., Haase, A., and Wain-Hobson, S., 1985, Nucleotide sequence of the visna lentivirus: relationship to the AIDS virus, Cell, 42:369.

Starcich, B., Hahn, B., Shaw, G. M., Modrow, S., Joseph, S. F., Wolf, H., Gallo, R. C., and Wong-Staal, F., Identification and characterization of conserved and divergent regions in the envelope gene of AIDS viruses, Cell, in press.

Wain-Hobson, S., Sonigo, P., Danos, O., Cole, S., and Alizon, M., 1985, Nucleotide sequence of the AIDS virus, LAV, Cell, 40:9.

Wong-Staal, F., and Gallo, R. C., 1985, Human T-lymphotropic retroviruses. Nature (London), 317:395.

Wong-Staal, F., Shaw, G. M., Hahn, B. H., Salahuddin, S. Z., Popovic, M., Markham, P. D., Redfield, R., and Gallo, R. C., 1985, Genomic diversity of human T-lymphotropic virus type III (HTLV-III), Science, 229:759.

ANTIVIRAL APPROACHES IN THE TREATMENT OF ACQUIRED IMMUNE DEFICIENCY SYNDROME

P. Chandra, A. Chandra, I. Demirhan and T. Gerber

Laboratory of Molecular Biology, Center of Biological Chemistry, University Medical School, Theodor-Stern-Kai 7 D-6000 Frankfurt/Main 70, FRG

1. Introduction

Acquired immune deficiency syndrome (AIDS) was first recognized in 1981 as an unexplained progressive immunodeficiency disorder associated with opportunistic infections (1) and Kaposi's sarcoma (2). Not knowing the nature of the etiological agent at that time, the therapeutic approaches were devised to treat the opportunistic infections and/or Kaposi's sarcoma; reconstitute the immunological status by bone marrow transplantation or adaptive transfer of immune competent cells; and through immunological enhancement using cytokines such as, Interleukin 2 and interferons, or immunogenic adjuvants. The identification of a virus associated to AIDS in the years 1983 (3) and 1984 (4,5,6), has provided important strategies in the diagnosis and treatment of this disease.

The virus, commonly termed as LAV (3) or HTLV-III (5,6) belongs to the class of retroviruses which contain RNA as their genetic element. The virus, HTLV-III/LAV, has a selective affinity for OKT4+ T-lymphocytes. In this respect, this virus resembles the other human retroviruses identified in lymphocytes of patients with adult T-cell leukemia (HTLV-I) and Hairy-cell leukemia (HTLV-II). These viruses were discovered in the years 1980 (7) and 1982 (8) by Gallo and his associates. For this reason, they have designated their AIDS-associated virus as HTLV-III. Other designations for the AIDS-associated virus include IDAV (Immunodeficiency-associated virus) and ARV (AIDS-related virus), discovered independently by Levy et al. (9). Adoption of an internationally acceptable name for this group of viruses has been recommended by the International Committee on the Taxonomy of Viruses (ICTV). The name proposed by this committee is HIV (10) to designate Human Immunodeficiency Virus. For the very reason that most of the cited literature has used the nomenclature HTLV-III/LAV, we will use the term HTLV-III/LAV to avoid confusion for the readers.

Unlike most of the diseases caused by viral infections, AIDS is a unique disease with a very wide spectrum of pathologic manifestations. Any consideration towards the development of a rational therapy for AIDS requires a complete knowledge about the molecular mechanisms involved in the manifestations of pathologic states involving different organs and different types of target cells. Although it is beyond the scope of this article to discuss the details of the diseases associated to AIDS, it is however, necessary to outline important features of some major pathologic manifestations in AIDS.

2. Pathologic manifestations associated to AIDS

The pathologic manifestations associated to AIDS can be divided into four major categories: i) Immunologic disorders; ii) Opportunistic infections; iii) associated neoplasms; and iv) neurologic disorders.

2.1 Immunologic disorders

The reported abnormalities of T-lymphocyte function include: selective T4-cell depletion (11-15), defective response to mitogen (15-17) and the mixed lymphocyte response (17) in bulk T-cell cultures, selective functional defects in the T4 helper/inducer lymphocyte subset (18), and defective T-cell-mediated cytotoxicity (19). Impaired production of lymphokines by T-lymphocytes from AIDS patients has been reported by Murray et al. (20). These authors have also reported a defect in the mitogen-induced production of γ-interferon in more than 70 % of AIDS patients. Other abnormalities of the immune system include: a preactivated state of monocytes manifested by elevated spontaneous interleukin-1 and prostaglandin E_2 production while lacking a normal stimulus-induced production of interleukin-1 (21), the presence of circulating (22) or inducible suppressor factors (23,24), elevated levels of α_1-thymosin (25), and decreased serum thymulin levels.

2.2 Infection in AIDS

AIDS patients are susceptible to a variety of protozoal (Peunomocystis carinii, Toxoplasma gondii, Cryptosporidium species, Entamoeba histolytica, Giardia lambia), fungal (Candida species, Cryptococcus neoformans, Histoplasma capsulatum), bacterial (Mycobacterium avium-intracellulare, Mycobacterium tuberculosis, Salmonella, Shigella), and viral (Cytomegalovirus, Herpes simplex, Herpes zoster, Epstein-Barr virus, Hepatitis viruses) infections. Although the spectrum of infections in AIDS patients from different risk groups is very similar, it is not identical. For example, infectious agents involved in members of all risk groups include P. carinii, Cryptococcus neoformans, Candida species, Toxoplasma gondii, M. avium-intracellulare, Cytomegalovirus, and Epstein-Barr virus. However, some infective agents appear to predominantly infect certain subgroups, probably due to unusual heavy exposure in the environment. Thus, M. tuberculosis and Salmonella infections are particularly common in AIDS patients from Haiti, and probably in patients from developing countries. Similarly, perirectal Herpes simplex and Cryptosporidium infections are more frequent among homosexual AIDS patients. Thus, the spectrum of infections in AIDS patients may be influenced by the environmental factors.

2.3 Associated Neoplasms

The association of neoplasms to this disease is not surprising since immunodeficiency diseases like Ataxia telangiectasia and Wiskott-Aldrich syndrome are known to have a high incidence of cancer (27). The most common type of malignancy observed in AIDS patients is the Kaposi's sarcoma, described by Moritz Kaposi in 1872 (28). Prior to the appearance of Kaposi's sarcoma in AIDS patients, this multifocal cutaneous vascular neoplasm was rare, and was found to be localized to the lower extremities of older men of mediterranean descent (29). Kaposi's sarcoma has also been reported in organ transplant recipients (30,31) and amoung young population in regions of equitorial Africa, where it is endemic (32,33).

Since the appearance of AIDS, there has been a dramatic rise in its occurence. Almost 40 % of the AIDS patients develop Kaposi's sarcoma as one of the symptoms that constitute the diagnosis of AIDS (34,35). It is interesting that one of the infective agents involved in AIDS has been shown to be associated to Kaposi's sarcoma. Geraldo and co-workers (36-38) and Boldogh et al. (39) have substantial evidence that cytomegalovirus is associated to Kaposi's sarcoma. Recently, CMV genome has been detected in Kaposi's sarcoma tissue from AIDS patients (40).

Recent reports constitute a rise in the incidence of non-Hodgkin's lymphoma among homosexuals (41), hemophiliacs, and other persons at risk for development of AIDS. The patients frequently exhibit lymphomas of B-cell origin in extranodal sites, particularly in the central nervous system. As reported by Ziegler et al. (41), 38 of 90 homosexual men with non-Hodgkin's lymphoma showed the involvement of central nervous system.

2.4 Neurologic disorders

In addition to the variety of complications resulting from the underlying immune dysfunction caused by the virus, central nervous system dysfunction occurs frequently in patients with AIDS (42). Resnick et al. (43) have reported the presence within the blood-brain barrier of a specific immunoglobulin to HTLV-III/LAV in patients with neurologic complications of AIDS, or the AIDS-related complex (ARC). Quantitative and qualitative analysis of the antibody present in the cerebrospinal fluid - especially the finding of unique oligoclonal IgG bands - indicates that the synthesis of IgG occured intrathecally, providing evidence that HTLV-III/LAV replication can take place in the brain. In addition, the AIDS-virus has been isolated from the cerebrospinal fluid (44-46). Shaw et al. (47) have been able to detect HLTV-III/LAV sequences in the brains of patients with AIDS and encephalopathy. The neuropathological abnormalities include subacute encephalitis (48), vacuolar degeneration of the spinal cord (49), an unexplained chronic meningitis or peripheral neuropathy (50,51).

3. Therapeutic Strategies

Although the major etiologic role can be assigned to HTLV-III/LAV, our knowledge about the cellular and molecular events leading to pathologic syndrome is incomplete. Some of the events involved in cellular and molecular pathogenesis, and their targeted role in designing therapeutic approaches will be described in this section.

Depending upon the type of manifested disease, treatment strategies have been developed and reported in literature. For example, attempts have been made to treat immunologic disorders in AIDS by interferons, interleukin-2, thymic hormones and factors, and several pharmacologic immunomodulators such as isoprinosine, azimexon, cimetidine and indomethacin. These studies have been recently reviewed by Lotze (52) and will not be discussed here.

The fact that most of the strategies directed to treat immunologic impairments were without mentionable success (52) indicates that an antiviral treatment is a primary requirement befory any immunomodulatory effect can be achieved.

3.1 Antiviral approaches in the control of AIDS

The overall genetic structure of HTLV-III/LAV is similar to that of other animal retroviruses. However, the sequence analysis of proviral DNA of HTLV-III/LAV genome predicts that, in addition to the conventional retroviral genes involved in virus replication, namely, gag, pol and env genes, the HTLV-III/LAV genome has two genes (53-56), termed sor (short open reading frame) and 3'orf (3' open reading frame). The presence of a third potential gene, termed tat (transactivation transcriptional gene), was also documented (57-60). Thus, HTLV-III/LAV has coding potential for three genes that are, so far, novel to this virus. Although tat gene has been shown to occur in other human lymphotropic viruses (57-61) and bovine leukosis virus (BLV), its organization in HTLV-III is unique in that, it is localized between sor and the env genes (58,60), whereas the tat gene of HTLV-I and BLV is located downstream from the env gene. All the three novel genes of HTLV-III/LAV contain introns and form polypeptides with molecular weights 24 - 25 kDa (sor), 14 - 15 kDa (tat) and 26 - 28 kDa (3'orf). These gene products display differential immune reactivity for HTLV-III positive human sera (62).

To study the role of viral genes controlling HTLV-III/LAV life cycle, the effects of site-directed mutations of the viral genome on replication and cytopathic potential of the virus have been studied. In particular, attempts have been made to define the role of three novel genes, described above. Mutations in two of the open reading frames, the sor gene and 3'orf gene, do not eliminate the ability of the virus to replicate in and kill T-lymphocytes (63). The transactivator (tat-III) gene encodes a 14 - 15 kDa protein that post-transcriptionally (64) stimulates HTLV-III/LAV long terminal repeat (LTR)-directed gene expression (57,58,60) via an interaction with specific target sequences in the leader of viral meassages (65). Mutations in the 5' portion of the first coding exon of the bipartite tat-III gene destroy the ability of the virus to efficiently synthesize structural proteins and to replicate (66). These mutations could be complemented in trans in cell lines that constitutively express the tat-III protein (66).

Further attempts led to an additional set of mutations near the two coding exons of the tat-III gene which dramatically attenuate the ability of the virus to express gag and env proteins. The observation that these mutations could not be complemented by the tat-III gene product led to the discovery of a second post-transcriptional gene (67) named art (anti-repressor translation) gene that partly overlaps the tat-III and env genes (Fig. 1). The protein product of art gene, 116 amino acids long, has no effect on viral mRNA concentrations but is required for expression of HTLV-III/LAV gag and env proteins. The ultimate conclusion of these experiments is that the art gene product specifically allows translation of gag and env mRNAs, thus unlocking the block imposed on virus expression.

In conclusion, HTLV-III/LAV has, in addition to gag, pol and env, four novel genes. Although no functional role has yet been assigned to sor and 3'orf genes, we know that these genes are not involved in controlling virus replication or the cytolytic activity of virus-infected T cells. The post-transcriptional regulation by tat-III and art genes offers attractive molecular targets to intervene in the replicative events of HTLV-III/LAV life cycle.

3.1.1 Replicative Cycle of HTLV-III/LAV

Although HTLV-III/LAV was discovered only three years ago, a dramatic progress has been made to the understanding of novel features of its replication. The replicative events in the life cycle of HTLV-III/LAV reflect many novel features which we do not find in the replicative cycle of other retroviruses. Some important features are shown in figure 2.

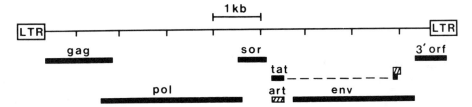

Figure 1. Location of the genes encoding the AIDS virus.

The two coding exons of the tat gene are indicated by black boxes, whereas hatched boxes designate exons of the art gene. Other gene designations appear in boxes.

The AIDS-virus, HTLV-III/LAV, is an exogenous virus without any cell-derived sequences as shown by the lack of hybridization of its genome to uninfected cell DNA (68). Unlike other retroviruses with transforming potential which have a site-specific integration of their proviral DNA, the HTLV-III/LAV DNA has two functional components in the infected cell (Fig. 2): one exists as polyclonally integrated provirus, the other as unintegrated superhelical or linear viral DNA. The persistence of a large amount of unintegrated viral DNA is unusual, but it has been shown for other cytopathic retroviruses, such as spleen necrosis virus (70), some strains of avian leukosis virus (71) and visna virus (72). However, some recent data indicate that the functional activity of the unintegrated species of HTLV-III/LAV proviral DNA is not directly related to the cytopathic activity of the virus (R.C. Gallo, personal communication).

The post-transcriptional regulation of gene expression by tat-III (see insert of Fig. 2) and art gene products offers an autonomous regulatory pathway in the life cycle of HTLV-III. Although the functional role of tat gene product has been elucidated in HTLV-I and HTLV-II, the regulation by the tat-III and art gene products involving both positive and negative elements is unique. Thus, it may serve as an important target in blocking the virus replication. We believe that an in vivo-modification of tat-III and art gene products may contribute to develop effective inhibitors of HTLV-III/LAV replication. Such an attempt has been made in our laboratory using D-Penicillamine, an amino-acid analogue to cysteine and valine. These studies will be described in a separate section.

Another novel feature of HTLV-III/LAV is its reverse transcriptase (RT) which catalyses the synthesis of HTLV-III/LAV proviral DNA. Studies on the biochemical and immunological characterization of RT from HTLV-III/LAV have been carried out in our laboratoy and will be briefly described.

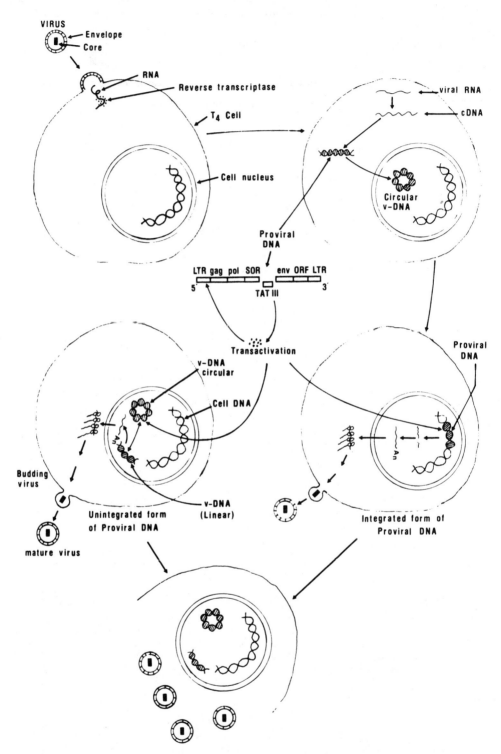

Figure 2. Replicative cycle of the AIDS-associated virus HTLV-III/LAV

3.1.2 Novel Features of HTLV-III/LAV Reverse Transcriptase

The enzyme from HTLV-III-infected H9 cells, or HTLV-III suspensions was purified by successive chromatography on DEAE-cellulose (DE23 and DE52) and phosphocellulose columns. A single peak of high reverse transcriptase activity was eluted from the phosphocellulose column at a salt concentration of 0.175 M KCl. This peak of activity was unable to transcribe the template primers $(dC)_n \cdot (dG)_{12}$ and $(dA)_n \cdot (dT)_{12}$, indicating that no DNA-dependent DNA polymerase activities were present in this fraction. A quantitative analysis of the enzyme at various purification steps showed a 120-140-fold purification, starting from the crude fraction obtained after centrifugation at 170 000 x g (73).

Table 1. Template primer requirements of the purified reverse transcriptase from HTLV-III

Template primer	^{3}H-labeled substrate	Enzyme activity per h(pmol/µg protein h)			
		Mg^{2+}	(mM)	Mn^{2+}	(mM)
$(rA)_n \cdot (dT)_{12}$	dTTP	15.45	(0.5)	0.23	(0.05)
$(rAm)_n \cdot (dT)_{12}$	dTTP	0.18	(3.0)	94.77	(0.05)
$(rC)_n \cdot (dG)_{18}$	dGTP	273.90	(15.0)	0.74	(0.075)
$(rCm)_n \cdot (dG)_{18}$	dGTP	38.17	(3.0)	89.03	(0.2)
70S RNA (SSAV)	dGTP	1.79	(10.0)	0	
70S RNA + $(dT)_{12}$	dGTP	4.78	(10.0)	0	
$(dT)_{12}$	dGTP	0.01	(0.5)	0	

Assays were performed with the same batch of enzyme using the same concentration. Viral RNA (70S) was tested in the presence of 3 other cold dNTPs with the indicated radioactive dNTP. DNA polymerase activity was measured by adding 20 µl of the enzyme fraction to a volume of 30 µl, which gave a final concentration of 50 mM Tris-HCl buffer (pH 8.0), 50 mM KCl, divalent cations (as indicated), 1 mM dithiothreitol, 20 µM each of the complementary dNTP, or ^{3}H-labeled substrate and 1.25 µg template primer (as indicated).

The purified and concentrated enzyme was tested with various template primers in the presence of different concentrations of divalent cations. These assays are useful in deriving information as to whether cellular DNA polymerases are present, and whether the enzyme has a type specificity; for example, mammalian viruses of the C-type have different ionic requirements from viruses belonging to the B-type. As follows from table 1, the purified enzyme transcribes $(rA)_n \cdot (dT)_{12}$, $(rAm)_n \cdot (dT)_{12}$, $(rC)_n \cdot (dG)_{12}$ and $(rCm)_n \cdot (dG)_{12}$; primer alone with either cation gives no activity. The pattern of template primer utilization by the enzyme distinguishes it from terminal deoxynucleotidyltransferase and host DNA polymerases. Another property unique to reverse transcriptase is its capacity to catalyze transcription of the viral RNA (70S RNA). This was confirmed for the HTLV-III enzyme using a purified 70S RNA from SSAV.

Except for the transcription of 2'-O-methylated templates, $(rAm)_n$ and $(rCm)_n$, all other template primers require Mg^{2+} for optimal activity. The fact that $(rC)_n \cdot (dG)_{12}$-dependent activity is several-fold higher than that catalyzed by $(rA)_n \cdot (dT)_{12}$ and is strictly magnesium-dependent constitutes a novel feature of the HTLV-III enzyme, compared to known properties of other C-type virus reverse transcriptases. In this respect, the enzyme resembles more closely reverse transcriptases from B- and D-type viruses.

To study the cross-reactivity between antigens of HTLV-III and other related viruses, we have recently generated hybridoma (mouse/mouse) clones which secrete monoclonal antibodies to p24, p31, gp41, gp120 and reverse transcriptase. Here, we will describe the serological analysis of reverse transcriptases purified from HTLV-I, HTLV-II and HTLV-III, using monoclonal antibodies to HTLV-III reverse transcriptase; the hybridoma clone used carries the designation 4F8.

The supernatant fluid obtained from the clone 4F8 was screened by ELISA using disrupted HTLV-III as antigen. Wells of a 96-well plastic tray were coated overnight with lysates of density-banded HTLV-I, HTLV-II and HTLV-III containing different amounts of antigen (table 2). The remaining protein-binding sites were saturated with 1 % BSA in PBS. 50 µl of the hybridoma supernatant was then added to each well and incubated for 1 h at 37 °C. The immuno-reactivity was measured by adding affinity purified goat anti-mouse IgM + M conjugated with ß-galactosidase, using p-nitrophenyl-ß D-galactopyranoside as substrate. The activities were measured at 405 nm in an ELISA reader.

The results shown in table 2 demonstrate that viral proteins from HTLV-I and HTLV-II do not bear any cross-reactive epitope to antibodies secreted by the clone 4F8. As a control in these experiments, BSA was used. On the other hand, a concentration-dependent cross-reactivity is exhibited by the viral antigens from HTLV-III against antibodies secreted by 4F8 clone.

Table 2. Serological relationship between human T-lymphotropic retroviruses defined by monoclonal antibodies

Experiment	$E_{405} \times 10^3$ at various virus (protein) concentrations						
	0(PBS/BSA)	15	30	60	125	250	500	1000
					(ng/well)			
HTLV-I	57	63	63	55	57	55	67	53
HTLV-II	64	69	64	51	63	77	60	50
HTLV-III	50	60	61	54	75	96	198	288

Supernatant fluid obtained from clone 4F8 was screened by ELISA using disrupted HTLV-III as antigen. Wells of a 96-well plastic tray were coated overnight with HTLV-I, HTLV-II and HTLV-III lysates containing different amounts of antigen. After saturation of the remaining protein-binding sites with 1 % BSA, 50 µl of culture supernatant from 4F8 was added to each well. The cross-reactivity was measured by adding anti-mouse IgG + M conjugated with ß-galactosidase, using p-nitrophenyl-ß-D-galactopyranoside as substrate. The activities were measured in an ELISA reader at 405 nm.

Following the procedure described above, wells of 96-well plastic trays were coated overnight with reverse transcriptases purified from HTLV-I, HTLV-II and HTLV-III. To document the specificity of 4F8 clone for reverse transcriptase, another hybridoma clone secreting monoclonal antibodies against the gp120 antigen of HTLV-III, designated 1A6, was included in this experiment. As follows from table 3, no cross-reaction between reverse transcriptases of HTLV-I and HTLV-II was observed towards antibodies to HTLV-III reverse transcriptase (clone 4F8). In the same experiment, the antibodies secreted by the clone 1A6 show no cross-reactivity towards the HTLV-III reverse transcriptase.

Table 3. Serological relationship between reverse transcriptases from human T-lymphotropic viruses defined by monoclonal antibodies to HTLV-III-RT

Experiment	$E_{405} \times 10^3$ at various RT concentrations (protein)						
	0(PBS/BSA)	15	30	60	125 (ng/well)	250	500	1000
HTLV-I RT								
(4F8)	51	56	44	48	47	68	48	39
(1A6)	45	44	41	45	56	56	53	39
HTLV-II RT								
(4F8)	45	45	56	46	54	61	51	39
(1A6)	49	42	42	48	53	55	50	42
HTLV-III RT								
(4F8)	55	44	49	70	82	154	231	287
(1A6)	44	50	56	52	50	53	55	42

Following the procedure described in the text and table 2, wells were coated overnight with various amounts of reverse transcriptases purified from HTLV-I, HTLV-II and HTLV-III. The clone 1A6, serving as negative control, secretes antibodies against glycoprotein gp120.

The specificity of antibodies secreted by the clone 4F8 for HTLV-III reverse transcriptase was further documented by immunoblotting. The electrophoretic separation of lysates from HTLV-I (lane 1), HTLV-II (lane 2) and HTLV-III (lane 3) are shown in figure 3A; lane 4 shows the separation of affinity-purified M_r markers (phosphorylase b, 94000; BSA, 67000; ovalbumin, 43000; carbonic anhydrase, 30000; soybean trypsin inhibitor, 20100 and α-lactalbumin, 14400). Figure 3B depicts the immunoblots of HTLV-I (lane 1), HTLV-II (lane 2) and HTLV-III (lane 3), developed against antibodies secreted by the clone 4F8. There is no cross-reacting band seen on HTLV-I and HTLV-II strips. The strip with HTLV-III shows two prominent bands in the region of M_r 53000 and 66 000. To separate the two reactivities, the clone 4F8 was subjected to repeated cycling. Even after the 4th cycling of 4F8 clone, the two reactivities could not be separated. This indicates that both the reactive antigens, 53 and 66 kDa proteins, have a common determinant recognized by the same epitope.

To elucidate the biochemical nature of 4F8-antibody interaction to HTLV-III reverse transcriptase, we have measured the catalytic activity

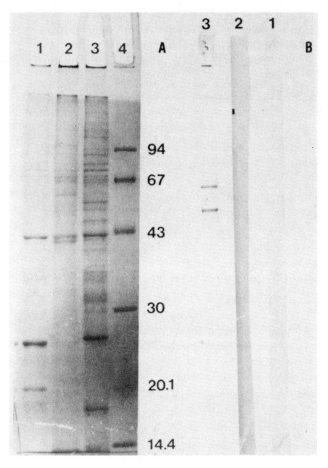

Figure 3. (A) SDS-polyacrylamide gel electrophoresis of virus lysates. 100 μl virus suspension containing 4 μl Tris-HCl (pH 6.8), 20 μl SDS and DTT (1 %) was incubated in a boiling water bath for 5 min. To this were added 80 μl glycerol and 10 μl bromophenyl blue (10 %); 15 μl of this suspension was used per slab. Lanes: 1, HTLV-I; 2, HTLV-II; 3, HTLV-III; 4, protein markers: phosphorylase b (94 kDa), bovine serum albumin (67 kDa), ovalbumin (43 kDa), carbonic anhydrase (30 kDa), soybean trypsin inhibitor (20.1 kDa), α-lactalbumin (14.4 kDa). (B) Immunoblot transfer analysis of antigenic cross-reactivities of HTLV-I, HTLV-II and HTLV-III towards antibodies secreted by hybridoma 4F8. Lanes: 1, HTLV-I; 2, HTLV-II; 3, HTLV-III.

of reverse transcriptase in the presence of antibodies. Figure 4 shows that antibodies secreted by the clone are unable to neutralize the enzymatic activity. Measurement of residual activity by immunoprecipitation of the antigen-antibody complex, however, shows a concentration-dependent inhibition of the enzyme activity. These data indicate that the antibodies secreted by 4F8 clone are not directed towards the active center of HTLV-III reverse transcriptase.

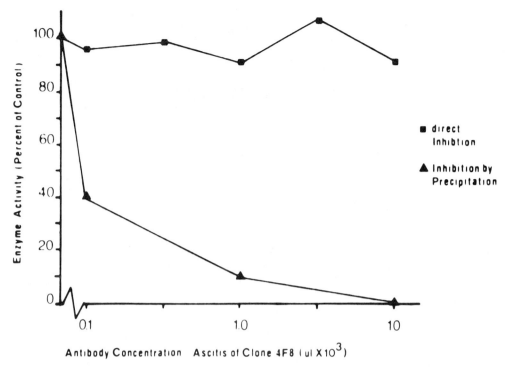

Figure 4. Catalytic activity of HTLV-III reverse transcriptase in the presence of antibodies secreted by the hybridoma clone 4F8.

Ascites prepared from clone 4F8 was used as the source of antibodies. Neutralizing activity was measured by preincubating 20 µl of the enzyme with 20 µl of the properly diluted ascites fluid overnight. 10 µl of this was used to assay the enzyme activity. To measure the residual activity, 10 µl of the enzyme was preincubated with 10 µl of the properly diluted ascites overnight at 4 °C. To this was added 100 µl of magnetic conjugate of goat anti-mouse IgG (Biomag M4400, Sebak, Aidenbach, FRG), and the tubes were placed on a magnetic separation device (Sebak). After 1 h, 30 µl aliquots were pipetted from the supernatant and assayed for the residual enzyme activity. All estimations were done in triplicate. In the control experiments, 1:100 dilution of ascites from another clone which showed no cross-reactivity to reverse transcriptase was used.

The biochemical and immunological data presented above suggest that there are two forms of reverse transcriptase activities having a common determinant for the antibodies secreted by clone 4F8. The antibodies secreted by clone 4F8 are non-neutralizing, and are specific for HTLV-III reverse transcriptase. Thus, the reverse transcriptase of HTLV-III is serologically different from those from HTLV-I and HTLV-II. Immunoblotting shows two forms of reverse transcriptase of M_r 53000 and 66000. There are two possible mechanisms to explain this heterogeneity of HTLV-III reverse transcriptase: one possibility is that the product of the pol gene is a polyprotein which is processed at different sites producing two forms of the molecule; the other is that the parent enzyme of M_r 66000 is processed by a protease associated to the HTLV-III genome. Experimental evidence is in favor of the second possibility. We have recently shown that HTLV-III reverse trasncriptase can be separated by isoelectric focusing into two peaks with isoelectric points of 5.75 and 6.25 (Fig. 5). These differences are probably due to cleavage of roughly 100 amino acids from the carboxyl end by a specific protease. The characterization of two enzymes separated by isoelectric focusing should help in answering these questions.

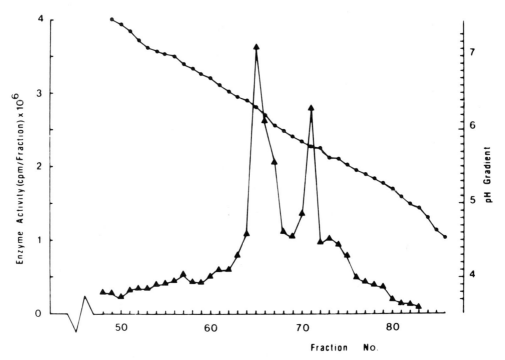

Figure 5. Profile of DNA polymerase activities after electrofocusing of the HTLV-III enzyme eluted from a phosphocellulose column. The enzyme activity of the eluted fractions was measured in the presence of $(rC)_n \cdot (dG)_{12}$ using Mg^{2+} (15 mM).

The studies reported in this section (74-76) conclude that HTLV-III/LAV has two catalytically active reverse transcriptases which can be biochemically separated on preparative isoelectric focusing columns. We believe that the presence of two catalytically active reverse transcriptases may account for the genomic heterogeneity of HTLV-III/LAV, especially in the env region.

3.1.3 Reverse Transcriptase Inhibitors

The strategical role of nucleic acids, or nucleic acid polymerases as targets, involves two distinct features. The investigational drug must have the capacity to recognize distinct bases or base-pair sequences, either by direct interaction between functional groupings on the base-pairs and the drug molecule, or indirectly via recognition of the conformational peculiarity of the nucleic acid molecule. The other alternative is the specific affinity of the drug for one or the other nucleic acid polymerizing enzymes. From the strategical standpoint this is a more specific approach, since a number of enzymes are involved in the polymerization of nucleic acids.

RNA directed DNA synthesis is unique to the life cycle of retroviruses. No similar reaction is known to occur in normal uninfected cells. Reverse transcriptase is a multifunctional enzyme, with at least four domains that mediate RNA dependent DNA polymerase activity, RNase H activity, and DNA dependent DNA polymerase activity and a site for nucleic acid binding. All these domain mediated functional activities offer attractive targets for inhibiting the synthesis of DNA in retroviruses. The final product of the reactions mediated by reverse transcriptase is a linear duplex DNA. This linear DNA is longer than the 35S subunit of genomic RNA and contains long terminal repeats (the composition of which can be written as 5'-U3-TR-U5-3') present at both ends (77,78).

The process of reverse transcription offers a unique target for drug design, and the number of compounds reported to inhibit this process has exceeded expectations. However, the critical analysis of the en bloc progress leaves a big gap. The in vitro assay systems used for reverse transcriptase determination reveal a variety of substrates, template-primers, and interacting compounds which can modulate the catalytic rate of DNA synthesis (79). Some examples of this type of modulation are: the detergent effect on the activity of rifamycins (80); influence of divalent cations (Mg^{2+} or Mn^{2+}) on the rate of DNA synthesis with different substrates and the role of chelating agents or cation binders in buffer (79,81); and the interaction of thiols with some potential inhibitors or their direct influence on the measured DNA synthesis. Thus slight variations in assay conditions may lead to wrong interpretations with respect to the specificity of a particular inhibitor in the viral DNA polymerase system.

The second problem is the interpretation of the enzymatic data with respect to the antiviral activity of these inhibitors (82). This is particularly the case with those compounds that exert their inhibitory action by complexing with one or more synthetic templates. Such an effect cannot be very specific for the viral enzyme. Drugs exhibiting cytotoxic effects to the extent of causing a delayed death or those which intervene in the replicative cycle of the host cell may give erroneous information about the specific antiviral activity of the compound. Thus antiviral studies in vitro should be carried out under conditions and at inhibitor concentrations which have little or no effect on the replicative cycle of the cell. Molecular manipulations of parent compounds have proved to be very useful in several instances in the development of inhibitors of viral DNA synthesis which exhibit a low cytotoxicity and at the same time a higher antiviral potential. This is evidenced by our earlier studies on distamycin derivatives (83,84), tilorone congeners (85-87), diamidine-phenylindole derivatives (88-90), and daunomycin derivatives (91-93). These studies have been reviewed extensively (94,95) and will not be discussed here.

Further efforts to develop compounds which inhibit viral DNA polymerases by interacting with templates may lead to the discovery of useful compounds exhibiting a higher therapeutic index, i.e., low cytotoxicity and high antiviral activity. However, this approach will not lead to the development of a specific inhibitor of the viral enzyme unless one finds a compound which binds specifically to 70S RNA. Although 70S RNA is a novel feature of retroviruses, from the chemical and physical standpoint it does not appear to offer any uniqueness that would distinguish it from cellular nucleic acids. Thus the strategical approach of developing an inhibitor of this type at the present state of our knowledge is unthinkable.

The second approach, which has proved to be more useful and relatively specific in developing such inhibitors, is to design compounds that bind to the viral enzyme. In searching for this type of inhibitor the enzymes chosen for comparison are important. Many studies have been done with either avian or mammalian reverse transcriptase, since these two types of enzymes have some different characteristics. A large number of studies claim specificity (or selectivity) of a compound by comparing an inhibitory response to a bacterial DNA polymerase with the viral reverse transcriptase. Such a comparison has no relevance to the selective nature of the compound. The most important approach to demonstrate selectivity of a compound is to compare inhibitory effects against various cellular DNA polymerases, such as DNA polymerases α, β and γ.

Our efforts to develop compounds that inhibit viral DNA polymerases by interacting directly with the enzyme led to the discovery of a polycytidylic acid analogue, containing 5-mercapto substituted cytosine bases, a partially thiolated polycytidylic acid. This compound, MPC, was found to inhibit the retroviral DNA polymerase in a very specific manner (82,94 100). Details on various aspects of MPC action are documented in literature (82,95,98,101). We will summarize only the main points relevant to its mode of action and selectivity as an inhibitor of retroviral reverse transcriptase. Partially thiolated polycytidylic acid preparations (MPC I, MPC II, and MPC III) containing 1.7, 3.5, and 8.6 % 5-mercaptocytidylate units inhibited the DNA polymerase activity of Friend leukemia virus and of other retroviruses in the endogenous reaction as well as in the presence of exogenous template-primers. The inhibitory activities were directly related to the percentage of thiolation. A maximum inhibition was observed with preparations containing 15 - 17 % of the thiolated cytosine bases. In these experiments nonthiolated samples of polycytidylic acid showed no inhibition of the reverse transcriptase reaction.

Polyuridylic acid thiolated by the same procedure (96) was also found to inhibit this reaction strongly (101,102); however, the nonthiolated sample of polyuridylic acid also inhibited the reaction moderately. Besides, the inhibitory effect of thiolated poly(U) samples was not strictly related to their degree of thiolation, as was observed for the thiolated samples of polycytidylic acid. The inhibitory effect of nonthiolated polyuridylic acid is presumably due to hydrogen bonding between poly(U) and the added template, $(rA)_n$. The inhibition of the endogenous reaction by nonthiolated poly(U) is explainable by the fact that genomic RNA contains large stretches of polyadenylic acid at its 3'-terminal. For this reason it was difficult to designate the role of thiolation on the inhibitory effect of 5-mercapto-polyuridylic acid samples.

The effect of 5-mercapto substituted pyrimidine bases of polynucleotides on the catalytic activities of reverse transcriptase and DNA polymerase ß purified from HTLV-III infected H9 cells is shown in Table 4. As follows from Table 4 aphidicolin, a specific inhibitor of eukaryotic DNA polymerase α, has no effect on the catalytic activity of DNA polymerase ß and the reverse transcriptase. In contrast, the catalytic activity of HTLV-III is strongly inhibited by modified poly(U) and poly(C), containing 5-mercapto substituted pyrimidines; at as small concentrations as 1 µg the HTLV-III reverse transcriptase loses almost 90 % of its catalytic activity in the presence of 5-mercaptopolyuridylic acid (SH = 4 %) or in the presence of MPC (SH = 15 and 30 %). The inhibitory effects of MPC are at low concentrations related to their degree of thiolation. This correlation is very specific if the MPC samples with lower degree of thiolation are used. Interestingly the DNA polymerase ß from HTLV-III infected H9 cells is not or is only slightly inhibited by MPC. On the other hand a poly(U) polymer containing 4-thiouridine moieties has no effect on the catalytic activities of reverse transcriptase and the DNA polymerase ß from HTLV-III infected H9 cells. This shows that 5-mercapto substitution of pyrimidines in polynucleotides is an effective modification for the development of specific inhibitors of retroviral reverse transcriptase.

Table 4. Effect of thiolated polynucleotides on DNA polymerase ß and reverse transcriptase activities from HTLV-III infected H9 cells

inhibitor	Concentration [µg/reaction mixture (50 µl)]	Enzyme activity (% of control)	
		$(dA)_n \cdot (dT)_{12}$	$(rC)_n \cdot (dG)_{18}$
None		100 (14.028)[a]	100 (20.130)[a]
Aphidicolin	10	92.4	123
Poly(U) (4-thio)	10	91.5	113
Poly(U) (5-thio)			
(SH = 4 %)	0.1	80.0	38.5
	1.0	64.0	12.6
Polycytidylic acid (5-thio)			
(SH = 15 %)	0.1	97.0	46.9
	1.0	90.0	16.1
(SH = 30 %)	0.1	90.0	38.5
	1.0	72.0	11.1

[a] pmol dNMP incorporation/µg protein/h

Mitsuya et al. (103) have recently reported the protection of T-cells in vitro against infectivity and cytopathic effect of HTLV-III by suramin (synonyms, Germanin, Bayer 205, Naganol, Antrypol), an antitrypanosomal drug. The development of Germanin goes back to the initial discovery of Paul Ehrlich (104) in 1904, who discovered that the sulfonate derivative of aminonaphthalene, Trypanrot, has a high therapeutic activity against trypanosomal infections. Paul Ehrlich was convinced, as a result of his discovery of salvarsan, that a therapeutic drug must have a color or chromophore for fixation to the cell: corpora non agunt nisi fixata. Due to its high therapeutic efficacy its aza derivative attracted the attention of several chemists who undertook various types of modifications. Two very important modifications have led to the development of Germanin; (a) Mesnill and Nicolle, the French chemists at the Institut Pasteur, showed that the urea derivative of Ehrlich's aza compound is by severalfold more active than is the parent compound (105); (b) Heymann (106), at the research laboratories of Farbenfabriken Bayer, discovered that the substitution of aza bonds by aminobenzoyl groups causes the color of active sulfonate derivate of naphthalene to disappear, but the therapeutic activity is retained. The marriage of these two observations resulted in the birth of Germanin in which both the chemical modifications were incorporated.

The effects of Germanin on the catalytic activities of DNA polymerase and HTLV-III reverse transcriptase are shown in Table 5. DNA polymerase α was a highly purified preparation from human placenta. As follows from the results Germanin inhibits both the enzymatic activities very strongly, but to the same extent; K_i values for their inhibition are very close to each other. As evident from its structure Germanin has a high binding capacity to proteins due to sulfonic acid moieties present in the molecule. Germanin has a very high binding affinity to serum albumin, which is perhaps responsible for the irreversible damage to the kidney.

Table 5. Inhibition of DNA polymerase α and HTLV-III reverse transcriptase by Germanin

System	[^{3}H]dNMP incorporation (% of control)	
	$(dA-dT)_n \cdot (dA-dT)_n$	$(rC)_n \cdot (dG)_n$
Complete	100 (9.414)^a	100 (67.719)^a
+ Germanin^b		
0.1 µg	33	27.8
1.0 µg	10.6	0.8
10.0 µg	0	0.2

a pmol dNMP incorporation/µg protein/h
b amount added to the reaction mixture (50 µl)

DNA polymerase α was purified from human placenta as described earlier (73). The reaction mixture contained 50 mM Tris-HCl buffer (pH 8.2); KCl (40 mM); magnesium ions (1.2 mM); dithiothreitol (1 mM); 20 µM concentrations each of the complementary deoxyribonucleoside triphosphates and the labeled substrate dTTP; and 1.25 µg of the copolymer $(dA-dT)_n \cdot (dA-dT)_n$.

Although the enzyme data on Germanin or Suramine effect do not indicate any selectivity towards the inhibition of reverse transcriptase (74), this compound was selected for clinical trials (107). Several recent reports have appeared to document the adverse effects of suramine in patients with AIDS and AIDS-related complex (108,109).

Another drug, 21'-tungsto-9-antimoniate (HPA23), based on the inhibition of reverse transcriptase, is now being clinically used to treat AIDS (110). We (111), and others (112), have found that DNA polymerase α is far more sensitive to HPA23 than reverse transcriptase. This compound is analogous to the previously published derivative, 5-tungsto-2-antimoniate, which was earlier reported by Sinoussi (113).

Other inhibitors of HTLV-III/LAV reverse transcriptase include: 3'-Azido-thymidine (114-116) and Foscarnet (117-119). Several new compounds are being developed which may act via reverse transcriptase inhibition. It should be emphasized that not the potency of their inhibitory response is important, but the selectivity which should be documented by careful experimentation using purified enzyme systems and optimal assay conditions.

3.1.4 Antiviral approach at post-transcriptional level

The recent knowledge about the post-transcriptional regulatory phenomenon operating in the life cycle of HTLV-III/LAV (see section 3.1) offers a very meaningful approach to develop selective antiviral compounds. Based on this approach, we have recently reported the inhibition of HTLV-III/LAV replication in cell culture by D-Penicillamine, an amino acid analogue of cysteine and valine (120). The effect of D- and L-penicillamine on the replication of HTLV-III in H9 cells was determined as a function of drug concentration by measuring the expression of viral proteins p15 (Fig. 6) and p24 (Fig. 7) in an immunofluorescence assay procedure using monoclonal antibodies. Fig. 6 shows a concentration dependent inhibition of p15 expression by both the compounds, L-penicillamine (filled circles) and D-penicillamine (open circles). At lower concentrations L-penicillamine is more effective than D-penicillamine. As follows from Fig. 6, at 5 µg/ml of drug, L-penicillamine inhibits the p15 expression to more than 80 % whereas D-penicillamine reduces the p15 expression to only about 40 %. However, at concentrations above 20 µg/ml the inhibitory response of both the isomers is similar. To obtain an inhibition of 98.5 to 99.4 % a drug concentration of 40 µg/ml was needed for both the isomers.

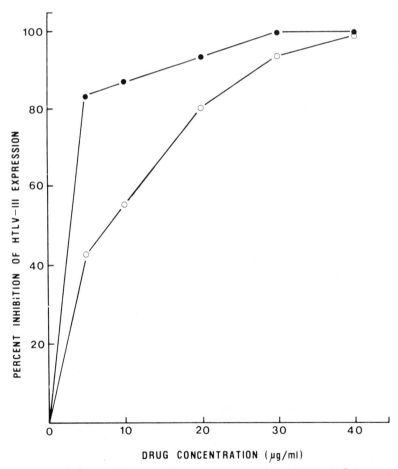

Figure 6. Inhibition of HTLV-III virus expression by D- and L-penicillamine as measured by monoclonal antibodies to HTLV-III p15. ●———●, L-penicillamine; o———o D-penicillamine.

Fig. 7. illustrates the inhibition of HTLV-III expression by D- and L-penicillamine measured by the immunofluorescence assay procedure using monoclonal antibodies to the viral protein p24. Both compounds inhibit p24 expression in a similar manner as that of p15. At lower concentrations the L-isomer is a far better inhibitor than the D-isomer of penicillamine. Thus, at a concentration of 5 µg/ml L-penicillamine inhibits the p24 expression to 75 % whereas the inhibition by D-penicillamine is only about 13 %. At drug concentration of 30 µg/ml or more the inhibitors response of both isomer is similar. To achieve a total inhibition 40 µg/ml was needed for both the isomers.

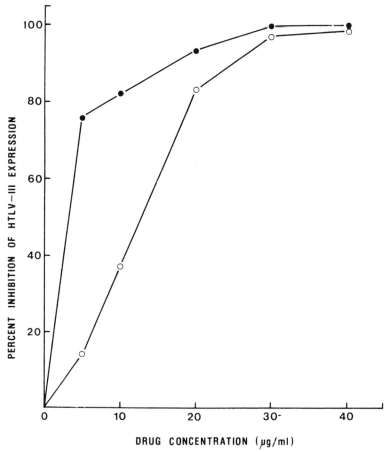

Figure 7. Inhibition of HTLV-III virus expression by D- and L-penicillamine as measured by monoclonal antibodies to HTLV-III p24. ●—●, L-penicillamine; o—o, D-penicillamine.

Clinical trials with D-Penicillamine in asymptomatic virus positive patients are in progress to evaluate the antiviral efficacy of the drug in vivo.

4. Future perspectives

Unlike other viral infections, HTLV-III/LAV infection involves a very wide spectrum of diseases. The association of HTLV-III/LAV to AIDS is now well established, but our knowledge regarding the pathogenic events at cellular and at molecular level induced or catalyzed by HTLV-III/LAV is incomplete. Many questions relevant to designing therapeutic strategies are still to be answered. As our knowledge on the molecular biology of this virus progresses, more selective targets will be available which may find application in designing potential compounds to block the replication of HTLV-III/LAV.

Based on the existing knowledge, we can design some strategies to chemically block the HTLV-III/LAV replication. For example, the self-regulatory mechanism of HTLV-III/LAV replication by the proteins coded by *tat*-III and *art* genes is an attractive and novel target. In vivo modification of these gene products will abolish this control mechanism resulting in the inhibition of virus replication. This approach may also lead to influence the cytopathic effect of the virus. It has been suggested that the C-terminal part of the trans-membrane protein (gp41) is probably involved in the cytopathic effect of HTLV-III. Therefore, the in vivo modification of this protein will be very interesting.

The classical approach of developing inhibitors of reverse transcriptase will be very rewarding for two reasons: the RT of HTLV-III/LAV has some novel features, as mentioned in this review; and the knowledge of this subject is available from the past (98). However, these studies should be carried out carefully with purified enzymes and well compared to homologous cellular enzymes α, β and γ. Without this type of comparative study, a reverse transcriptase inhibitor does not make any sense, and clinical trials with such compounds may end up in a great disappointment.

References

1. Pneumocystis pneumonia - Los Angeles. Morbid Mortal Weekly Rep., 30: 250-252 (1981).
2. Kaposi's sarcoma and Pneumocystis pneumonia among homosexual men - New York City and California. Morbid Mortal Weekly Rep., 30: 305-308 (1981).
3. F.Barre-Sinoussi, J.C. Chermann, F. Rey, M.T. Nugeyne, S. Chamaret, J. Gruest, C. Dauguet, C. Axler-Blin, F. Vezinet-Brun, C. Rouzioux, W. Rozenbaum, and L. Montagnier, Isolation of a T-lymphotropic retrovirus from a patient at risk for acquired immune deficiency syndrome (AIDS), Science, 220: 868-871 (1983).
4. P.M. Feorino, V.S. Kalyanaraman, H.W. Haverkos, C.D. Cabradilla, D.T. Warfield, H.W. Jaffe, A.K. Harrison, M.S. Gottlieb, D. Goldfinger, J.C. Chermann, F. Barre-Sinoussi, T.T. Speia, J.S. McDougal, J.W. Curran, L. Montagnier, P.A. Murphy, and D. Francis, Lymphadenopathy associated virus infection of a blood donor-recipient pair with acquired immunodeficiency syndrome, Science, 225: 69-72 (1984).
5. M.Robert-Guroff, M. Brown, and R.C. Gallo, HTLV-III neutralizing antibody in patients with AIDS and ARC, Nature, 316: 72-74 (1985).

6. M. Popovic, M.G. Sarngadharan, E. Reed, and R.C. Gallo, Detection, isolation, and continuous production of cytopathic human T lymphotropic retrovirus (HTLV-III) from patients with AIDS and pre-AIDS, Science, 224: 497-500 (1984).
7. R.C. Gallo, Detection and isolation of type C retrovirus particles from fresh and cultured lymphocytes of a patient with cutaneous T-cell lymphoma, Proc. Natl. Acad. Sci. USA, 77: 7415 (1980).
8. V.S. Kalyanaraman, M.G. Sarngadharan, M. Robert-Guroff, I. Miyoshi, D. Blayney, D. Golde, and R.C. Gallo, A new subtype of human T-cell leukemia virus (HTLV-II) associated with a T-cell variant of hairy cell leukemia, Science, 218: 571-573 (1982).
9. J.A. Levy, A.D. Hoffman, S.M. Kramer, J.A. Landis, and J.M. Shimabukuro, Isolation of lymphocytopathic retroviruses from San Francisco patients with AIDS, Science, 225: 840-842 (1984).
10. ICTV, What to call the AIDS Virus? Nature, 321, 10 (1986).
11. F.P. Siegal, C. Lopez, G.S. Hamer et al., Severe acquired immunodeficiency in homosexual males, manifested by chronic perianal lesions, New Engl. J. Med., 305: 1439 (1981).
12. M.S. Gottlieb, R. Schroff, H.M. Schander et al., Pneumocystis carinii pneumonia and mucosal candidiasis in previously healthy homosexual men: Evidence of a new acquired cellular immunodeficiency, New Engl. J. Med., 305: 1425 (1981).
13. D. Mildvan, U. Mathur, R.W. Enlow et al., Opportunistic infections and immunodeficiency in homosexual men, Ann. Intern. Med., 96: 700 (1982).
14. M.S. Gottlieb, J.E. Groopman, W.M. Weinstein et al., The acquired immunodeficiency syndrome, Ann. Intern. Med., 99: 208 (1983).
15. A.S. Fauci, A.M. Macher, D.L. Longo et al., Acquired Immunodeficiency syndrome: Epidemiologic, clinical, immunologic, and therapeutic considerations, Ann. Intern. Med., 100: 92 (1983).
16. N. Ciobanu, K. Welk, G. Kruger et al., Defective T-cell responses to PHA and mitogenic monoclonal antibodies in male homosexuals with acquired immunodeficiency syndrome and its in vitro correction by interleukin-2, J. Clin. Invest., 3: 332 (1983).
17. S. Gupta, and B. Safai, Deficient autologous mixed lymphocyte reaction in Kaposi's sarcoma associated with deficiency of Leu-3 positive responder cells, J. Clin. Invest., 71: 296 (1983).
18. H.C. Lane, H. Masur, L.C. Edgar et al., Abnormalities of B lymphocyte activation and immunoregulation in patients with the acquired immunodeficiency syndrome, New Engl. J. Med., 309: 453 (1983).
19. A.H. Rock, H. Masur, H.C. Lane et al., Interleukin-2 enhances the depressed natural killer and CMV-specific cytoxic activation of lymphocytes from patients with the acquired immunodeficiency syndrome, J. Clin. Invest., 72: 398 (1983).
20. H.W. Murray, B.Y. Rubin, H. Masur et al., Impaired production of lymphokines and immune (gamma) interferon in the acquired immunodeficiency syndrome, New Engl. J. Med., 310: 883 (1984).
21. P. Smith, K. Ohura, H. Masur et al., Monocyte function in the acquired immune deficiency syndrome: Defective chemotaxis, J. Clin. Invest., in press
22. S. Cunningham-Rundles, M.A. Michelis, and H. Masur, Serum suppression of lymphocyte activation in vitro in acquired immunodeficiency disease, J. Clin. Immunol., 3: 156 (1983).
23. J. Laurence, and H.G. Kunkel, Soluble suppressor factors in patients with acquired immune deficiency syndrome, Clin. Res., 31: 347A (1983).
24. J. Laurence, and L. Mayer, Immunoregulatory lymphokines of T hybridomas from AIDS patients: Constitutive and inducible suppressor factors, Science, 225: 66 (1984).

25. E.M. Hersh, J.M. Reuben, A. Rios et al., Elevated serum thymosin alpha$_1$ levels associated with evidence of immune dysregulation in male homosexuals with a history of infectious diseases or Kaposi's sarcoma, New Engl. J. Med., 308: 45 (1983).
26. M.Dardenne, J.-F. Bach, and B. Safai, Low serum thymic hormone levels in patients with acquired immunodeficiency syndrome, New Engl. J. Med., 309: 48 (1983).
27. B.D. Spector, G.S. Perry, and J.H. Kersey, Genetically determined immunodeficiency disease (GDID) and malignancy: Report from the immunodeficiency cancer registry, Clin. Immunol. Immunopathol., 11: 12 (1978).
28. M.Kaposi, Idiopathisches multiples Pigment Sarcom der Haut, Arch. Dermatol., 4: 465 (1872).
29. W.D. McCarthy, and G.T. Pack, Malignant blood vessel tumors. A report of 56 cases of angiosarcoma and Kaposi's sarcoma, Surg. Gynecol. Obstet., 91: 465 (1950).
30. B.D. Myers, E. Kessler, J. Levi et al., Kaposi's sarcoma in kidney transplant recipients, Arch. Intern. Med., 133: 307 (1974).
31. I.Penn, Kaposi's sarcoma in organ transplant recipients: Report of 20 cases, Transplant., 27: 8 (1979).
32. J.F. Taylor, A.C. Templeton, C.L. Vogel et al., Kaposi's sarcoma in Uganda: A clinicopathological study, Int. J. Cancer, 8: 122 (1971).
33. H.W. Haverkos, and J.W. Curran, The current outbreak of Kaposi's sarcoma and opportunistic infections, Cancer, 32: 330 (1982).
34. Centers for Disease Control: Update on acquired immunodeficiency syndrome (AIDS), U.S. MMWR, 31: 507 (1982).
35. R.L. Modlin, J.T. Crissey, and T.H. Tea, Kaposi's sarcoma, Int. J. Dermatol., 22: 443 (1983).
36. G.Giraldo, E. Beth, and F. Haguenau, Herpes-type virus particles in tissue culture of Kaposi's sarcoma from different geographic regions, J. natl. Cancer Inst., 49: 1509 (1972).
37. G.Giraldo, E. Beth, F.M. Kourilsky et al., Antibody patterns to herpesviruses in Kaposi's sarcoma, Serological association of European Kaposi's sarcoma with cytomegalovirus, Int. J. Cancer, 15: 839 (1975).
38. G.Giraldo, E. Beth, W. Henle et al., Antibody patterns to herpes-viruses in Kaposi's sarcoma. II. Serological association of American Kaposi's sarcoma with cytomegalovirus, Int. J. Cancer, 22: 126 (1978).
39. I.Boldogh, E. Beth, E.S. Huang et al., Kaposi's sarcoma IV. Detection of CMV DNA, CMV RNA, and CMNA in tumour biopsies, Int. J. Cancer, 28: 469 (1981).
40. W.L. Drew, M.A. Conant, R.C. Miner et al., Cytomegalovirus and Kaposi's sarcoma in young homosexual men, Lancet, 2: 125 (1982).
41. J.L. Ziegler, J.A. Beckstead, P.A. Volberding et al., Non-Hodgkin's lymphoma in 90 homosexual men, New Engl. J. Med., 311: 565 (1984).
42. B.D. Jordan, B. Navia, C. Petito, E. Cho, and R. Price, Neurological syndromes complicating AIDS, Front. Radiat. Ther. Oncol., 19: 82-87 (1985).
43. L.Resnick, F. DiMarzo-Veronese, J. Schüpbach et al., Intra-blood-brain-barrier synthesis of HTLV-III-specific IgG in patients with neurologic symptoms associated with AIDS or AIDS-related complex, New Engl. J. Med., 313: 1498-1504 (1985).
44. D.C. Gajdusek, H.L. Amyx, C.J. Gibbs et al., Infection of chimpanzees by human T-lymphotropic retroviruses in brain and other tissues from AIDS patients, Lancet, 1: 55-56 (1985).
45. D.D. Ho, T.R. Rota, R.T. Schooley et al., Isolation of HTLV-III from cerebrospinal fluid and neural tissues of patients with neurologic syndromes related to the acquired immunodeficiency syndrome, New Engl. J. Med., 313: 1493-1497 (1985).

46. J.A. Levy, J. Shimabukuro, H. Hollander, J. Mills, and L. Kaminsky, Isolation of AIDS-associated retroviruses from cerebrospinal fluid and brain of patients with neurological symptoms, Lancet, 2: 586-588 (1985).
47. G.M. Shaw, M.E. Harper, B.H. Hahn et al., HTLV-III infection in brains of children and adults with AIDS encephalopathy, Science, 277: 177-182 (1985).
48. S.Nielsen, C.K. Peito, C.D. Urmacher, and J.B. Posner, Subacute encephalitis in acquired immune deficiency syndrome: a postmortem study, Am. J. Clin. Pathol., 82: 678-682 (1984).
49. C.K. Petito, B.A. Navia, E.-S. Cho, B.D. Jordan, D.C. George, and R.W. Price, Vacuolar myelopathy pathologically resembling subacute combined degeneration in patients with the acquired immunodeficiency syndrome, New Engl. J. Med., 312: 874-879 (1985).
50. W.D. Snider, D.M. Simpson, S. Nielsen, J.W.M. Gold, C.E. Metroka, and J.B. Posner, Neurological complications of acquired immune deficiency syndrome: analysis of 50 patients, Ann. Neurol., 14: 403-418 (1983).
51. R.M. Levy, D.E. Bredesen, and M.L. Rosenblum, Neurological manifestations of the acquired immunodeficiency syndrome (AIDS): experience at UCSF and review of the literature, J. Neurosurg., 62: 475-495 (1985).
52. M.T. Lotze, Treatment of immunologic disorders in AIDS, in: "AIDS" (V.T. deVita, S. Hellman, and S.A. Rosenberg, Eds.) J.B. Lippincott Comp. USA, pp. 235-264 (1985).
53. R.Sanchez, M.D. Power, P.J. Barr, K.S. Stoimer, M.M. Stempien, S.L. Brown-Shimer, W.W. Gee, A. Renard, A. Randolph, J.A. Levy, D. Dina, and P.A. Luciw, Nucleotide sequence and expression of an AIDS-associated retrovirus (ARV-2), Science, 227: 484-492 (1985).
54. L.Ratner, W. Haseltine, R. Patarca, K.J. Livak, B. Starcich, S.F. Josephs, E.R. Doran, J.A. Rafalski, E.A. Whitehorn, K. Baumeister, L. Ivanoff, S.R. Petteway, M.L. Pearson,, J.A. Lautenberger, T.S. Papas, J. Ghrayeb, N.T. Chang, R.C. Gallo, and F. Wong-Staal, Complete nucleotide sequence of the AIDS virus, HTLV-III, Nature, 313: 277-284 (1985).
55. S.Wain-Hobson, P. Sonigo, O. Danos, S. Cole, and M. Alizon, Nucleotide sequence of the AIDS virus, LAV, Cell, 40: 9-17 (1985).
56. M.A. Muesing, D.H. Smith, C.D. Cabradilla, C.V. Benton, L.A. Lasky, and D.J. Capon, Nucleic acid structure and expression of the human AIDS/lymphadenopathy retrovirus, Nature, 313: 450-458 (1985).
57. J.Sodroski, C. Rosen, F. Wong-Staal, S.Z. Salahuddin, M. Popovic, S.K. Arya, R.C. Gallo, and W.A. Haseltine, Trans-acting transcriptional regulation of T-cell leukemia virus type III long terminal repeat, Science, 227: 171-173 (1985).
58. S.K. Arya, C. Guo, S.F. Josephs, and F. Wong-Staal, Trans-activator gene of T-lymphotropic virus type III (HTLV-III), Science, 229: 69-73 (1985).
59. S.K. Arya, R.C. Gallo, B.H. Hahn, G.M. Shaw, M. Popovic, S.Z. Salahuddin, and F. Wong-Staal, Homology of genome of AIDS-associated virus with genomes of human T-cell leukemia viruses, Science, 225: 927-930 (1984).
60. J.Sodroski, R. Patarca, C. Rosen, F. Wong-Staal, and W. Haseltine, Location of the trans-activating region on the genome of human T-cell lymphotropic virus type III, Science, 229: 74-77 (1985).
61. J.G. Sodroski, C.A. Rosen, and W.A. Haseltine, Trans-acting transcriptional activation of the long terminal repeat of human T-lymphotropic viruses in infected cells, Science, 225: 381-385 (1984).
62. S.K. Arya, and R.C. Gallo, Three novel genes of human T-lymphotropic

virus type III: Immune reactivity of their products with sera from acquired immune deficiency syndrome patients, Proc. Natl. Acad. Sci. USA, 83: 2209-2213 (1986).
63. J.G. Sodroski, W.C. Goh, C.A. Rosen, and W.A. Haseltine, Replicative and cytopathic potential of HTLV-III/LAV with sor gene deletions, Science, in press.
64. C.A. Rosen, J.G. Sodroski, W.C. Goh, A.I. Dayton, J. Lippke, and W.A. Haseltine, Post-transcriptional regulation accounts for the trans-activation of the human T-lymphotropic virus type III, Nature, 319, 555-559 (1986).
65. C.A.Rosen, J.G. Sodroski, and W.A. Haseltine, The location of cis-acting regulatory sequences in the human T-cell lymphotropic virus type III (HTLV-III/LAV) long terminal repeat, Cell, 41: 813-823 (1985).
66. A.I. Dayton, J.G. Sodroski, C.A. Rosen, W.C. Goh, and W.A. Haseltine, The Trans-activator gene of the human T cell lymphotropic virus type III is required for replication, Cell, 44:941-947 (1986).
67. J.Sodroski, W.C. Goh, C. Rosen, A. Dayton, E. Terwilliger, and W. Haseltine, A second post-transcriptional trans-activator gene required for HTLV-III replication, Nature, 321: 412-417 (1986).
68. B.H. Hahn, G.M. Shaw, S.K. Arya, M. Popovic, R.C. Gallo, and F. Wong-Staal, Molecular cloning and characterization of the virus associated with AIDS (HTLV-III), Nature, 312: 166-169 (1984).
69. F.Wong-Staal, B. Hahn, G.M. Shaw, S.K. Arya, M. Harper, M. Gonda, R. Gilden, L. Ratner, B. Starcich, T. Okamoto, S.F. Josephs, and R.C. Gallo, Molecular characterization of human T-lymphotropic leukemia virus type III associated with AIDS, in: "Retroviruses in human lymphoma/leukemia" (M. Miwa et al., Eds.) VNU Science Press, Utrecht, pp. 291-300 (1985).
70. E.Keshet, and H.M. Temin, Cell killing by spleen necrosis virus is correlated with a transient accumulation of spleen necrosis virus DNA, J. Virol., 31: 376-388 (1979).
71. S.K. Weller, A.E. Joy, and H.M. Temin, Correlation between cell killing and massive second round superinfection by members of some subgroups of avian leukosis virus, J. Virol., 33: 494-506 (1980).
72. J.E. Clements, and O. Narayan, A physical map of the linear un-integrated DNA of Visna virus, Virology, 113: 412-415 (1981).
73. A.Vogel, and P. Chandra, Evidence for two forms of reverse transcriptase in human placenta of a patient with breast cancer, Biochem. J., 197: 553-563 (1981).
74. P.Chandra, A. Vogel, and T. Gerber, Inhibitors of retroviral DNA polymerases: Their implication in the treatment of AIDS, Cancer Res. (Suppl.), 45: 4677S-4684S (1985).
75. A.Chandra, T. Gerber, and P. Chandra, Biochemical heterogeneity of reverse transcriptase purified from AIDS virus, HTLV-III. FEBS Letters, 197: 84-88 (1986).
76. A.Chandra, T. Gerber, S. Kaul, C. Wolf, I. Demirhan, and P. Chandra, Serological relationship between reverse transcriptases from human T-cell lymphotropic viruses defined by monoclonal antibodies, FEBS Letters, 200: 327-332 (1986).
77. T.W. Hsu, J.L. Sabran, G.E. Mark, R.V. Guntaka, and J.M. Taylor, Analysis of unintegrated avian RNA tumor virus double-stranded DNA intermediates, J. Virol., 28: 810-818 (1978).
78. P.R. Shank, and H.E. Varmus, Virus specific DNA in the cytoplasm of ASV-infected cells is a precursor to covalently closed circular DNA in the nucleus, J. Virol., 25: 104-114 (1978).
79. P.S. Sarin, and R.C. Gallo, RNA-directed DNA polymerase, Int. Rev. Sci., 6/8: 219-254 (1974).

80. F.M. Thompson, L.J. Libertini, U.R. Joss, and M. Calvin, Detergent effects on reverse transcriptase activity and on inhibition by rifamycin derivatives, Science, 178: 505-507 (1972).
81. H.M. Temin, and D. Baltimore, RNA-directed DNA synthesis and RNA tumor viruses, Virus Res., 17: 129-186 (1972).
82. P.Chandra, U. Ebener, L.K. Steel, H. Laube, D. Gericke, B. Mildner, T.J. Bardos, Y.K. Ho, and A. Götz, A molecular approach to inhibit oncogenesis by RNA tumor viruses, Ann. New York Acad. Sci., 284: 444-462 (1977).
83. P.Chandra, F. Zunino, A. Götz, A. Wacker, D. Gericke, A. Di Marco, A.M. Casazza, and F. Giuliani, Template specific inhibition of DNA polymerases from RNA tumor viruses by distamycin A and its structural analogues, FEBS Letters, 21: 154-158 (1972).
84. P.Chandra, A. Di Marco, F. Zunino, A.M. Casazza, D. Gericke, F. Giuliani, C. Soranzo, R. Thorbeck, A. Götz, F. Arcamone, and M. Thione, The role of molecular structure on the inhibition of DNA-polymerases from RNA tumor viruses, viral multiplication and tumor growth by some antitumor antibiotics, Naturwissenschaften, 59: 448-455 (1972).
85. P.Chandra, F. Zunino, and A. Götz, Bis-DEAE-fluorenone: a specific inhibitor of DNA polymerases from RNA tumor viruses, FEBS Letters, 22: 161-164 (1972).
86. P.Chandra, Molecular approaches for designing antiviral and antitumor compounds, Top. Curr. Chem., 52: 99-139 (1974).
87. P.Chandra, G. Will, D. Gericke, and A. Götz, Inhibition of DNA polymerases from RNA tumor viruses by tilorone and congeners: site of action, Biochem. Pharmacol., 23: 3259-3265 (1974).
88. P.Chandra, and B. Mildner, Zur molekularen Wirkungsweise von Di-amidinphenylindol (DAPI): I. Physikochemische Untersuchungen zur Charakterisierung der Bindung von DAPI an Nukleinsäuren, Cell. Mol. Biol., 25: 137-146 (1979).
89. B.Mildner, and P. Chandra, Zur molekularen Wirkungsweise von DAPI. II. Einwirkung von DAPI auf die Matrizenfunktion der DNA und der Polydesoxynukleotide in dem DNA Polymerase System bei Bakterien, Eukaryonten und RNA-Tumorviren, Cell. Mol. Biol., 25: 399-407 (1979).
90. P.Chandra, and B. Mildner, Zur molekularen Wirkungsweise von Di-amidinphenylindol (DAPI): III. Physikochemische Untersuchungen zur Bindung von DAPI-Derivaten an die DNA bzw. Polydesoxynukleotide und ihre Auswirkung auf die Matrizenfunktion der Nukleinsäuren, Cell. Mol. Biol., 25: 429-433 (1979).
91. P.Chandra, F. Zunino, A. Götz, D. Gericke, R. Thorbeck, and A. Di Marco, Specific inhibition of DNA polymerases from RNA tumor viruses by some new daunomycin derivatives, FEBS Letters, 21: 264-268 (1972).
92. P.Chandra, A. Di Marco, F. Zunino, A.M. Casazza, D. Gericke, F. Giuliani, C. Soranzo, R. Thorbeck, A. Götz, F. Arcamone, and M. Ghione, The role of molecular structure on the inhibition of DNA polymerases from RNA tumor viruses, viral multiplication and tumor growth by some antitumor antibiotics, Naturwissenschaften, 59: 448-455 (1972).
93. P.Chandra, D. Gericke, F. Zunino, and B. Kornhuber, The role of chemical structure to cytostatic activity of some daunomycin derivatives, Pharmacol. Res. Commun., 4: 269-272 (1972).
94. P.Chandra, Molecular approaches for designing antiviral and antitumor compounds, Top. Curr. Chem., 52: 99-139 (1974).
95. P.Chandra, L.K. Steel, U. Ebener, M. Woltersdorf, H. Laube, B. Kornhuber, B. Mildner, and A. Götz, Chemical inhibitors of oncorna-viral DNA polymerases, in: "International Encylcopaedia of

Pharmacology and Therapeutics (Section 103)", Pergamon Press, Oxford, pp. 47-89 (1980).
96. P.Chandra, and T.J. Bardos, Inhibition of DNA polymerases from RNA tumor viruses by novel template analogues: partially thiolated polycytidylic acid, Res. Commun. Chem. Pathol. Pharmacol., 4: 615-622 (1972).
97. P.Chandra, U. Ebener, and A. Götz, Inhibition of oncornaviral DNA polymerase by 5-mercapto polycytidylic acid: mode of action, FEBS Letters, 53: 10-14 (1975).
98. P.Chandra, L.K. Steel, U. Ebener, M. Woltersdorf, H. Laube, B. Kornhuber, M. Mildner, and A. Götz, Chemical inhibitors of oncornaviral DNA polymerases: biological implications and their mode of action, Pharmacol. Ther. A, 1: 231-287 (1977).
99. P.Chandra, U. Ebener, T.J. Bardos, D. Gericke, B. Kornhuber, and A. Götz, Inhibition of viral reverse transcriptase by modified nucleic acids: biological implications and their mode of action, in: "Fogarty International Center Proceedings No. 28", United States Government Printing Office, Washington, DC, pp. 169-186 (1977).
100. P.Chandra, Selective inhibition of oncornaviral functions (a molecular approach), in: "Antimetabolites in Biochemistry and Biology and Medicine" (J. Skoda, and P. Langen, Eds.) Pergamon Press, Oxford, pp. 249-261 (1979).
101. U.Ebener, Hemmung der viralen Reverse Transkriptase und Leukämogenese durch modifizierte Nukleinsäuren, PhD. Dissertation, Johann Wolfgang Goethe-Universität, Frankfurt (1977).
102. P.Chandra, I. Demirhan, U. Ebener, and B. Kornhuber, Virus-associated DNA polymerizing activities: their role in designing antiviral and antitumor drugs, in: "Targets for the Design of Antiviral Agents" (E. de Clercq, and R.T. Walker, Eds.) Plenum Publishing Corp., New York, pp. 307-335 (1984).
103. H.Mitsuya, M. Popovic, R. Yarchoan, S. Matsushita, R.C. Gallo, and S. Broder, Suramin protection of T-cells in vitro against infectivity and cytopathic effect of HTLV-III, Science, 226: 172-174 (1984).
104. P.Ehrlich, and K. Shiga, Berl. Klin. Wochschr., 41:329-362 (1904).
105. F.Mesnil, and M. Nicolle, Ann. Inst. Pasteur, 20: 417 (1906).
106. B.Z. Heymann, Angew. Chem., 37: 585 (1924).
107. S.Broder, J.M. Collins, P.D. Markham et al., Effects of suramin on HTLV-III/LAV infection presenting as Kaposi's sarcoma or AIDS-related complex. Clinical pharmacology and suppression of virus replication in vivo, Lancet, ii: 627-630 (1985).
108. W.Busch, R. Brodt, A. Ganser, E.B. Helm, and W. Stille, Suramin treatment for AIDS, Lancet, iii: 1247 (1985).
109. A.Fauci, and H.C. Lane, Therapeutic approaches to the underlying immune defect in patients with AIDS, in: "II. Internat. Conf. AIDS", Paris, Commun. SP8 (1986).
110. W.Rozenbaum, D. Dormont, B. Spire, E. Vilmer, M. Gentilini, C. Griselli, L. Montagnier, F. Barre-Sinoussi, and J.-C. Chermann, Antimoniotungstenate (HPA23) treatment of three patients with AIDS and one with prodrome, Lancet, i: 450-451 (1985).
111. I.Demirhan, and P. Chandra, unpublished observations.
112. K.Ono, H. Nakane, T. Matsumoto, F. Barre-Sinoussi, and J.-C. Chermann, Inhibition of DNA polymerase α activity by ammonium-21-tungsto-9-antimonate (HPA23), Nucleic Acids Res., 15: 169-172 (1984).
113. F.Sinoussi, Inhibition of RNA-dependent DNA polymerase of oncornaviruses by 5-tungsto-2-antimoniate, Cancer Res., 16:201 (1975).
114. P.A. Furman, M. StClair, K. Weinhold, J.A. Fyfe, S. Nusinoff-

Lehrmann, and D.W. Barry, Selective inhibition of HTLV-3 by BW A509U, ICAAC, Abstr. 440 (1985).
115. H.Mitsuya, D.W. Barry, S. Nusinoff-Lehrmann, and S. Broder, BW A509U blocks HTLV-3 infection of T lymphocytes in culture, ICAAC, Abstr. 437 (1985).
116. M.H. StClair, K. Weinhold, C.A. Richards, D.W. Barry, and P.A. Furman, Characterization of HTLV-3 reverse transcriptase and inhibition by the triphosphate of BW A509U, ICAAC, Abstr. (1985).
117. E.G. Sandström, R.E. Byington, J.C. Kaplan, and M.S. Hirsch, Inhibition of human T-cell lymphotropic virus type III in vitro by phosphonoformate, Lancet, i: 1480-1482 (1985).
118. P.S. Sarin, Y. Taguchi, D. Sun, A. Thornton, R.C. Gallo, and B. Öberg, Inhibition of HTLV-III/LAV replication by foscarnet, Biochemical Pharmacol., 34: 4075-4079 (1985).
119. L.Vrang, and B. Öberg, Pyrophosphate analogs as inhibitors of human T-lymphotropic virus type III reverse transcriptase, Antimicrobial Agents and Chemotherapy, in press (1986).
120. P.Chandra, and P.S. Sarin, Selective inhibition of replication of the AIDS-associated virus, HTLV-III/LAV, by synthetic D-penicillamine, Arzneimittelforschg. (Drug Res.), 36: 184-186 (1986).

INHIBITION OF HTLV-III REPLICATION IN CELL CULTURES

P. S. Sarin, D. Sun, A. Thornton and Y. Taguchi

Laboratory of Tumor Cell Biology
National Cancer Institute
Bethesda, Maryland 20892, USA

A human T-lymphotropic retrovirus (HTLV-III) has been identified as the etiological agent for acquired immune deficieny syndrome (AIDS) and AIDS related complex (ARC) (1-4). Various therapeutic approaches are currently being investigated to control the disease either with inhibitors of reverse transcriptase and virus replication or with a vaccine. HTLV-III is a cytopathic retrovirus which selectively infects T-helper cells and kills OKT4+ T helper cells resulting in immune suppression (1-5). HTLV-III contains an RNA directed DNA polymerase (reverse transcriptase) and buds from the cell membrane like other animal retroviruses (6,7). The replication of virus in the infected cells and further infection of uninfected cells with the newly produced virus can be interfered by chemotherapeutic agents that can attack the various steps in the replication cycle (Fig. 1) including virus attachment, reverse transcription, and DNA integration.

The virus identified as the etiological agent of AIDS infects T-helper cells by binding to the cell through a receptor site identified as the T4 receptor or CD4 (8,9). The binding of HTLV-III to the T4 cells could be blocked by antibodies to the T4 receptor (8,9) suggesting that the virus binds to the cells through this receptor or a receptor adjacent to or overlapping with the T4 receptor. Most of the AIDS or ARC patients contain antibodies to the viral antigens that are either not neutralizing or have very low virus neutralizing activity and hence not protective against virus infection of uninfected T cells. The virus after attachment to the T4 cells through a receptor on the cell surface, enters the cell followed by uncoating of the virus, synthesis of DNA from the viral RNA and subsequent integration of the viral DNA into the host genome. The viral DNA can also be present in the unintegrated form and can replicate in the cytoplasm. Agents such as amantadine have been used in the past to block infection with influenza virus (10). Whether similar agents will be useful in preventing HTLV-III infection of T cells and other susceptible cells such as macrophages, B cells and monocytes which can also be infected with HTLV-III to a lesser extent, remains to be determined.

To control HTLV-III infection and to prevent replication in the AIDS and ARC patients, it is important to examine the various steps involved in the HTLV-III infection and replication in T cells. HTLV-III infects the target cells by attaching through a specific receptor, entry into the

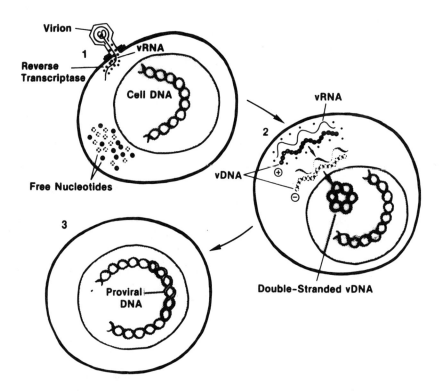

Figure 1. Various stages in the infection of T cells with HTLV-III and virus replication in infected cells

cell, uncoating and exposing reverse transcriptase and viral RNA, followed by transcription of the RNA into complementary DNA (cDNA), conversion to proviral DNA (double stranded DNA, dsDNA) followed by circularization and integration into host DNA. The transcription of DNA to messenger RNA and protein synthesis in the infected cell results in the synthesis of viral RNA and gag and envelope proteins which then attach to the cell membrane in the form of a nucleoid followed by budding and release of more infectious HTLV-III virus particles.

Interference with virus attachment

Since HTLV-III binds to the T cells through the T4 receptor (8,9) or a receptor adjacent or overlapping with the T4 receptor, it should be possible to block virus attachment to the target cells by blocking virus attachment to the receptor by using monoclonal antibodies which specifically block the receptor site and prevent virus attachment. Monoclonal antibodies made against the T4 receptor (especially 4a and 4i) have recently been shown to block the infection of OKT4+ positive T-helper cells (Krohn et al., unpublished results). Synthetic peptides specific for the receptor site to which the virus binds may also be effective in blocking virus infection, as seen in the case of myxovirus and paramyxovirus infections (9). Another approach would be to use agents that will destroy the integrity of the viral envelope, thus poking holes in the virus envelope making it inactive or less infectious. One such agent (AL721) was recently shown (12) to interfere with HTLV-III infection and replication by extracting cholesterol from the viral envelope, which is composed of glycoproteins, phospholipid and cholesterol (Sarin and Crews, unpublished results). Another compound that has been found to block HTLV-III replication in cell culture is D-Penicillamine (13). Preliminary results with D-Penicillamine in the clinic have been encouraging and this compound is currently undergoing clinical trials against asymptomatic homosexuals with lymphadenopathy and ARC patients. D-Penicillamine has been used in the past for the treatment of Wilson's disease, chronic hepatitis and rheumatoid arthritis, and has shown some immunosuppressive activity. Whether any of these compounds will be effective in treatment of AIDS or ARC remains to be seen.

Interference with reverse transcription

Inhibition of reverse transcriptase activity of retroviruses has been a major target for the development of antiviral agents against replication of animal retroviruses. Rifamycins (14) and streptovaricins (15) have been the earliest compounds that were used to inhibit replication of Rauscher murine leukemia virus (RLV) and feline leukemia virus (FeLV) both in vitro and in vivo. Some of the rifamycins (Table 1) analogs found to be active against RLV did not inhibit HTLV-III replication in H9 cells. Whether the rifamycin analogs found to be active in early studies with animal retroviruses and inactive in blocking HTLV-III replication was due to the fact that the compounds were synthesized in the 1970's and hence may have degraded over the years or whether they are really inactive against HTLV-III is difficult to discern. Synthetic polynucleotides (16) were also reported to be effective in blocking murine virus replication in vivo and in vitro by interfering with the binding of template primer to murine retrovirus reverse transcriptase. Several compounds including 5-mercapto-poly C have recently been reported by Chandra and co-workers to block HTLV-III reverse transcriptase (17). A potent inhibitor of reverse transcriptase is foscarnet (phosphonoformate) (18,19). The chemical structure of

Table 1. Rifamycin Derivatives Inactive in HTLV-III Replication System

No.	Code	Chemical Structure
1	AF/AMP (Rifampicin)	3-(4-methylpiperazinoiminomethyl) rifamycin SV
2	AF/AP (N-demethyl rifampicin)	3-(piperazinoiminomethyl) rifamycin SV
3	AF/ABDMP (DMB)	3-(4-benzyl-2,6-dimethylpiperazinoiminomethyl) rifamycin SV
4	AF/DNFI	3-(2,4-dinitrophenylhydrazonomethyl) rifamycin SV
5	AF/DPI	3-(dipropylhydrazonomethyl) rifamycin SV
6	AF-013	3-formyl rifamycin SV: O-n-octyloxime
7	AF-05	3-formyl rifamycin SV: O-(diphenylmethyl) oxime
8	PR/14	3'-acetyl-1,2'-dimethylpyrrolo[3,2-c]-4-deoxy rifamycin SV
9	PR/19	3'-acetyl-1'-benzyl-2'-methylpyrrolo[3,2-c]-deoxyrifamycin SV

foscarnet and the mechanism of action of this compound is shown in Figure 2. This compound was first identified for treatment of cytomegalovirus (CMV) infections and has been used in the clinic to control CMV infections (20). Foscarnet completely inhibits reverse transcriptase activity at a concentration of 150 µM and the concentration for complete inhibition of HTLV-III replication in H9 cell cultures is between 150 - 300 µM (Fig. 3). The drug does not show any toxicity to the cells in culture up to a concentration of 750 µM. Foscarnet is currently undergoing clinical trials in AIDS and ARC patients in various countries. This compound is relatively non-toxic and it is given to the patients by continuous infusions. It will be of interest to produce a slow release oral form of this compound which could be prescribed to the AIDS/ARC patients as outpatients in the clinic or hospitals.

Suramin (21,22) and 3'-azidothymidine (AZT) (23,24) are two other reverse transcriptase inhibitors that are being actively evaluated both in the laboratory and the clinic. In clinical trials both Suramin and Azidothymidine suppress HTLV-III during the course of treatment, virus reappears after discontinuation of treatment suggesting that the virus suppression is transient. Unfortunately, Suramin has also shown toxic side effects in AIDS/ARC patients and hence cannot be administered for extended periods of time. No significant clinical benefit was observed in KS and AIDS patients treated with Suramin (Table 2), and in some patients increased incidence of KS lesions was observed.

Ribavarin has been used to block replication of murine retroviruses both in vitro and in vivo and more recently, it was shown to block the replication by HTLV-III/LAV in vitro (25). This compound is undergoing clinical trials in AIDS and ARC patients in France and other countries but the results so far look less than promising.

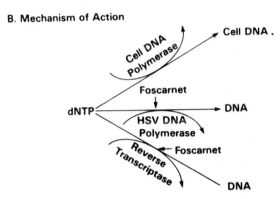

Figure 2. Structure and mechanism of action of foscarnet.
(A) Chemical structure of foscarnet.
(B) Mechanism of action of foscarnet against cellular and viral DNA polymerases.

Ammonium antimony tungstate (HPA-23) is another inhibitor of HTLV-III/LAV reverse transcriptase which has recently been used in clinical trials in France on patients with AIDS and ARC (26). Transient reduction in HTLV-III/LAV circulating in peripheral blood lymphocytes was observed, but the compound has been used at doses showing toxic side effects. In our studies (Sarin et al., unpublished results) HPA-23 did not show any significant reduction in HTLV-III/LAV replication in cell culture system at concentrations that are not toxic to the cultured cells. Further expanded clinical trials on HPA-23 are currently in progress and will show whether this compound will eventually be useful in the treatment of AIDS or ARC.

Cyclosporin is another compound that was recently publicized by French workers in the treatment of AIDS patients. The structure of cyclosporin (Fig. 4) and its possible mechanism of action (Fig. 5) show that this

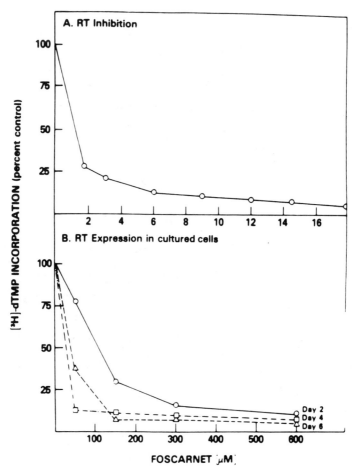

Figure 3. Effect of foscarnet on HTLV-III reverse transcriptase and virus replication
(A) Inhibition of HTLV-III reverse transcriptase.
(B) Inhibition of HTLV-III replication in cell culture.

Table 2. Treatment of AIDS, ARC, and Kaposi Sarcoma Patients with Suramin[18]

Diagnosis	TH/TS* BT	AT	Clinical Status
A. Early KS			
Patient No. 1	0.62	0.54	No change
Patient No. 2	1.1	0.5	New KS lesions
Patient No. 3	0.35	0.5	No change
B. Advanced KS			
Patient No. 4	0.4	0.17	New KS lesions
Patient No. 5	0.5	0.38	No change
C. ARC			
Patient No. 6 (LAS)	0.66	1.0	No change
Patient No. 7 (LAS)	1.0	1.0	No change
Patient No. 8 (oral candidiasis)	0.22	0.25	No change
Patient No. 9 (oral candidiasis)	0.33	0.5	PCP Pneumonia

*TH = T-helper cells; TS = T-suppressor cells; BT = before treatment, AT = after treatment

compound is immunosuppressive. In our studies on the inhibition of HTLV-III replication in H9 cells, we found that this compound does not inhibit replication of HTLV-III nor that of HTLV-I or HTLV-II (Fig. 6). HTLV-I is a retrovirus that is associated with the cause of adult T cell leukemia (ATL) and HTLV-II is a related virus which has been isolated from patients with hairy cell leukemia (1,3,27).

A list of other compounds that we have examined in our laboratory is given in Table 3. Majority of these compounds have an ID-50 of around 300 µg/ml except foscarnet, doxorubicine-HCl and HPA-23.

Inhibitors of DNA and RNA Transcription

Another approach that has been explored for inhibition of retrovirus replication is the use of antisense oligonucleotide inhibitors (synthetic oligonucleotides) designed to bind to specific target sites of the viral genome. Zamecnik and co-workers (28) used synthetic oligonucleotides (chain length 13-15) in the inhibition of Rous sarcoma virus in cell culture. These studies prompted us to examine oligonucleotides specific

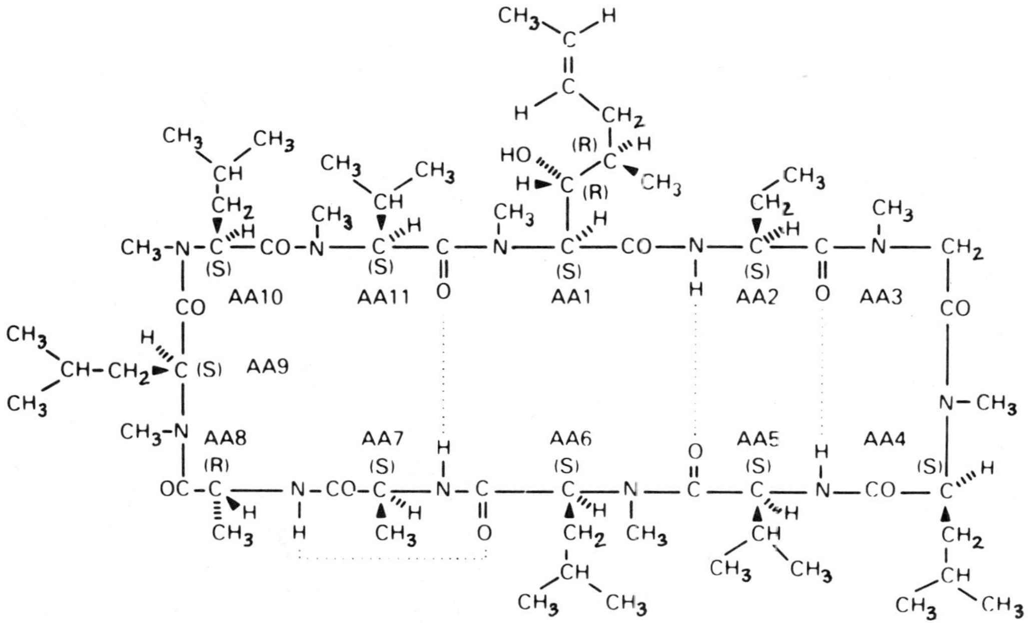

Figure 4. Structure of cyclosporin.

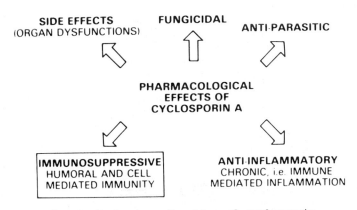

Fig. 5. Mechanism of action of cyclosporin.

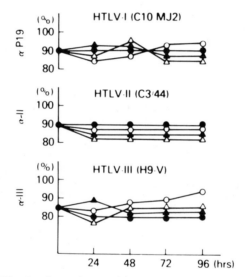

Figure 6. Effect of cyclosporin on HTLV-I, HTLV-II or HTLV-III replication in cell culture.
(A) HTLV-I; (B) HTLV-II; (C) HTLV-III.

Table 3. HTLV-III Reverse Transcriptase Inhibition by Various Drugs

NSC No.	Drug	ID50 (µg/ml)
1. 103-627	Azacytidine	> 300
2. 253-272	Caracemide	> 300
3. 241-240	Carboplatin	> 300
4. 145-668	Cyclocytidine HCl	> 300
5. 126-849	Deazuridine	> 300
6. 261-036	Desmethylmisonidazole	> 300
7. 132-313	Diandydrogalactitol	> 300
8. 118-994	Diglycoaldehyde	> 300
9. 264-880	Dihydroazacytidine HCl	> 300
10.	Doxorubicin HCl	> 20
11. 134-490	Emofolin	> 300
12. 296-961	Ethiofos	> 300
13. 312-887	Fludarabine Phosphate	> 300
14. PFA	Foscarnet	> 10
15.	HPA-23	> 100
16. 301-467	Hydroxyethyl-nitro imidazole acetamide	> 300
17. 169-780	ICRF-187	> 300
18. 129-943	ICRF-159	> 300
19. 132-319	Indicine N-oxide	> 300
20. 8806	Melphalan	> 300
21. 261-037	Misonidazole	> 300
22. 224-131	PALA-Disodium	> 300
23. 118-742	Pentamethylamine-2 HCl	> 300
24. 218-321	Pentostatin	> 300
25. 135-758	Piperazinedione-bis chlorpiperidyl HCl	> 300
26. 192-965	Spirogermanium HCl	> 300
27. 314-055	SR-2555	> 300
28. 148-958	Tegafut	> 300
29. 286-193	Tiazofurin	> 300
30. 281-272	5-Azacytosine arabinoside	> 300

for certain regions of the HTLV-III genome. Since the complete nucleotide sequence of the HTLV-III genome is known, we picked regions adjacent to the primer binding site and tat-III gene splice acceptor and donor sites (Table 4). As shown in Table 4, oligonucleotides of chain length twenty were found to be most active in inhibiting virus replication (29). The greatest inhibition of virus replication was observed by oligonucleotides specific for the tat-III gene splice acceptor and donor sites. Whether these compounds will prove to be useful in the treatment of AIDS or ARC patients remains to be determined.

Table 4. Effect of synthetic oligonucleotides on HTLV-III replication in cell culture

No.	Oligonucleotide Sequence	Chain Length	Conc. µg/ml	HTLV-III Binding Site	Percent Control RT	p15	p24
1	None	-					
2	CCCCAACTGTGTACT	15	5	None	100	100	100
3	CTGCTAGAGATddT	12	5	5'-vicinal to PBS	90	85	65
			10	5'-vicinal to PBS	83	85	50
4	CTGCTAGAGATddT	12	10	5'-vicinal to PBS	100	90	88
			20	5'-vicinal to PBS	100	72	62
5	CTGCTAGAGATTTTCCACAC	20	50	5'-vicinal to PBS	50	50	50
			10x3	5'-vicinal to PBS	50	25	25
6	TTCAAGTCCCTGTTC-GGGCGCCAAAA	26	50	at PBS	20	96	92
7	GCGTACTCACCAGTCGCCGC	20	50	splice donor site (tat-III)	15	60	40
8	ACACCCAATTCTGAAAATGG	20	50	splice acceptor site (tat-III)	33	5	12

PBS = primer binding site

REFERENCES

1. R.C. Gallo, S.Z. Salahuddin, M. Popovic, G.M. Shearer, M. Kaplan, B.F. Haynes, T.J. Palker, R. Redfield, J. Oleske, B. Safai, G. White, P. Foster, and P.D. Markham, Frequent detection and isolation of cytopathic retrovirus (HTLV-III) from patients with AIDS and at risk for AIDS. Science, 224: 500-503 (1984)
2. P.S. Sarin, and R.C. Gallo, The involvement of human T-lymphotropic retroviruses in T cell leukemia and immune deficiency. Cancer Reviews, 1: 1-17 (1986).
3. P.S. Sarin, and R.C. Gallo, Human T-lymphotropic retroviruses in adult T-cell leukemia-lymphoma and acquired immune deficiency syndrome. J. Clin. Immunolog., 4: 415-423 (1984)
4. R.C. Gallo, The human T cell leukemia/lymphotropic retroviruses

(HTLV) family: past, present and future. Cancer Res. (Suppl.) 45: 4524- 4533 (1985)
5. M.Popovic, M.G. Sarngadharan, E. Read, and R.C. Gallo, Detection, isolation and continuous production of cytopathic retroviruses (HTLV-III) from patients with AIDS and preAIDS. Science, 224: 497-500 (1974)
6. P.S. Sarin, and R.C. Gallo, RNA directed DNA polymerase, in: "MTP International Rev. Science" (Burton, K., edt.) Butterworths,London, vol. 6. pp. 219-254 (1974)
7. H.Temin, RNA directed DNA synthesis. Sci. American, 226:25-33 (1972)
8. A.B. Dalgleish, P.C. Beverley, P.R. Clapham, D. Crawford, M.F. Greaves, and R.A. Weiss, The CD4 (T4) antigen is an essential component of the receptor for the AIDS retrovirus. Nature, 312: 763-767 (1984)
9. D.Klatzmann, E. Champagne, and S. Chamaret, T-lymphocyte T4 molecule behaves as the receptor for human retrovirus LAV. Nature, 312: 767-768 (1984)
10. R.Dolin, R.C. Reichman, H. Madore, R. Maynard, P. Linton, and J. Weber-Jones, A controlled trial of amantadine and rimantadine in the prophylaxis of influenza A infection. New Engl. J. Med., 307: 508-509 (1982)
11. C.D. Richardson, A. Scheid, and P.W. Choppin, Specific inhibition of paramyxovirus and myxovirus replication by oligonucleotides with amino acid sequences similar to those at the N-termini of the F1 or HA2 viral polypeptides. Virology, 105:205-222 (1985)
12. P.S. Sarin, R.C. Gallo, D.I. Scheer, F. Crews, and A.S. Lippa, Effects of AL 721 on HTLV-III infectivity: a novel approach for the potential treatment of AIDS. New Engl. J. Med., 313:1289-1290 (1985)
13. P.Chandra, and P.S. Sarin, Selective inhibition of replication of the AIDS associated virus HTLV-III/LAV by synthetic D-penicillamine. Arzneimittelforschung (Drug Research) 36: 184-186 (1986)
14. C.Gurgo, Rifamycins as inhibitors of RNA and DNA polymerases, in: "Inhibitors of RNA and DNA polymerases. Intl. Encyl. of Pharmacol. Therapeutics" (P.S. Sarin, R.C. Gallo, edts.) Pergamon Press, New York, Section 103, pp. 235-247 (1980)
15. B.I.Milavetz, and W.A. Carter, Streptovaricins, in: "Inhibitors of RNA and DNA polymerases. Intl. Encycl. of Pharmacol. Therapeutics" (P.S. Sarin, R.C. Gallo, edts.) Pergamon Press, New York, Section 103, pp. 191-206 (1980)
16. P.M. Pitha, and J. Pitha, Polynucleotide analogs as inhibitors of RNA and DNA polymerases, in: "Intl. Encycl. of Pharmacol. and Therapeutics" (P.S. Sarin, R.C. Gallo, edts.) Pergamon Press, New York, Section 103, pp. 235-247 (1980)
17. P.Chandra, A. Vogel, and T. Gerber, Inhibitors of retroviral DNA polymerase. Their implication in treatment of AIDS. Cancer Res. (Suppl.), 45: 4677-4684 (1985)
18. P.S. Sarin, Y. Taguchi, D. Sun, A. Thornton, R.C. Gallo, and B. Oberg, Inhibition of HTLV-III/LAV replication by foscarnet. Biochem. Pharmacol., 34: 4075-4079 (1985)
19. E.G. Sandstrom, J.C. Kaplan, R.E. Byington, and M.S. Hirsch, Inhibition of human T-cell lymphotropic virus type III in vitro by phosphonoformate. Lancet, 1: 1480-1482 (1985)
20. B.Oberg, Control of cytomegalovirus infections with foscarnet, personal communication
21. H.Mitsuya, M. Popovic, R. Yarchoan, S. Matsushita, R.C. Gallo, and S. Broder, Suramin protection of T cells in vitro against infectivity and cytopathic effect of HTLV-III. Science, 226:172-174 (1984)

22. S.Broder, R. Yarchoan, J.M. Collins, H.C. Lane, P.D. Markham, R.W. Klecker, R.R. Redfield, H. Mitsuya, D.F. Hoth, E. Gelmann, J.E. Groopman, L. Resnick, R.C. Gallo, C.E. Myers, and A.S. Fauci, Effects of suramin on HTLV-III/LAV infection presenting as Kaposi sarcoma or AIDS related complex: clinical pharmacology and suppression of virus replication in vivo. Lancet, 2:627-630 (1985)
23. H.Mitsuya, K.J. Weinhold, P.A. Furman, M.H. Clair, S.N. Lehrman, R.C. Gallo, D. Bolognesi, and S. Broder, 3'-azido-3'-deoxythymidine (BW A509U): an antiviral agent that inhibits the infectivity and cytopathic effect of human T-lymphotropic retrovirus type III/lymphadenopathy associated virus in vitro. Proc. Natl. Acad. Sci. (USA), 82: 7096-7100 (1985)
24. R.Yarchoan, R.W. Klecker, K.J. Weinhold, P.D. Markham, H.K. Lyerly, D.T. Durack, E. Gelmann, S.N. Lehrmann, R.M. Blum, D.W. Barry, G.M. Shearer, M.A. Fischl, H. Mitsuya, R.C. Gallo, J.M. Collins, D. Bolognesi, C.E. Myers, and S. Broder, Administration of 3'-azido-3'-deoxythymidine, an inhibitor of HTLV-III/LAV replication to patients with AIDS or AIDS related complex. Lancet, 1: 575-580 (1986)
25. J.B. McCormick, J.P. Getchell, S.W. Mitchell, and D.R. Hicks, Ribavarin suppressed replication of lymphadenopathy associated virus in cultures of human adult T-lymphocytes. Lancet, 2:1367-1369 (1984)
26. W.Rozenbaum, D. Dormont, B. Spire, E. Vilmer, M. Gentilini, C. Griscelli, L. Montagnier, F. Barre-Sinoussi, and J.C. Chermann, Antimoniotungstate (HPA23) treatment of three patients with AIDS and one with prodrome. Lancet, 1: 450-451 (1985)
27. P.S. Sarin, and R.C. Gallo, Human T-cell leukemia-lymphoma virus (HTLV). Prog. Hematol., 13: 149-161 (1983)
28. P.C. Zamecnik, and M.L. Stephenson, Inhibition of Rous sarcoma virus replication and transformation by a specific oligonucleotide. Proc. Natl. Acad. Sci. (USA), 75: 280-284 (1978)
29. P.C. Zamecnik, J. Goodchild, Y. Taguchi, and P.S. Sarin, Inhibition of replication and expression of the human T-cell lymphotropic virus (HTLV-III) in cultured cells by exogenous synthetic oligonucleotides complementary to viral RNA. Proc. Natl. Acad. Sci. (USA), in press

PROTEINS ENCODED BY THE HUMAN T-LYMPHOTROPIC VIRUS Type III/
LYMPHADENOPATHY ASSOCIATED VIRUS (HTLV-III/LAV) GENES

M.G. Sarngadharan[*], Fulvia diMarzo Veronese[*], and Robert C. Gallo[+]

[*] Department of Cell Biology, Bionetics Research, Inc. Rockville, MD 20850

[+] Laboratory of Tumor Cell Biology
National Cancer Institute, Bethesda, MD 20892

Introduction

Human T-lymphotropic virus-type III/lymphadenopathy associated virus (HTLV-III/LAV) has been identified as the primary etiologic agent of human acquired immune deficiency syndrome (AIDS) (1-4). Among members of epidemiologically identified risk groups of AIDS, there has been a high prevalence of HTLV-III/LAV infection irrespective of whether they have clinical AIDS or have only some signs and symptoms associated with AIDS [AIDS-related complex (ARC)], or are completely free of any disease symptoms (4-6). Thus a significant proportion (up to 50 or 60 %), depending on where they live) of homosexual males at risk for AIDS have serum antibodies to HTLV-III/LAV proteins (6,7). Infectious virus has been isolated from lymphocyte cultures derived from many of these asymptomatic antibody-positive individuals (2,8). HTLV-III/LAV-infection through contaminated blood has been identified as the cause of AIDS in transfusion recipients with no other risk factor for AIDS (9,10).

Serological and virus isolation studies have clearly identified an asymptomatic, contagious carrier state as part of the clinical spectrum of HTLV-III/LAV infection (2,8-11). Serological assays currently in use, for instance, the ELISA and Western blot assays for HTLV-III/LAV antibodies (4,5) are designed to identify HTLV-III/LAV infection. Interpretation of these assays, especially the Western blot patterns, requires an understanding of the various HTLV-III/LAV antigens. Characterization of the viral antigens and identification of the viral genes encoding these antigens might help in diagnosing infection and in devising strategies for developing antiviral vaccines. Further, the recognition of the most immunogenic HTLV-III/LAV antigens during the natural infection will be extremely valuable in formulating more sensitive and specific antibody tests.

Gene organization of HTLV-III/LAV

All replication competent retroviruses have three essential genes: gag, pol and env which code for the internal structural proteins, the reverse transcriptase (RT) and the envelope proteins respectively. HTLV-III/LAV also has these three genes. In addition it has at least three other functional genes identified as sor, tat and 3'orf (12-14). Figure 1 shows a diagrammatic representation of the viral genome, indicating organization of the individual genes. A great deal of information is known about the products of the gag, pol and the env genes and will be presented in some detail in the following sections. We know that sor, tat and 3'orf are also functional genes, and a short description of these will be presented towards the end.

Fig. 1. Gene organization of HTLV-III/LAV

Products of HTLV-III/LAV gag gene

Retroviral gag gene generally codes for a polyprotein which, during virus maturation, is processed into many smaller proteins. These gag proteins constitute the major protein fraction in any mature virion. HTLV-III/LAV is no exception. Like in HTLV-I and -II, the major immunogenic gag protein in HTLV-III/LAV is a 24,000 dalton protein (p24) (4,5,15,16) and is the middle peptide released during the processive cleavage of the precursor molecule (Fig. 2). The amino terminal peptide (p17) is smaller than the homologous protein of HTLV-I (4,17). The carboxy terminal peptide of HTLV-III/LAV is p15 which may be further processed into two smaller peptides, a p7 and a p9.

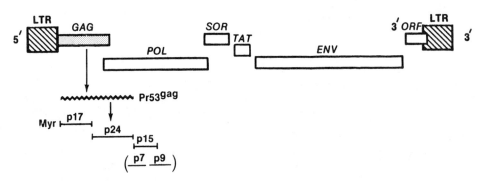

Fig. 2. Processing of the HTLV-III/LAV gag gene product

Much of the structural correlations between the precursor molecule in the infected cell and the processed proteins in the mature virion were made with the help of specific monoclonal antibodies raised against individual proteins (Ref. 18 and our unpublished data). For instance, monoclonal antibodies to both p17 and p24 precipitated a common 53,000 dalton-protein from extracts of cells producing HTLV-III/LAV. Therefore, the specific antigenic epitopes recognized by anti-p17 and anti-p24 are both present on the p53 molecule. The cellular p53 is thus the precursor of both p24 and p17. When HTLV-III/LAV-producing H9 cells were metabolically labeled with [^{3}H]myristic acid and the labeled cell extract was incubated with anti-p17 and anti-p24 in separate experiments, both antibodies precipitated radioactive p53. In addition, the p17 precipitated by anti-p17 was also radioactive. No radioactivity was detected in p24 precipitated by anti-p24. These experiments demonstrate that HTLV-III/LAV, like HTLV-I and other mammalian retroviruses, incorporates a myristate-radical in an acyl linkage at the amino terminal end (19,20). As expected, the amino terminal amino acid of p17 was blocked and was inaccessible to Edman degradation.

Both p17 and p24 are immunogenic in the natural host, although p24 is more immunogenic than p17. Sera of most HTLV-III/LAV-infected people react with these antigens in immunological assays. The carboxy terminal gag protein is much less immunogenic than p24 and p17 and antibodies to this, or its further processed smaller derivatives, are not recognized in the sera of many virus-infected individuals.

Immunologically HTLV-III/LAV is a unique retrovirus with very little cross-reactivity with other human or animal retroviruses. A rabbit anti-serum to HTLV-II reacted with HTLV-III/LAV p24 on the Western blot but there was no reactivity in the reciprocal test (15). Mouse monoclonal antibodies reactive with p24 of HTLV-I, -II and -III have been raised and these antibodies reveal no cross-reactivities between HTLV-III/LAV and either of the other two viruses (18). A one-way cross reactivity was recognized between HTLV-III/LAV p24 and equine infectious anemia virus (EIAV) p26. Horse sera with antibodies to EIAV p26 reacted with HTLV-III/LAV p24 on Western blot immunoassay (15) but antibodies to HTLV-III did not react with the EIAV core protein under the same conditions. The extent of this cross-reactivity was limited because it could not be detected in conventional competition radio immunoassays (15).

Products of the HTLV-III/LAV pol gene

HTLV-III/LAV pol gene codes for a reverse transcriptase (RT) and an endonuclease (Fig. 3) (12). The RT of HTLV-III/LAV shares many biochemical features with the RT of HTLV-I. It prefers Mg^{2+} over Mn^{2+} as the cofactor for the enzyme activity. Also like HTLV-I RT, the enzyme from HTLV-III/LAV has a higher apparent molecular weight than its counterpart in other mammalian retroviruses (21). It does, however, have many unique properties as described below.

Approximately 80 % of all HTLV-III/LAV-seropositive individuals have a serological reaction to two peptides at 66 and 51 kilodaltons in Western blot analysis. A mouse hybridoma producing a monoclonal antibody (M3364) against these antigens was obtained, which even after repeated cloning retained its specificity to both peptides, indicating that p66 and p51 shared peptide sequences (22) (Figure 4). Similar findings are obtained in another independent study (23). These proteins were purified from an extract of HTLV-III/LAV by an immunoaffinity procedure employing the immobilized monoclonal antibody. The isolated protein was homogeneous and contained two immunoreactive peptides at 66 and 55 kilodaltons (Figure 5). Amino terminal Edman degradation of the p66/51 generated single amino acid residues in 17 successive cycles. A comparison of the amino acid sequence observed with the sequence predicted from the known nucleotide sequence of HTLV-III/LAV, clone BH10 and BH5 showed a perfect match for a product of the pol gene segment between nucleotides 2130 and 2180 (12). The sequence date indicate that p66 and p51 have common sequence at the amino terminus. p51 therefore must be derived from p66 by a specific processive cleavage near the carboxy terminus.

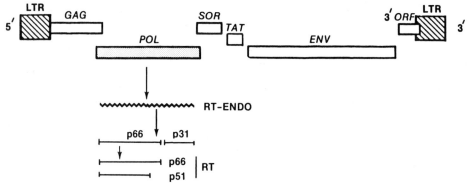

Fig. 3. <u>Pol</u> gene products

The immunoaffinity-purified p66/p51 was enzymatically active reverse transcriptase. Whether both the peptides are necessary for the enzyme activity or either peptide is independently active is not known. The presence of two subunits in the purified RT molecule makes it unique among well-analyzed mammalian retroviruses. A similar structural feature was known for avian retroviral RT (24), but all the type C viral RTs have been known to be single subunit enzymes (25). The monoclonal antibody M3364 did not neutralize the enzyme activity and so was not directed against the active site of the enzyme. However, the antibody specifically precipitated the enzyme protein as demonstrated by a concentration dependent reduction of activity when the antigen-antibody complex was removed by centrifugation (Fig. 6).

Nucleotide sequences 3' to the RT portion of the <u>pol</u> gene code for a 31 kilodalton endonuclease molecule (12,26). This part of the <u>pol</u> gene was cloned into bacterial expression vectors for direct expression of the endonuclease and for expression as a fusion protein with human superoxide dismutase. The endonuclease and the fusion protein contained in the bacterial extracts were found to be immunoreactive when analyzed by Western blot immunoassay using sera of AIDS patients (26). From initial results, the endonuclease appears to correlate well with RT in eliciting antibodies in the infected host (Ref. 26 and our unpublished data).

Products of HTLV-III/LAV env gene

Sera of HTLV-III/LAV-infected people generally react with three glycoproteins (gp160, gp120 and gp41) present in virus-infected cells (27,28), two of which (gp120 and gp41) are also present in the mature virion (29). Their relationship with each other and with the virus itself was analyzed by structural studies on radiolabeled proteins, synthesized <u>de</u> <u>novo</u> in the infected cell and fractionated with the help of either antibody-positive patient sera or specific mouse monoclonal antibodies (28,30).

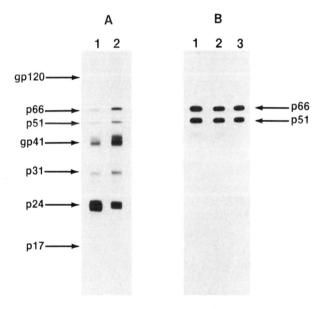

Fig. 4. Specificity of the monoclonal antibody to HTLV-III/LAV reverse transcriptase. Western blot immunoassay of HTLV-III/LAV reverse transcriptase using two antibody-positive human sera (Panel A) and clones of a mouse monoclonal antibody (Panel B) was done as described (22). (Reprinted from Reference 22).

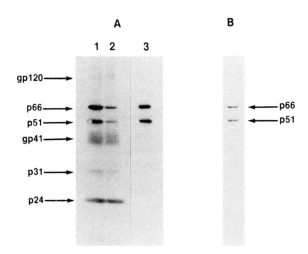

Fig. 5. Immunoaffinity purification of HTLV-III/LAV reverse transcriptase. An extract of HTLV-IIIB was chromatographed on an immunoaffinity column containing purified IgG from a monoclonal antibody to HTLV-III/LAV reverse transcriptase (M3364) covalently attached to activated CH-Sepharose. Fractions were analyzed by Western blot immunoassay using an antibody-positive human serum (Panel A). Lane 1, virus extract before chromatography. Lane 2, unbound flow through fraction from the antibody column. The reaction with p66 and p51 indicates that the column was overloaded and that all the p66 and p51 were not retained by the column. Lane 3, reverse transcriptase eluted from the antibody column. Panel B shows the Coomassie blue stained pattern of the protein represented in lane 3 of Panel A. (Reprinted from Reference 22).

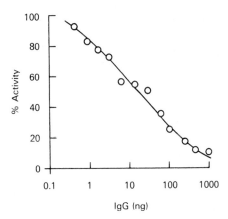

Fig. 6. Immunoprecipitation of reverse transcriptase with M3364 IgG. 10 µl of a HTLV-IIIB extract (1 mg) were incubated with the indicated concentrations of IgG from the monoclonal antibody for 60 minutes at room temperature in 0.1 ml of PBS containing 0.016 ml of Protein A-Sepharose that had been previously treated with rabbit antiserum to mouse κ light chain. Aliquots (0.055 ml) of clarified incubation mixture were withdrawn and assayed for the remaining reverse transcriptase activity. (Reprinted from Reference 22).

Allan et al. (30) demonstrated that when HTLV-III/LAV-producing cells were labeled in culture with [^{3}H]leucine, [^{35}S]cysteine or [^{3}H]valine and the radioactive proteins in the cell extract were immunoprecipitated with HTLV-III/LAV-antibody positive human sera, the immunoprecipitates contained, among other virus-related antigens, gp160 and gp120. These were fractionated by electrophoresis on SDS-polyacrylamide gels and recovered by elution from separated gel bands. NH$_2$-terminal amino acid sequencing by Edman degradation yielded identical radiolabeled amino acid sequences for both gp160 and gp120 in the initial 40 degradation cycles. A comparison of the experimentally observed positions of leucine, cysteine and valine with the predicted amino acid sequence of HTLV-III/LAV gene products (12) gave a perfect match for gp160 and gp120 to be products of the env gene of HTLV-III/LAV. Since the NH$_2$-terminal amino acid sequence of both these proteins is identical, gp160 is considered to be a precursor of gp120. This hypothesis is further strengthened by the results of actual pulse-chase experiments (30).

Primary retroviral envelope glycoproteins as synthesized in the cell are known to be processed during virus maturation into a larger external component and a smaller transmembrane portion (31). By analogy gp120 should be considered the external envelope glycoprotein of HTLV-III/LAV. The transmembrane protein should, therefore, be the carboxy terminal fragment of the processive cleavage of gp160. During Western blot analysis of HTLV-III/LAV proteins two glycoproteins gp120 and gp41 were recognized as viral components (29). In fact, gp41 was the most correlative antigenic band recognized in Western blots by sera of HTLV-III/LAV-infected individuals (4,5).

To understand the nature of gp41 and to verify whether this is the transmembrane protein of HTLV-III/LAV, we decided to determine its amino acid sequence as follows: A mouse hybridoma secreting a specific monoclonal antibody was developed. H9 cells producing HTLV-III/LAV were labeled with either [^{3}H]leucine or [^{3}H]isoleucine and the extracts of these labeled cells were immunoprecipitated with the monoclonal antibody to HTLV-III/LAV gp41. The immunoprecipitates were fractionated by SDS-polyacrylamide gel electrophoresis, the radioactive gp41 bands were cut out and the protein eluted. These were then subjected to automated Edman degradation. Radioactive isoleucine was recovered in cycle 4 among 24 cycles analyzed and radioactive leucine was recovered in cycles 7, 9, 12, 26, 33 and 34 out of 40 cycles analyzed (Figure 7). The amino acid sequence determined is a perfect match with the predicted sequence and it precisely locates gp41 in the env gene of HTLV-III/LAV provirus clone BH 10 and BH 8 (12). The sequencing data identifies gp41 as the carboxyterminal fragment generated by cleavage of gp160 between Arg^{518} and Ala^{519} of the primary env gene product (Fig. 8). The NH$_2$-terminal gp120 should be the external glycoprotein and gp41 the transmembrane protein in the mature virion. Figure 9 provides a diagrammatic representation of the processing of the env gene product.

Products of tat, sor, and 3'orf genes of HTLV-III/LAV

In addition to gag, pol and env, the conventional retroviral genes involved in replication, nucleotide sequence analysis (12) and functional analysis of these sequences (13) have identified at least three new genes in HTLV-III/LAV. These are termed short open reading frame (sor), 3'-open reading frame (3'orf) and transactivator gene (tat). These have recently been structurally identified and functionally characterized by way of c-DNA cloning. DNA sequence analysis shows that tat and 3'orf genes contain three exons and their transcription into RNA involves two splicing events and that the sor gene contains at least two exons (32).

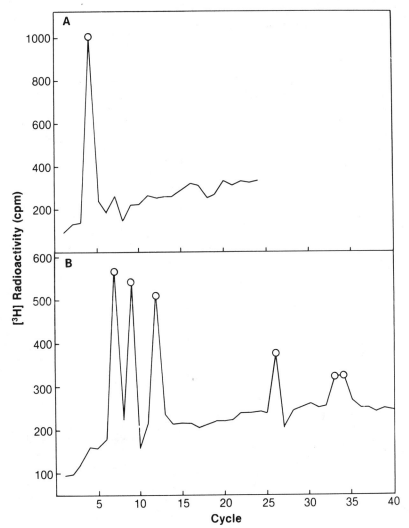

Figure 7. Amino terminal sequence analysis of p41 labeled with (A) [^{3}H]isoleucine and (B) [^{3}H]leucine. H9/HTLV-III cells were labeled with the radioactive amino acids and the labeled p41 was isolated from cell extract by immunoprecipitation with a monoclonal antibody to HTLV-III/LAV gp41. The radioactive p41 band was identified by autoradiography and was sliced out of the gel and eluted with water. The dialyzed proteins were subjected to semiautomated Edman degradation. The recovery of radioactivity in each cycle is given in the figure. Positive identifications are indicated by the open circles. Isoleucine was found in position 4 while leucine was found in positions 7, 9, 12, 26, 33, and 34. (Reprinted from Reference 28).

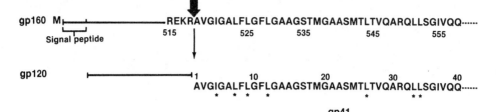

Fig. 8. Diagram indicating the cleavage of gp160 to gp120 and gp41. Numbers below the amino acid sequence of gp160 denote the positions in the deduced amino acid sequence for the primary gene product. Bold arrow indicates the identified cleavage site. Amino acid residues identified by asterisks are those determined by radiolabel sequence analysis (see Fig. 7). Numbers above the amino acids in gp41 denote the degradation cycle of the sequencing procedure. (Reprinted from Reference 28).

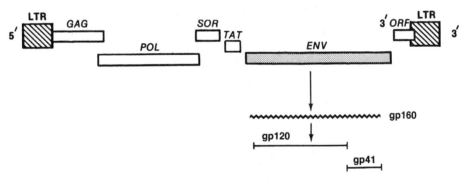

Fig. 9. Diagramatic representation of the processing of HTLV-III/LAV env gene product

In vitro transcription and translation of the cloned spliced sequences show that the sor, tat and 3'orf genes code for polypeptides with apparent molecular weights of 24-25,000, 14-15,000 and 26-28,000, respectively. All three peptides are immunoreactive and immunogenic in the natural host (32). None of them, however, is as frequently recognized by human sera as the gag, pol, or env proteins.

Role of gag, pol, and env gene proteins in serological detection of antibodies to HTLV-III/LAV

Currently available methods for the serological detection of HTLV-III/LAV infection are based on a simple ELISA procedure followed by a confirmatory Western blot immunoassay (4,5). In principle, the ELISA involves incubating the test human sera with antigens in extracts of HTLV-III/LAV fixed on surfaces of microtiter plate wells and measuring the amount of human immunoglobulins bound to the wells by means of a second antibody with specificity to human immunoglobulins. The second antibody is prelabeled with an appropriate enzyme that is capable of generating stable colored products when exposed to specific substrates. The color is measured using specially designed colorimeters called ELISA readers. The assay system is optimized so that the color yield in the final step is proportional to the amount of human antibodies bound to the antigen-coated plates in the initial step.

The success of the test system depends on having an antigen preparation representing all the HTLV-III/LAV proteins. A low antigen concentration on the plates or an antigen preparation lacking some key viral proteins might lead to false negative results with sera having low antibody titer or sera that have antibodies only to few HTLV-III/LAV proteins. On the other hand, too much antigen on the plates or too many contaminants in the antigen preparation increase the background and lead to false positive results.

The commonly confirmatory test for HTLV-III/LAV antibodies is a Western blot immunoassay. It involves the electrophoretic fractionation of the virus antigens on an SDS-polyacrylamide slab gel and electro-blotting the antigens from the gel onto a nitrocellulose sheet and performing the immunoassay on the nitrocellulose sheet or vertical strips cut from the sheet (4,5,10). This test allows for the identification of the antigen bands that react in the immunoassay. Since all the major HTLV-III/LAV-antigens have been well characterized, a judicial examination of the blots will reveal whether HTLV-III/LAV antigens are involved in the antibody reaction, and thus help to identify the true antibody reaction to HTLV-III/LAV. Although all of the functional HTLV-III/LAV genes so far identified have been shown to be immunogenic in the natural host (4,5,32), the products of _gag_, _pol_, and _env_ genes are the most commonly identified antigens in immune reactions with sera of HTLV-III/LAV-infected people (4,5). Among them, gp41 is the most commonly recognized antigen in the Western blot assay (4,5). In immuno-precipitation (RIP) assays using metabolically labeled HTLV-III/LAV proteins, all of the _env_ gene proteins (pg160, gp120, and gp41) are identified in the immunoprecipitates (27,28). Next to gp41, the _gag_ protein p24 and the RT peptides p66 and p51 are the most immunogenic antigens on the Western blots. Occasionally only p24 is identified on a Western blot by sera of individuals recently seroconverting to HTLV-III/LAV. This is probably because of lower antibody titers to other viral antigens early in seroconversion. RIP assays are probably more sensitive to identify other antigens at this stage of seroconversion. Subsequent bleeds of these seroconverters reveal the usual serological patterns when tested on the Western blots (33).

Concluding Comments

HTLV-III/LAV is a unique retrovirus which contains, in addition to the _gag_, _pol_, and _env_ genes that all replication competent retroviruses have, at least three new functional genes, namely _sor_, _tat_ and 3'_orf_. Although _tat_ is present in HTLV-I and -II and bovine leukemia virus, its location

5' to the env gene in HTLV-III still makes this virus different from HTLV-I/BLV group of viruses in which tat is located 3' to the env gene (34,35). The identification and functional characterization of the various gene products and their relative immunogenicity in the host during natural infection helps to develop more sensitive and specific assays for the detection of antibodies in sera of infected individuals. Qualitative and quantitative differences exist in the antibody pattern. Identification of the most immunoreactive and immunodominant antigens recognized by sera of infected individuals with various clinical diagnoses will help in developing new assays. These antigens may be purified from the virus or more conveniently and economically obtained through recombinant DNA technology. For instance, an assay employing a combination of peptides representing certain regions of the env gene with the major gag protein p24, and the pol gene products will be sensitive enough to detect even low titer serum antibodies and will certainly be highly specific.

References

1. M. Popovic, M.G. Sarngadharan, E. Read, and R.C. Gallo, Detection, isolation, and continuous production of cytopathic retroviruses (HTLV-III) from patients with AIDS and pre-AIDS. Science, 224: 497-500 (1984).
2. R.C. Gallo, S.Z. Salahuddin, M. Popovic, G.M. Shearer, M. Kaplan, B.F. Haynes, T.J. Palker, R. Redfield, J. Oleske, B. Safai, G. White, P. Foster, and P.D. Markham, Frequent detection and isolation of cytopathic retroviruses (HTLV-III) from patients with AIDS and at risk for AIDS. Science, 224: 500-503 (1984).
3. F. Barre-Sinoussi, J.-C. Chermann, F. Rey, M.T. Nugeyre, S. Chamaret, J. Gruest, C. Dauguet, F. Axler-Bline, F. Brun-Vezinet, C. Rouzioux, W. Rozenbaum, and L. Montagnier, Isolation of a T-lymphotropic retrovirus from a patient at risk for acquired immune deficiency syndrome (AIDS). Science, 220: 868-871 (1983).
4. M.G. Sarngadharan, M. Popovic, L. Bruch, J. Schuepbach, and R.C. Gallo, Antibodies reactive with human T-lymphotropic retroviruses (HTLV-III) in the serum of patients with AIDS. Science, 224:506-508 (1984).
5. B. Safai, M.G. Sarngadharan, J.E. Groopman, K. Arnett, M. Popovic, A. Sliski, J. Schuepbach, and R.C. Gallo, Seroepidemiological studies of human T-lymphotropic retrovirus type III in acquired immunodeficiency syndrome. Lancet, i:1438-1440 (1984).
6. M.G. Sarngadharan, P.D. Markham, and R.C. Gallo, Human T-cell leukemia viruses, in: "Virology" (B.N. Fields, D.M. Knipe, R.M. Chanock, J.L. Melnick, and R.E. Shope, Eds.), Raven Press, New York, pp. 1345-1371 (1985).
7. J.E. Groopman, K.H. Mayer, M.G. Sarngadharan, D. Ayotte, A.L. DeVico, R. Finberg, A.H. Sliski, J.D. Allan, and R.C. Gallo, Seroepidemiology of HTLV-III among homosexual men with acquired immunodeficiency syndrome, generalized lymphadenopathy and asymptomatic case controls in Boston. Ann. Int. Med., 102:334-337 (1985).
8. S.Z. Salahuddin, P.D. Markham, M. Popovic, M.G. Sarngadharan, S. Orndorff, A. Fladagar, A. Patel, J. Gold, and R.C. Gallo, Isolation of infectious T-cell leukemia/lymphotropic virus type III (HTLV-III) from patients with acquired immunodeficiency syndrome (AIDS) or AIDS-related complex (ARC) and from healthy carriers. A Study of risk groups and tissue sources. Proc. Natl. Acad. Sci. USA, 82:5530-5534 (1985).

9. J.E. Groopman, S.Z. Salahuddin, M.G. Sarngadharan, J.I. Mullins, J.L. Sullivan, C. Mulder, C.J. O'Hara, S.H. Cheeseman, H. Haverkos, P. Forgacs, N. Riedel, M.F. McLane, M. Essex, and R.C. Gallo, Virologic studies in a case of transfusion-associated AIDS. N.Engl. J. Med., 311:1419-1422 (1984).
10. H.W. Jaffe, M.G. Sarngadharan, A.L. DeVico, L. Bruch, J.P. Getchell, V.S. Kalyanaraman, H.W. Harverkos, R.L. Stoneburner, R.C. Gallo, and J.W. Curran, Infection with HTLV-III/LAV and transfusion-associated acquired immunodeficiency syndrome. J. Amer. Med. Assoc. 253: 770-773 (1985).
11. J.E. Groopman, M.G. Sarngadharan, S.Z. Salahuddin, R. Buxbaum, M.S. Huberman, J. Kinniburgh, A. Sliski, M.F. McLane, M. Essex, and R.C. Gallo, Apparent transmission of human T-cell leukemia virus type III to a heterosexual women with the acquired immunodeficiency syndrome. Ann. Int. Med., 102:63-66 (1985).
12. L.Ratner, W. Haseltine, R. Patarca, K.J. Livak, B. Starcich, S.F. Josephs, E.R. Doran, J.A. Rafalski, E.A. Whitehorn, K. Baumeister, L. Ivanoff, S.R. Petteway, M.L. Pearson, J.A. Lautenberger, T.S. Papas, J. Ghrayeb, N.T. Chang, R.C. Gallo, and F. Wong-Staal, Complete nucleotide sequence of the AIDS virus, HTLV-III. Nature, 313: 277-284 (1985).
13. S.K. Arya, C. Guo, S.F. Josephs, and F. Wong-Staal, Trans-activator gene of human T-lymphotropic virus type III (HTLV-III), Science, 229: 69-73 (1985).
14. J.Sodroski, R. Patarca, C. Rosen, F. Wong-Staal, and W. Haseltine, Location of the trans-activating region on the genome of human T-cell lymphotropic virus type III. Science, 229: 74-77 (1985).
15. M.G. Sarngadharan, L. Bruch, M. Popovic, and R.C. Gallo, Immunological properties of the gag protein p24 of the acquired immunodeficiency syndrome retrovirus (human T-cell leukemia virus type III). Proc. Natl. Acad. Sci. USA, 82: 3481-3484 (1985).
16. V.S. Kalyanaraman, C.D. Cabradilla, J.P. Getchell, R. Narayanan, E.H. Braff, J.-C. Chermann, F. Barre-Sinoussi, L. Montagnier, T.J. Spira, J. Kaplan, D. Fishbein, H.W. Jaffe, J.W. Curran, and D.P. Francis, Antibodies to the core protein of lymphadenopathy-associated virus (LAV) in patients with AIDS. Science, 225: 321-323 (1984).
17. V.S. Kalyanaraman, M. Jarvis-Morar, M.G. Sarngadharan, and R.C.Gallo, Immunological characterization of the low molecular weight gag gene products p19 and p15 of human T-cell leukemia-lymphoma virus (HTLV) and demonstration of human neutral antibodies to them. Virology, 132: 61-70 (1984).
18. F.D. Veronese, M.G. Sarngadharan, R. Rahman, P.D. Markham, M. Popovic, A.J. Bodner, and R.C. Gallo, Monoclonal antibodies specific for p24, the major core protein of human T-cell leukemia virus type III. Proc. Natl. Acad. Sci. USA, 82: 5199-5202 (1985).
19. A.M. Schultz, and S. Oroszlan, In vivo modification of retroviral gag gene-encoded polyproteins by myristic acid. J. Virol., 46: 355-361 (1983).
20. S.Oroszlan, T.D. Copeland, V.S. Kalyanaraman, M.G. Sarngadharan, A.M. Schultz, and R.C. Gallo, Chemical analysis of human T-cell leukemia virus structural proteins. In: "Human T-Cell Leukemia/Lymphoma Virus" (R.C. Gallo, M. Essex, and L. Gross, Eds.) Cold Spring Harbor Laboratory, New York, pp. 91-100 (1984).
21. A.Chandra, T. Gerber, and P. Chandra, Biochemical heterogeneity of reverse transcriptase purified from the AIDS virus, HTLV-III. FEBS Letters, 197: 84-88 (1986).
22. F.D. Veronese, T.D. Copeland, A.L. DeVico, R. Rahman, S. Orszlan, R.C. Gallo, and M.G. Sarngadharan, Characterization of highly immunogenic p66/p51 as the reverse transcriptase of HTLV-III/LAV. Science, 231: 1289-1291 (1986).

23. A.Chandra, T. Gerber, S. Kaul, C. Wolf, I. Demirhan, and P. Chandra, Serological relationship between reverse transcriptases from human T-cell lymphotropic viruses defined by monoclonal antibodies. FEBS Letters, 200: 327-332 (1986).
24. D.P. Grandgenett, G.F. Gerard, and M. Green, A single subunit from avian myeloblastosis virus with both RNA-directed DNA polymerase and ribonuclease H activity. Proc. Natl. Acad. Sci. USA, 70: 230-234 (1973).
25. G.F. Gerard, and D. P. Grandgenett, Purification and characterization of the DNA polymerase and RNAse H activities in Moloney murine sarcoma-leukemia virus. J. Virol., 15: 785-797 (1975).
26. K.S. Steimer, K.W. Higgins, M.A. Powers, J.C. Stephans, Y. Gyenes, C. George-Nascimento, P.A. Luciw, P.J. Barr, R.A. Hallewell, and R. Sanchez-Pescador, Recombinant polypeptide from the endonuclease region of the acquired immune deficiency syndrome retrovirus polymerase (pol) gene detects serum antibodies in most infected individuals. J. Virol., 58: 9-16 (1986).
27. L.W. Kitchen, F. Barin, J.L. Sullivan, M.F. McLane, D.B. Brettler, P.H.Levine, and M. Essex, Aetiology of AIDS-antibodies to human T-cell leukemia virus (type III) in haemophiliacs. Nature, 312: 367-369 (1984).
28. F.D. Veronese, A.L. DeVico, T.D. Copeland, S. Oroszlan, R.C. Gallo, and M.G. Sarngadharan, Characterization of gp41 as the transmembrane protein coded by the HTLV-III/LAV envelope gene. Science, 229: 1402-1405 (1985).
29. M.G. Sarngadharan, F.D. Veronese, S. Lee, and R.C. Gallo, Immunological properties of HTLV-III antigens recognized by sera of patients with AIDS and AIDS related complex and of asymptomatic carriers of HTLV-III infection. Cancer Res., 45: 4574s-4577s (1985)
30. J.S. Allan, J.E. Colligan, F. Barin, M.F. McLane, J.G. Sodroski, C.A. Rosen, W.A. Haseltine, T.H. Lee, and M. Essex, Major glycoprotein antigens that induce antibodies in AIDS patients are encoded by HTLV-III. Science, 228, 1091-1094 (1985).
31. D.P. Bolognesi, R.C. Montelaro, H. Frank, and W. Schaefer, Assembly of type C oncornaviruses: A model. Science, 199: 183-186 (1978).
32. S.K. Arya, and R.C. Gallo, Three novel genes of human T-lymphotropic virus type III: Immune reactivity of their products with sera from acquired immune deficiency syndrome patients. Proc. Natl. Acad. Scie. USA, 83: 2209-2213 (1986).
33. N.Manca, F.D. Veronese, D.D. Ho, M.G. Sarngadharan, and R.C. Gallo, Variable immunological anti-viral profiles in sera from HTLV-III-infected patients, submitted for publication.
34. J.G. Sodroski, C.A. Rosen, and W.A. Haseltine, Trans-acting transcriptional activation of the long terminal repeat of human T-lymphotorpic viruses in infected cells. Science, 225: 381-385 (1986).
35. C.A. Rosen, J.G. Sodroski, R. Kettman, A. Burny, and W.A. Haseltine, Transactivation of the bovine leukemia virus long terminal repeat in BLV-infected cells. Science, 227, 320-322 (1985).

CYTOPATHOGENIC MECHANISMS WHICH LEAD TO

CELL DEATH OF HTLV III INFECTED T CELLS

D. Zagury[*], R. Cheynier[*,**], J. Bernard[*], R. Leonard[*,***],
M. Feldman[**], P. Sarin[***] and R. C. Gallo[***]

[*] Université P. et M. Curie (Paris) and Institut J. Godinot
Reims, [**] Weizmann Institute of Science (Rehovot),
[***] Laboratory of Tumor Cell Biology N.I.H. (Bethesda)

Cytopathogenic processes of HTLV III-LAV infected T4 cells which lead to T4 depletion in AIDS patients, has been experimentally investigated in long term culture of AIDS T cells (A-CTC) and of HTLV III-LAV infected T cells from normal donors (N-CTC).

MATERIAL AND METHODS

Long term culture of normal and AIDS CTC

In order to obtain long term normal and AIDS CTC, conditions previously employed in our laboratory for long term growth of normal T cell clones were used. Culture conditions were different from primary PBL cultures. As indicated in Figure 1, the modifications consisted of dilution of the cell number from $10^5 - 10^6$ cells per ml (usual culture conditions) to $10^3 - 10^4$ cells per ml and the addition of a feeder cell layer of $10^5 - 10^6$ irradiated (4000 rads) peripheral blood mononuclear cells pooled from 10 - 20 donors.

RESULTS

AIDS CTC secrete IL2 and express HTLV III after PHA activation

A-CTC when cultured in presence of exogenous IL2, proliferate for long term periods, up to two months. As shown in Table I, as well as N-CTC, A-CTC secrete IL2 only after a PHA stimulation in presence of macrophages and B cells. IL2 effector function is not performed in non activated CTC. In the same line, as indicated in Table II, column 4, HTLV III-LAV is not expressed in viral infected A-CTC, growing in presence of exogenous IL2. However, when A-CTC are activated by PHA, retroviral expression is manifested by both Reverse Transcriptase (RT) in culture supernatants and indirect immunofluorescence in presence of HTLV III-LAV p24 and p15 monoclonal antibodies.

Normal CTC may be infectet by HTLV III-LAV containing culture supernatants

N-CTC, whether stimulated by PHA in presence of macrophages and

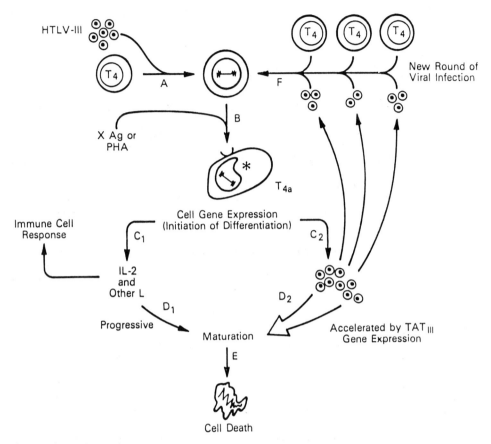

Figure 1. Schematic representation of biological events leading to HTLV-III/LAV infection, viral expression, and cell death of T_4 cells.
  A. T_4 lymphocytes, either activated or not, are preferential target cells for HTLV-III/LAV.
  B. Infected T_4 cells, as non-infected ones, are activated either specifically by corresponding antigen (XAg) or polyclonally by PHA (T_4a = activated T_4).
  C. Activation initiates effector differentiation by expression of cellular genes. Differentiation of infected T_4 cells promotes IL_2 secretion (physiological pathway) and HTLV-III/LAV expression (pathological pathway). (L = lymphokines).
  D. In activated T_4 cells maturation occurs progressively; in infected cells viral expression is accelerated through TAT-III gene expression which leads to maturation by over 1000-fold.
  E. Terminal differentiation leads to cell death.

Table 1. IL-2 production and reverse transcriptase activity (RT) in long-term T cell cultures from peripheral blood of AIDS (A-CTC) and normal donors (N-CTC) before and after immune activation.

CTC Origin	HTLV-III Serology	PHA Activation	IL-2 Production[1] (IL-2 U/2x10^4 cell)	RT Activity[2] (cpm/experiment x 10^{-3})
Experiment I				
AIDS (RK)[3]	+	−	< 260	< 1
		+	>1400	28
AIDS (JMB)	+	−	< 260	< 1
		+	>1400	6.6
Normal$_1$ (MF)	−	−	< 260	< 1
		+	>1400	< 1
Experiment II				
AIDS (BS)	+	−	< 260	< 1
		+	>1400	17
AIDS (AC)	+	−	< 260	< 1
		+	>1400	3.2
Normal$_2$ (BR)	−	−	< 260	< 1
		+	>1400	< 1
Experiment III				
AIDS (AD)	+	−	< 260	< 1
		+	>1400	8.4
AIDS (MP)	+	−	< 260	< 1
		+	>1400	31
Normal$_1$ (MF)	−	−	< 260	< 1
		+	>1400	< 1

B cells, or not, may be infected by H9-HTLV III culture supernatants. After activation, HTLV III-LAV infected N-CTC cultured with exogenous IL2, hydrocortisone and anti human αINF, expressed HTLV III. As indicated in Table II, RT was shown in culture supernatant on day 5 to 9. Infected cultures degenerate after viral expression. Table II shows too that only T4 cells containing subpopulation are infected in CTC (experiment II).

As indicated in Table III, infection of non stimulated N-CTC do not promote viral expression. However, when infected cells are secondarily activated, HTLV III-LAV is expressed after 5-6 days of culture in presence of hydrocortisone and human anti αIFN. Viral expression disappears after 2-4 days and is followed by cell death.

DISCUSSION

From these results, we can infer the following data:
1) N-CTC may be infected by HTLV III-LAV without activation.
2) Infection concerns T4 cells containing subpopulation.

3) A-CTC, which proliferate in presence of exogenous IL2 for long term periods, and infected N-CTC, secrete endogenous IL2 and espress HTLV III-LAV only after polyclonal PHA activation.
4) Viral expression is followed by cell death.

From these results, a cytopathogenic process for HTLV III-LAV infection of T4 cells can be drawn in a scheme (Fig. 1). This representation emphasizes that multiple round of viral infection may promote, in vitro, T4 cell depletion, and that physiological effector pathways (IL2 production) may be impaired in infected T cells. T4 cell depletion and immune function impairment may in turn lead, progressively after a variable latency period, to acute immunodeficiency syndrome (AIDS).

Table II. Effect of immune activation on IL-2 and virus production in long term T-cell cultures from peripheral blood of normal donors (N-CTC) infected in vitro with HTLV-III.

CTC	HTLV-III Serology	PHA Activation	IL-2 production (IL-2 U/2×10^4 cells)	RT activity (cpm/experiment $\times 10^{-3}$)
Normal donor (mixed cell population)	−	−	< 260	< 1
		+	>1400	< 1
HTLV-III infected normal donor (mixed cell population)	−	−	< 260	< 1
		+	1400	46
HTLV-III infected normal donor (Non-T_4 subpopulation)	−	−	< 260	< 1
		+	< 260	< 1
HTLV-III infected normal donor (Non-T_8 subpopulation)	−	−	< 260	< 1
		+	>1400	26.5

Twenty days old activated and non-activated T-cell cultures (CTC) derived from the same normal donor were assayed for IL-2 production and RT activity. After activation by PHA for 24 hours, the CTC from the donor, were treated with polybrene (2 μg/ml), incubated in the presence of HTLV-III containing culture supernatant fluid, obtained from the H9/HTLV-III-B2 cell culture, for 1 hour at 37° and cultured in RPMI-1640/FCS medium containing exogenous IL-2, anti αIFN and hydrocortisone. Non-T_4 and non-T_8 subpopulations were obtained after treatment of the total CTC by complement dependent cytoxicity in the presence of OKT4 and OKT8 monoclonal antibodies, respectively. Non-activated cells were used as controls.

Table III. HTLV-III expression in long-term cultured normal T-cells (N-CTC).

Days after infection	RT activity[2]	
	Non-activated	PHA-activated
3	0	0
5	0	4
7	0	39
9	0	23

[1] 25 day old T-cell cultures from normal donors were infected with HTLV-III/LAV. Cells were treated with polybrene (2 µg/ml), incubated in the presence of HTLV-III/LAV containing culture supernatants for one hour at 37° and cultured again in the presence of IL-2, cortisone and sheep antihuman αIFN. After 1-2 hours, infected CTC were washed twice and divided in two cultures. One was maintained in the same medium containing IL-2 (non-activated) and the other (PHA-activated) was stimulated by PHA in presence of macrophages and B-cells, as described in Table I.

[2] Viral expression was measured by RT activity in culture supernatants as described in Table I, and the results are expressed in cpm/experiment $\times 10^{-3}$.

CLINICAL ASPECTS OF THYMIC FACTORS IN THE TREATMENT OF

IMMUNODEFICIENCY DISEASES AND NEOPLASIA: ACHIEVEMENTS AND FAILURES

John R. Hobbs

Department of Chemical Immunology, Charing Cross and
Westminster Medical School, Westminster Hospital
Page Street, London, SW1P 2AR, G.B.

INTRODUCTION

Since Jacques Miller showed the importance of the thymus and the term "T-Lymphocyte" was introduced, there followed much basic research, and for the past 8 years, a variety of thymic factors have become available for therapeutic studies in man. All those currently available have to be given parenterally, and fall into two main groups, (see Tables 1 and 2). Their modes of action (1) will be briefly mentioned, but primarily to stress the importance of the sequential action of a series of factors.

It was natural to first apply these to the genetic (primary) immunodeficiencies but, of course, it would be foolish to expect too much from such studies for, indeed, if the cells capable of responding or the capacity to release the mediators expected during the response are genetically impaired or absent, it is expecting a lot for any hormone to evoke activity in useless targets. On the other hand, where the T-lymphocyte deficiency has been acquired (secondary) (2) subsequent to environmental influences upon a previously normal population of cells, then more satisfactory targets may be able to respond much better to an increase in thymic factors, and this certainly has been the most rewarding area in therapeutic studies.

T-lymphocyte deficiency itself, predisposes to neoplasia to a certain extent (3), and undoubtely aggravates its complications and those of the treatment methods used for cancer (4), so that in neoplasia there are preventative and resurrective roles for thymic therapy. Nevertheless, the most desirable action of such factors would be from their lymphopoietic and clonal expansion capacity, in the hope that their use could generate new clones of T-lymphocytes that might become effective in eradicating the neoplasm, and it is important in designing such protocols to avoid elimination of any such clones by cycles of cytoreductive therapy.

THYMIC FACTORS

It is not proposed to even consider the use of whole extracts of thymus containing intact proteins, for whether such extracts are animal (xenogeneic) or denatured in their preparation, they are likely to be highly antigenic, so that apart from the first shot of therapy, they are very likely to become immunologically neutralized in a short time. In their repeated use, there can be anaphylactic reactions dangerous to the patient.

Table 1. Thymic Extracts

Thymosin Fraction 5	(TF5)	Hoffman LaRoche
Thymostimulin	(TS, TP1)	Serono
Thymic Humoral Factor	(THF)	Trainin
Polish	(TFX)	Polfa Gorski
Suppressin A		Osband
Chinese Porcine	(PTI)	Jin

The two main groups that have been used safely in man comprise either extracts of whole animal thymus purified to peptide level (see Table 1), or some of the actual individual peptides, either isolated or synthesized for therapy (see Table 2). The pure peptides listed all have their own distinct pI of isoelectric focussing in PAG, pH 3.5 - 9.5, and different amino-acid sequences. There are at least six different peptides with thymic activity which can be extracted from bovine thymus. The three synthetic variations come from natural parent peptides, present in bovine thymus and those and the other three [i] have all been isolated from thymus; [ii] 5 have been localised in thymic biopsies using immuno-localisation methods; [iii] 2 are known to circulate in human plasma; [iv] Thymulin-Zn cannot be assayed in athymic plasma; [v] 4 have been shown to bind to T-lymphocytes with Kds of $10^{-7} - 10^{-10}$; [vi] examples of their modes of action are summarized in Table 3 and Figure 1.

Table 2. Therapeutic Thymic Polypeptides

				pI
Synthetic	α_1-Thymosin	28aa(α_1T)	A. Goldstein	4.2
	Thymulin-Zinc	9aa(FTS-Zn)	J.-F. Bach	7.5
	Thymopoietin	5aa(TP5)	G. Goldstein	5.5
Purified	α_7-Thymosin	2,500 D	A. Goldstein	3.5
	β_4-Thymosin	4,982 D	A. Goldstein	5.4
	THF	3,200 D	N. Trainin	5.7

From such studies, it is clear that there is a sequential action of different peptides of thymic origin acting at various stages during the maturation of the T-lymphocyte. Since in secondary T-cell deficiency and in neoplasia, the mechanisms of impairment are not entirely clear and may be multiple, at this stage of our art, a strong case can be made for using the thymic extracts comprising mixtures of these active polypeptides on the grounds that the one prescription is likely to find a site of action or provide a complementary activity from multiple actions. Of course, the drawback of such "blunderbuss" therapy is that inhibitory actions, perhaps unwanted, might also come into play as well as the desired stimulatory ones. This same risk, however, applies to the single peptide, for it has been shown that thymic factors, like most drugs, show a bell-shaped curve in their responses, with usually an optimal concentration in a given situation above which the effect can be the opposite.

Table 3. Modes of Action of Thymic Factors

(a) Induction of T-lymphocyte markers or receptors

Marker/Receptor	Factor	System	Ref.
Terminal Deoxynucleotidyl Transferase	TF5 β_4-Thymosin	T-Cell depletion by hydrocortisone	7
E-Rosettes	TF5	Melanoma PBM	8
HLA(OKT8)	TS	Atopic PBM	9
H2	Suppressin A	Atopic OKT4	10
DR(OKT4)	TS	AIDS PBM	2
CON-A	TS	PBM	1

(b) Increased Lymphokine Production

Target		Presumed Lymphokine		
Human PBM	TS	ConA-induced	Inhibition of Oncovirus RD114	11
Mouse Spleen	TS	Tumor induced Granulocyte Monocyte Colony Stimulating Factor	Mouse bone marrow culture	11
S-RBC Human T-Cells	TFX	B-Cell Growth Factor B-Cell Differentiation Factor	Pokeweed B-Cell Ig Synthesis	12
Human PBM	TF5	Migration Inhibitory Factor	Macrophage	13

Our own group have, for many years, undertaken in vitro studies of peripheral blood mononuclear (PBM) fractions from our patients, carefully excluding foetal calf serum which could cloud the issue. It has sometimes been possible to show that a single peptide has had no effect and the whole extract has produced an induction of new surface markers in the PBM, indicating that for that patient, the mixture is better than the single peptide. Until more is known about the actions of the individual peptides, a case can be made for using the mixture as extracted with its components, presumably balanced in the way in which they were in the original donor animal. As our knowledge advances, it may be possible to select individual peptides for individual patients. However, at the present time, it is much more practical to use the extract mixture.

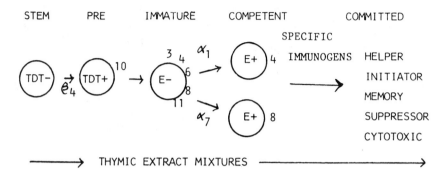

Fig. 1. To show some known steps in T-lymphocyte maturation facilitated by thymic factors. Committed lymphokine release can affect other cells, B-cell growth, etc. The free numbers refer to OKT antigens; those with Greek symbols, to pure thymosins; 'E' indicates SRBC-rosettes without foetal calf serum.

While we, ourselves, have drawn attention to the importance of the dose level (5), there is not always time or, indeed, enough cells to test a whole range of concentrations in vitro and for the purposes of therapeutic trials it is more convenient to choose the single dose level for all the subjects, which can later be increased in the non-responders. We have adopted a loading dose schedule, shown (6) to be more effective than no loading dose or a lower loading dose. A maintenance dose at least twice a week is then preferred because, in vivo, the actions of the extract seem to last only 30 - 48 hours. There is also the added advantage that regular dosage tends to prevent the emergence of reaginic antibodies to a level where they could cause anaphylaxis.

SKIN TESTING

Since T-cell deficiency itself predisposes to atopy, and there is always the possibility that the biological peptide extract could be contaminated with a few molecules of proteins (e.g. bovine), intra-dermal skin testing is always used just before we begin a course of treatment. A small portion of the solution prepared for injection (e.g. 50 mg in 1 ml) is diluted to 0.1 mg/ml and 0,1 ml is injected intra-dermally. The patient should not be on antihistamines, and because of the other factors in thymic extracts, a minor flare is not uncommon. A wheal exceeding 10 mm, at 20 minutes has ab initio been observed in less than 1 % of subjects, but unless the treatment is essential, would contra-indicate its use. Sometimes the need for treatment can justify its use after full cover with prior antihistamine has been given to the patient with intravenous hydrocortisone and adrenalin at standby. Where there has been a break in treatment, or any form of reaction that occurs with eosinophilia, we repeat such a skin test before recommencing the therapy and in that situation, about 4 % of the subjects will develop a wheal. With these precautions, over 1.000 patients have been treated without any anaphylaxis developing.

NORMAL CONTROLS

Reliable and normal ranges (2,14) with the correct statistical analyses (mostly Log-normal) on data from healthy normal subjects up to the age of 75 years are summarised in Table 4.

Table 4. Laboratory Criteria for T-cell Deficiencies. PBL = PB Lymphocytes. No foetal calf serum is used in tests 2,3,4,6, and 7.

1.	PBL Total	$< 1.35 \times 10^9/L$
2.	PBL E-Rosettes	$< 0.47 \times 10^9/L$
3.	PBL Thymidine-uptake vs Candida	< 4500 CPM
4.	PBL Thymidine-uptake vs 5-Donor-Pool	< 8000 CPM
5.	Skin delayed hypersensitivity vs Candida	Subnormal
6.	PBL OKT4	$< 0.375 \times 10^9/L$
7.	PBL OKT8	$< 0.28 \times 10^9/L$
8.	Serum IgE	Raised
9.	Culture supernatant from 3, 4, etc.	Subnormal Lymphokines

In table 4, the 2.5 % lower normal limit has been used; it has worked well in over 15,000 studies of patients. In such tests, we tend to progress from 1 down to 7, only undertaking lymphokine assays where indicated; e.g. absent delayed hypersensitivity in the skin in the presence of clear lymphocyte transformation to the immunogen. High serum IgE levels when found, can give early warning of atopic complications.

PRIMARY IMMUNODEFICIENCIES

Failure of the thymus on an inborn basis is best typified in the Di George syndrome where, however, the defect can be partial and show spontaneous recovery or, worse, if unrecognised and untreated before the age of six months, can progress to an irreversible state, where there is no response to thymic factor therapy; in such cases, bone marrow transplantation is required to re-establish adequate target lymphocytes on which thymic factor therapy can then act. The first choice in Di George patients is always a trial of thymic factor therapy as it does not carry the risks of bone marrow transplantation or, indeed, thymus grafting, where donor lymphocytes can survive and set up graft-versus-host disease. In the few patients who have been tested, there usually has been a good correlation between the in vitro response of the lymphocytes of the patient and the effects of thymic factor therapy in vivo, but there can be exceptions. Often, where the therapy has restored an infant to a healthy condition over six months, it can be discontinued and careful follow-up can show that the lymphocyte function of a patient has reached a state where it seems that autocrine feedback loops render it self-supporting. One of our patients has now needed no further therapy for 13 years and is a very fit, healthy teenager who plays rugby. In other cases it may be found that continuous courses of therapy are needed. Secondary failure of the thymus is a feature of adenosine deaminase and 5-purine phosphorylase deficiencies if these are not treated within the first six weeks of birth, and again, a combination of thymic factor therapy with irradiated blood cell therapy can sometimes by themselves restore the immune competence without the need for the risks of GVHD. Such cases, however, seem to be rare and most of the patients in this category are better treated by a displacement bone marrow transplant (15). There are many other causes of T-cell deficiency which appear to have a genetic basis and which, perhaps, are not as rapidly fatal as is usual for the Swiss-type of severe combined immune deficiency (SCID). Most workers are agreed that the Swiss-type only accounts for about 20 % of all varieties of SCID, but that it never responds to thymic factor therapy and the only effective treatment is bone marrow transplantation (15) and this is probably because the severe lymphopaenia of both T- and B-cells from birth implies there are no satisfactory targets on which thymic factors could act. It is in the other forms of SCID that there are occasionally surprising responses to thymic factor therapy, which therefore can justify a trial of such therapy, especially for those patients who have near normal numbers of total lymphocytes. As the other varieties of SCID are more clearly defined, it may be possible to identify which types best respond to such thymic therapy.

There are also milder forms of pure T-cell deficiency; again, mostly ob obscure origin and later onset (such as the Nezelof and Matsaniotis syndromes) of which some 50 % may respond to thymic factor therapy (16). Frequently such patients relapse when the therapy is withdrawn, so that careful monitoring of their progress is indicated.

Ther are T-cell deficiencies which follow on other fundamental genetic defects; in Wiskott-Aldrich syndrome there is a failure of the correct interaction between antigen-presenting-, T- and B-cells due to a membrane protein abnormality which probably also causes the defect in platelet function; in one of Fanconi's syndromes a failure of DNA repair results in increasing chromosomal aberrations within high turnover populations such as the T-lymphocytes, and similarly occurs in ataxia telangiectasia. For all three of these conditions, it would be fair to say that thymic factor therapy has not had any sustained success and that the first two are best treated by displacement bone marrow transplantation (15).

The atopic state undoubtedly has a genetic basis, affecting 8 - 12 % of the population and an important component of it is T-cell deficiency with (a) reduced numbers of total lymphocytes (9), (b) fewer E-Rosetting T-cells (9), (c) excess of B-cells bearing IgE, (d) deficiencies of T-cell subsets bearing Fc-γ receptors (17) or OKT8 markers, and it is in these subsets that the specific IgE-regulating T-cells are found (18).

Preliminary studies of the effect of thymic factor therapy in severely atopic subjects indicate that it is possible to increase the number of cells bearing E-rosettes and the number of PBL in the OKT8 subclass, and this can be associated with reductions of the total serum IgE level. In children there has been recording of remarkable improvement in symptoms, but this does tend to happen anyway, so that it will need carefully monitored long-standing double-blind trials to establish whether there is a place for thymic factor therapy. In adults with severe atopic dermatitis, a similar study is continuing and it would be premature to make any recommendation. However, it would be fair to state that such studies seem to be rewarding and should be continued.

SECONDARY T-CELL DEFICIENCIES

T-lymphocyte deficiency can be called secondary when [i] the past history or previous tests indicate that the subject's T-cell function was previously normal; [ii] it is associated with a condition in which it is known that T-cell deficiency occurs and can return to normal when that condition is corrected (releasing, however, that for some it can remain irreversible). Using the criteria in table 4, over 15,000 hospital patients have been tested by my department and some 20 % have been found to have deficiencies. For many of those patients, the deficiencies have contributed towards their illness, but in 4 % of all patients it was severe enough to be the main cause of their admission. The causal associations are reviewed in depth elsewhere (2) and are summarised in table 5.

Table 5. Secondary T-Lymphocyte Deficiencies

1. Age	(< 2 yrs.; > 40 yrs.)	
2. Post-Viral	Rubella, Cytomegalovirus, Epstein-Barr, Varicella, Herpes Simplex, Mumps, Coxsackie, Influenza, Adenoviruses, Hepatitis B, Hepatitis A, Rota Virus, HTLV, AIDS-Agent, Others (Post Chlamydia, Mycoplasma, etc.)	
3. Malnutrition	Protein-Calorie; Vitamins A,D,E; Fe,Zn,Cu.	
4. Neoplasia	T-Cell Tumours suppress normal T-cells; Others	
5. Iatrogenic	Irradiation, Cytotoxics; Corticosteroids, Sex-Steroids; Anti-Inflammatories, etc.	
6. Other Conditions	Lepromatous Leprosy; Sarcoidosis, Renal Disease, Others	

It can result in wasting, diarrhoea, disseminated infections (fungal, viral or mycobacterial), complicated infections (giant-cell pneumonia, pneumocystis, slow resolution, recurrent, haemorrhagic or encephalitic), aberrant immunity (auto-immunity or GVHD, especially after unirradiated transfusions) and acquired atopy.

<u>Age</u>. Children, especially under the age of 3 years, have large null populations of PBM, presumably making strong demands on a diminishing thymus, so that a relative deficiency can often be detected and, of course would predispose to the occurrence and spread of neoplasia, known to be commoner in that particular age group of children. After the age of 40 years, assay of α_1-Thymosin and TZn have independently shown declining values of circulating levels in the plasma and it is after this age that again neoplasia becomes more common and dysregulation presumably permits autoimmune disease to occur. After a preliminary report (19) double-blind trials are under way to find out if thymic factor therapy can reduce the number of infections and possibly auto-immune and neoplastic complications in the older age group.

<u>Post-viral</u>. When a patient first presents with T-cell deficiency and a viral infection, it can be difficult to decide which came first. T-cell impairment can be found before the onset of symptoms and immediately after many viral infections (20); and this has been proven for at least 18 viruses that affect man. It is probable that some subjects are more susceptible than others to a prolonged effect; e.g. following acute anterior uveitis, HLA-B-27 subjects continue much longer than their spouses (21). This also seems true for the X-linked proliferative syndrome following EB virus infection (22) and for some forms of chronic active hepatitis (23). Possible mechanisms have been reviewed elsewhere (20).

Treatment with thymic factors has been successful in diminishing further spread of the virus (across the body fluids) and in accelerating the healing of associated lesions in varicella zoster, haemorrhagic chicken pox, and herpes simplex (20). In some series the results have been so spectacular as compared to historical controls, that it has not been considered ethical to undertake a double-blind trial (20).

It is likely that the immunodeficiency acquired following multiple transfusions for aplastic anaemia (24) or in association with Kaposi's sarcoma as long ago as 1973 (25) were due to viruses transmitted to those patients. The retrovirus of AIDS is, of course, today's proven example. Herein, it has been reported (2) that thymostimulin therapy can have a transient lymphopoietic effect which is very impressive with lymphocyte counts rising from 0.05 to 1.00 x 10^9/L, but they have then been observed to fall away again despite continuing treatment. I believe this could be due to the fact that the new young lymphocytes will all go through a phase where they are simultaneously expressing the receptors recognised by the antibodies to OKT4, OKT8 and OKT6 prior to the further differentiation of the cell to commit itself to one or other major subsets. This means that all young lymphocytes will expose an OKT4 receptor at some time, and this is known to be the site of entry for the AIDS virus. When it was first isolated it was called the lymph-adenopathy-associated virus (LAV) (26) and we have used TS treatment in patients in what is considered to be a prodromal stage for AIDS and, again, with only limited success. There has been some increase in antibody titres and in a few patients they seem to have clinically improved and not progressed to AIDS, but our numbers of patients studied do not permit reliable statistical judgements. Nevertheless, as the disease advances, it is clear that thymic factor therapy alone is a failure, but it could well be improved by the use of Interferon therapy at the time of the lymphopoiesis, hopefully reducing the amount of free virus that might be able to penetrate the new young lymphocytes, and when they have appeared it would seem appropriate to add Inteleukin-2 therapy to accelerate their maturation and allow them to lose their OKT4 marker and become competent OKT8 cytotoxic lymphocytes. Such combination therapy deserves fuller and further study.

Thymic factor therapy has also been disappointing in cytomegalovirus and EB virus infections, but has, perhaps been a little more hopeful in severe measles (27).

NEOPLASIA

In some 4 % of patients with pre-existing T-cell deficiencies, neoplasia has subsequently occured and since most of these were children, this must be considered a very excessive rate (3). It must be remembered, however, that some of the DNA repair defects as in ataxia telangiectasia, Fancon's syndrome and xeroderma pigmentosa, acquire T-cell deficiency at the same time as they acquire the vast increase in somatic mutations which result in neoplasia in over two-thirds of the patients. It could also be said to be true that T-cell impairment would permit EB virus or Hepatitis B virus to produce more proliferation within a given subject and, thus lead to the increased incidence of lymphomas, nasopharyngeal and hepatic cancers which are associated with those viruses. From preliminary studies it would seem that the use of thymic factor therapy probably would not prevent EB virus induced neoplasia, whereas the preliminary results in chronic active hepatitis do suggest that thymic factor therapy might be helpful in preventing complications (28). It is likely that the prophylactic role of thymic factor therapy in such special subgroups and perhaps in the aged, will become important after adequate studies have been completed. Among established neoplasias it is known that tumours of T-lymphocytes, themselves, can produce marked depression of the normal T-cells (4) but, of course, they are rare, representing only a few percent of leukaemias and lymphomas. Hodgkin's disease also causes T-cell deficiency.

Among other kinds of neoplasia, T-cell deficiency is more common with the advancing stage of the tumour and can be general and non-specific or related to specific anti-tumour responses (29). Furthermore, most kinds of anti-tumour treatments such as radiotherapy, cytotoxic drugs (with the exception of Vinca and Cis-platinum derivatives), surgery and anaesthesia themselves suppress T-cell functions (2). Thymic factor therapy can ressurect the acquired T-cell deficiency for all these conditions and thereby reduce the infectious complications and perhaps aggravate less the unlimited spread of the tumour. It is known that among the non-specific depressions, one of the earliest to fail is in the cytotoxic function, mostly expressed in the OKT8 subset and reflected in tests 4 (30) and 7 (2) of table 4. The above, therefore, provide resurrective roles for thymic factor treatments in cancer patients.

Specific immunotherapy of tumours has, in general, had very little success, despite numerous attempts and, of course, the use of thymic factors would imply specific active immunotherapy for a patient. Many of man's naturally occurring tumours enjoy antigen-deletion, bearing very little in the way of a useful immunological target for specific responses. While about half of human tumours can be shown to be antigenic, it is usually other species that see any immunological difference better than do human beings and among the latter it is allogeneic members who can do better than the patient himself. This is illustrated in the marked difference in final cure rate after bone marrow transplantation for leukaemia between allogeneic survivors (< 40 % relapse rate) and twin survivors (>70 % relapse rate). To best exploit whatever antigenicity may underline such successes, it is important to reduce the tumour to a minimal residual state, as has been well demonstrated in choriocarcinoma (31). What is far less productive is the use of thymic factor therapy in recurrent tumours or in combination schedules where it alternates with the cytotoxic therapy (32). Thymic factors have a special place in the attempts to induce active specific immunotherapy because of their capacity to induce lymphopoiesis. It is in the generation and expansion of new clones of lymphocytes that there could arise the possibility of a line sufficiently cytotoxic to have an effect against microscopic metastases remaining in a patient in a minimal residual tumour state. The previous growth of the primary neoplasm within the patient can be considered as a declaration that any effective cytotoxic reaction of the patient is being overcome by the growth of that tumour, possibly the production of free circulating antigenic moieties or minor immune complexes which will block lymphocyte proliferation by not properly crosslinking any specific receptors that may be present. With the reduction of the tumour mass to low levels, and perhaps the added impetus after plasma exchange (29) to remove any remaining such blocking factors, there is the possibility that new clones could emerge, migrate to the region of the tumour and be stimulated to proliferate and overcome it. It follows that this would be more likely to happen in childhood tumours, particularly at the end of cytoreductive therapy but certainly not mixed with it (32) when, of course, any new dividing clones would be eliminated by the cytoreductive therapy.

Malignant melanoma is another tumour that has a wide agespread and certainly in the younger patients, who tend to have a worse prognosis, there is a case for, after the removal of the primary tumour, conducting trials of thymic factor treatments. This has been done in three centres (33,34,35), and for all three, it does appear in double-blind studies that those patients who have received the active drug, Thymostimulin, have had a lower recurrence rate than those who did not. It is a little to early to fully assess survival curves which in melanoma at large, require 5 - 10 years observation, but in the study confined to primary melanoma of the trunk, with no detectable lymph node involvement, it is known that such patients have a very high relapse rate within two years. The preliminary results do suggest that there has been a significant reduction in relapse rate and are shown in Figure 2.

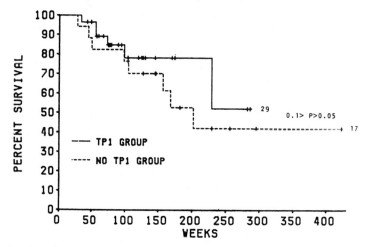

Fig. 2. Survival curves after primary surgery for Stage I truncal melanoma (combined data from 33,34).

Although the survival curves do not yet show a significant difference between the two groups, the recurrence rate by Chi-square testing yielded P < 0.01. Before surgery, 19 % of the 46 patients showed lymphopaenia, which has more prognostic value than even Clarke's thickness, a favourite parameter of histopathologists (33), and 46 % had subnormal numbers of OKT8 cells. In the placebo limb, these parameters worsened, whereas in those who were treated, 75 % corrected to a normal level of either total lymphocytes or OKT8 cells. In this comparison, P was < 0,001.

Studies of localised non-small cell lung cancer (no metastases detected) controlled by local radiotherapy showed a favourable response to α_1-Thymosin (6), with relapse-free survival 31 % at 1 year compared to 0 % in the control limb (P = 0.03). It was noted that the best results were in the patients with the smallest tumour masses, i.e. the closest to the minimal residual tumour state.

SUMMARY

Table 6 summarises the clinical situation in which Tymic Factor treatments can be considered to have achieved successes.

Table 6. Successful clinical applications of Thymic Factor therapies.

PHASE I STUDIES

Di George syndrome
Other genetic T-cell deficiencies
Histiocytosis-X
Disseminated Varicella
Herpes zoster
Complicated Herpes simplex
Severe measles
Guillain-Barre, subacute sclerosing panencephalitis
Hepatitis-B-associated chronic hepatitis
Lepromatous Leprosy

PHASE II TRIALS

Hospitalised measles
Recurrent Herpes simplex
Herpes keratitis
Disseminated Varicella
Severe atopy
Non-small cell lung cancer
Malignant melanoma

It is important to recognise and prove T-lymphocyte deficiency, and the secondary forms respond much better to such treatments than do the primary where there will always be the problem of inadequate targets.

The most obvious failures have been for the two viruses, AIDS and EB, which are known to actually infect lymphocytes and also for cytomegalovirus. In the cancer field, thymic factors have prophylactic and resurrective roles and evidence is accumulating that in minimal residual tumour states they may have specific active immunotherapeutic roles. Here again, failures can be expected where there is too much residual tumour capable of abrogating or blocking any active response by the patient, or where, because of antigen deletion in the tumour or an absence in the repertoire of the patient, it is not possible to establish an immune reaction. Other failures are summarised in Table 7.

Table 7. Contraindications to Thymic Extract Therapy

1. Wheal >12 mm. 20 mins. after 10 µg intradermally

2. Autoimmune and rheumatoid diseases

3. Graft-versus-Host disease

4. <4 weeks before cytoreductive treatments

Thymic factor treatments can also reduce the incidence of infection in the older age group (19) and during cytoreductive treatments (32) and in animal studies have been shown to reduce the incidence of subsequent tumours (11), so that there could be an important prophylactic area with regard to cancer in man and such preventative studies seem worthy of further exploration.

ACKNOWLEDGEMENTS

Most of the T-cell analyses in my own laboratory were undertaken by Dr. M. Kuppner and Miss N. Nagvekar. Financial help has come from the Fane, Bostic, and Dobson Trusts and also from Serono (UK) and Hoffman LaRoche, USA. I am grateful to all my clinical colleagues who co-operated in the studies of the patients.

REFERENCES

1. J. R. Hobbs, N. Byrom, and N. Nagvekar, The Induction of Lymphocyte receptors by Thymic Factors, in: "Peptide Hormones as Mediators in Immunology and Oncology", R.-D. Hesch, and M. J. Anderson, eds., Raven Press, New York (1985).
2. J. R. Hobbs, N. A. Byrom, J. D. Chambers, S. A. Williamson, and N. Nagvekar, Secondary T-Lymphocyte Deficiencies, in: "Thymic Factor Therapy", N. A. Byrom, and J. R. Hobbs, eds., Raven Press, New York (1984).
3. F. Aiuti and L. Businco, "The Immune System: Functions and Therapy of Dysfunction". G. Doria, and E. Estakol, eds., Academic Press, New York (1980).
4. J. R. Hobbs, Thymic factors to treat cancer patients, especially those with T-cell deficiencies, in: "Active biological Substances in Cancer Therapy", in press (1985).
5. N. Byrom, S. Retsas, A. J. Dean, and J. R. Hobbs, The Importance of Dose of Thymosin for the in vitro Induction of T-lymphocytes from Patients with Solid Tumours, Clin. Oncol., 4:34 (1978).
6. R. Schulof, M. Lloyd, T. Chorba, J. Cox, S. Palaszynski, and A. Goldstein, Immunorestorative and therapeutic effects of synthetic Thymosin-α_1 in patients with lung cancer, in: "Thymic Factor Therapy", N. A. Byrom, and J. R. Hobbs, eds., Raven Press, New York, (1984).
7. S.-K. Hu, T. L. K. Low, and A. L. Goldstein, Modulation of terminal deoxynucleotidyl transferase activity by thymosin, in: "Thymic Factor Therapy", N. A. Byrom, and J. R. Hobbs, eds., Raven Press, New York (1984).
8. N. A. Byrom, N. M. Nagvekar, and J. R. Hobbs, Thymic Factor Modulation of Lymphocyte Surface Markers, in: "Thymic Factor Therapy", N. A. Byrom, and J. R. Hobbs, eds., Raven Press, New York (1984)
9. R. C. D. Staughton, N. A. Byrom, N. M. Nagvekar, and J. R. Hobbs, TP1 Therapy of adult atopic eczema, in: " Thymic Factor Therapy" N. A. Byrom, and J. R. hobbs, eds., Raven Press, New York (1984)
10. M. E. Osband, E. B. Cohen, D. Hamilton, Z. sen Ho, D. L. Shipman, C. S. Panse, and R. A. Laursen, Suppressin, a calf thymus-derived inducer of histamine H2 receptor-bearing suppressor T-cells: biochemical and biological characteristics, in: "Thymic Factor Therapy", N. A. Byrom, and J. R. Hobbs, eds., Raven Press, New York (1984).
11. J. Shoham and A. S. Klein, The Potential of Thymic Factors in the Treatment of Cancer, in: "Thymic Factor Therapy", N. A. Byrom, and J. R. Hobbs, eds., Raven Press, New York (1984).
12. A. Gorski, I. Podobinska, M. Nowaczyk, and G. Korczak-Kowalska,

Immunomodulatory effects of thymic factor (TFX) on the interactions between T- and B-cells and their sensitivity to immunosuppressive agents, in: "Thymic Factor Therapy", N. Y. Byrom, and J. R. Hobbs, eds., Raven Press, New York (1984).
13. A. L. Goldstein, T. L. K. Low, N. R. Hall, P. H. Naylor, and M. M. Zatz, Thymosin and thymosin-like preparations, in: "Thymic Factor Therapy", N. A. Byrom, and J. R. Hobbs, eds., Raven Press, New York (1984).
14. J. R. Hobbs, S. Malka, and N. A. Byrom, Density and inducibility of T-cell receptors in relation to age of the lymphocyte donor and conditions of culture, Protides Biol. Fluids, 25:605-610 (1977).
15. J. R. Hobbs, Correction of 34 Genetic Diseases by Displacement Bone Marrow Transplantation, Plasma Ther. Transfus. Technol., 6: 221-246 (1985).
16. F. Aiuti, G. Russo, M. Carbonari, O. Pontesilli, and M. Fiorilli, A rational approach for the use of thymic hormones in viral infection and primary immunodeficiencies, in: "Peptide hormones as mediators in immunology and oncology", R. D. Hesch, and M. J. Atkinson, eds., Raven Press, New York (1985).
17. M. A. Campbell, Lymphocytes with receptors for IgG and IgM in atopic eczema and their relationship to serum IgE levels, Clinical Allergy, 11:509-513 (1981).
18. K. Ishizaka, Twenty years with IgE: From the identification of IgE to regulatory factors for the IgE response, J. Immunol., 135: i-x (1985).
19. F. Pandolfi, I. Quinti, G. Bonomo, A. Morrone, and M. Cherchi, Cellular Immunity and Thymic Hormone Therapy in Aged Humans and Asymptomatic Homosexuals, in: "Thymic Factor Therapy", N. A. Byrom, and J. R. Hobbs, eds., Raven Press, New York (1984).
20. J. R. Hobbs, N. Byrom, N. Nagvekar, C. M. E. Rowland Payne, and J. K. Oates, Post-viral T-cell Deficiencies and their Treatment, in: "Progress in Immunodeficiency Research and Therapy 1", C. Griscelli, and J. Vossen, eds., Elsevier Science Publishers, B. V., Amsterdam (1984).
21. N. A. Byrom, J. R. Hobbs, D. M. Timlin, M. A. Campbell, A. J. Dean, M. Webley, and D. A. Brewerton, T and B Lymphocytes in patients with acute anterior uveitis and ankylosing spondylitis, and in their house-hold contacts, Lancet, 2:601-603 (1979).
22. T. Lindsten, J. K. Seeley, M. Ballow, K. Sakamoto, S. St. Onge, J. Yetz, P. Aman, and D. Purtilo, Immune Deficiency in the X-linked lymphoproliferative syndrome, II. Immunoregulatory T Cell Defects, J. Immunol., 129:2536-2540 (1982).
23. J. Lindberg, A. Lindholm, P. Lundiz, and S. Iwarson, Trigger factors and HL-A antigens in chronic active hepatitis, Brit. Med. J., 4:77-79 (1975).
24. J. R. Hobbs, Disturbance of the immunoglobulins, in: "Scient. Basis Med. Ann. Rev.", Athlone, London, p. 114 (1966).
25. J. F. Taylor, Lymphocyte transformation in Kaposi's sarcoma, Lancet, 1:883-884 (1973).
26. F. Barre-Sinoussi, J. C. Chermann, F. Rey, M. T. Nugeyre, S. Chamaret, J. Gruest, C. Dauguet, C. Axler-Blin, F. Vezinet-Brun, C. Rouzioux, W. Rozenbaum, and L. Montagnier, Isolation of a T-Lymphotropic Retrovirus from a Patient at Risk for Acquired Immune Deficiency Syndrome (AIDS), Science, 220: 868-871 (1983).
27. N. Trainin, Z. T. Handzel, M. Pecht, Y. Varsano, D. W. Beatty, and R. Zaizov, Therapeutic effects of THF in viral diseases, in: "Thymic Factor Therapy", N. A. Byrom, and J. R. Hobbs, eds., Raven Press, New York (1984).

28. A. P. De Felici, M. A. Longo, E. Giradi, and G. Visco, Thymostimulin therapy in chronic active hepatitis, in: "Thymic Factor Therapy", N. A. Byrom, and J. R. Hobbs, eds., Raven Press, New York (1984).
29. J. R. Hobbs, N. Byrom, P. Elliott, C.-J. Oon, and S. Retsas, Cell Separators in Cancer Immunotherapy, Exp. Hematol., 5:95-103 (1977).
30. C. Butterworth, C.-J. Oon, G. Westbury, and J. R. Hobbs, T-Lymphocyte Responses in Patients with malignant Melanoma, Europ. J. Cancer, 10:639-646 (1974).
31. K. D. Bagshawe, Choriocarcinoma, Arnold, London (1969).
32. B. De Bernardi, G. Pastore, A. Garaventa, C. Rosanda, M. Carli, L. Cordero di Montezemolo, F. Bassani, P. F. Biddau, G. Calculli, C. Guazzelli, G. Loiacono, A. Mancini, C. Pianca, and A. Russo, Thymic Hormone Immunotherapy in Advanced Neuroblastoma, in: "Thymic Factor Therapy", N. A. Byrom, and J. R. Hobbs, eds., Raven Press, New York (1984).
33. M. G. Bernengo, G. C. Doveil, F. Lisa, M. Meregalli, M. Novelli, and G. Zina, The immunological profile of melanoma and the role of adjuvant thymostimulin immunotherapy in Stage I patients, in: "Thymic Factor Therapy", N. A. Byrom, and J. R. Hobbs, eds., Raven Press, New York (1984).
34. R. W. Norris, N. A. Byrom, N. M. Nagvekar, A. J. Dean, P. Manhaffey, and J. R. Hobbs, Thymostimulin plus Surgery in the Treatment of Primary Truncal Malignant Melanoma: Preliminary results of a U.K. Multi-centre Clinical Trial, in: "Thymic Factor Therapy", N. A. Byrom, and J. R. Hobbs, eds., Raven Press, New York (1984).
35. E. Azizi, H. J. Brenner, and J. Shoham, Postsurgical Adjuvant Treatment of Malignant Melanoma Patients by the Thymic Factor Thymostimulin, Arzneim.-Forsch., 34:1043-1046 (1984).

AN OVERVIEW OF THE CURRENT UNDERSTANDING AND

MANAGEMENT OF THE NON-HODGKIN'S LYMPHOMAS

Ellen R. Gaynor, and John E. Ultmann

Department of Medicine, University of Chicago and
the University of Chicago Cancer Research Center
Chicago, IL

INTRODUCTION

The hematologic malignancies have always provided the model tumor system for the oncologist. These tumors traditionally have been among the most responsive to chemotherapy and hence chemotherapeutic developments stemmed largely from knowledge gained from the management of these diseases. What has been lacking is an understanding of the malignant process itself and so while we have been able to control and often cure these diseases, a void has existed in our understanding of basic pathophysiology. The past several years have brought tremendous developments in the fields of cytogenetics, immunology, molecular biology, and virology. Whereas previously we have been able to speak of these diseases in merely descriptive terms, we are now quickly gaining an understanding of the events leading to the phenomenon of malignant transformation. It is apparent that once the process is understood, it will be merely a matter of time before we are better able to control the process. While the gap between basic pathophysiology information and therapeutic application is a reality, it probably will not be an insurmountable hurdle.

In this brief review we will attempt to summarize the current approaches to the classification and therapy and our current understanding of the diverse group of diseases which we group together as the non-Hodgkin's lymphomas (NHL).

CLASSIFICATIONS

In his now famous description of the disease which bears his name, Thomas Hodgkin made the insightful observeration that this was a "primary affection of the lymph glands" rather than a disease process which had been propagated to the lymphatic system (1). As the microscope gained wider usage, it became clear that in fact there were many morphologically distinct pathologic processes which involved the lymphatic system. Because the choice of appropriate therapy rests upon accurate diagnosis and classification, there has been intense interest over the years in the development of a meaningful classification of these diverse diseases. One of the most successful and most widely used classification systems was that proposed by Rappaport in which the lymphomas were grouped according to the predominant cell type and the presence or absence of a nodular pattern in the involved lymph node (2). The system proved to be valuable because it was not only relatively simple and reproducible but it provided a grouping of the lymphomas into prognostic categories. In general, lymphomas with a nodular architecture and those composed predominantly of small cells had a better prognosis than the diffuse varieties.

With developments in the field of immunology, it became apparent that the Rappaport system was inadequate. Normal T and B lymphocytes were known to be cells which passed through various stages of differentiation in the process of maturation. Lymphomas, the malignant counterparts of normal T and B cells, thus represented clonal expansions at various stages in this differentiation process. Because the Rappaport system failed to incorporate these key concepts, various other classification systems were developed including the Lukes-Collins (3) and Kiel (4) systems. These systems attempted to clearly delineate the lymphomas as the malignant expressions of various stages of normal lymphocyte development. Unfortunately the proliferation of classification systems led to confusion and problems with communication among pathologists. In an attempt to provide a common working scheme which would incorporate the key concepts of the existing classification systems and which would provide a common language for pathologists, the International Working Formulation for Clinical Usage was recently proposed (5). In this system lymphomas are basically grouped into one of three categories, low grade, intermediate grade, and high grade histologies, correlating with the clinical behavior of these diseases. Though a very useful system, the Working Formulation is not entirely reproducible from one pathologist to another und thus shares a limitation seen with all other attempts at classification.

With the knowledge that normal lymphocytes express definite surface antigens at various stages in their development and with the production of monoclonal antibodies directed against these surface markers, the problem of reproducibility is being minimized. Through the use of monoclonal antibodies, all of the common T and B cell malignancies can now be classified as to the stage of differentiation of the malignant cells. In the case of lymphomas which cannot be adequately classified microscopically, the malignant cells can be tested against a panel of monoclonal antibodies and through the use of such immunological probes can be precisely classified.

In those rare situations where precise T cell or B cell lineage cannot be determined immunologically or in situations where the lymphoid origin of a neoplastic process cannot be determined, it is now possible through the techniques of molecular genetics to examine immunoglobulin gene rearrangements within the neoplastic cells to determine whether the neoplasia is malignant or benign and whether the cell of origin is lymphoid (6,7). In all cells except the committed B cell, immunoglobulin genes are encoded by discontinous segments of genetic material. Such an arrangement (germline state) is altered very early in B cell development as the genetic material rearranges prior to the production of immunoglobulin. Furthermore, since a B cell lymphoma represents a clonal expansion of a single transformed cell, one would expect to find identical DNA rearrangements in all cells of the malignant clone. Through the use of molecular genetic analysis, it is now possible to accurately distinguish the cell of origin as being B cell or non-B cell and, furthermore, to establish clonality in those situations where this may be histologically impossible.

It is obvious that advances have been made in the past several years in our ability to diagnose accurately the non-Hodgkin's lymphomas. There is no doubt that such precision has enabled us to intelligently choose the best mode of therapy for the individual patient.

PATHOPHYSIOLOGY

Immunology

As noted above, we now appreciate the fact that the malignant lymphomas represent clonal expansions at a specific stage of normal lymphocyte differentiation. With the recognition of the immune system as a network of cells originating from a primitive stem cell, it is now apparent that the neoplastic process might be more appropriately viewed from the perspective of differentiation. Whereas previously our attention has focused on malignant proliferation, we must now focus on a more fundamental question, namely, What is the target cell for malignant transformation? (8).

Chronic myelogenous leukemia (CML) is known to be a stem cell disorder in which all progeny lines - erythrocytes, granulocytes, and platelets - are affected by the malignant process. Clinically the disease manifests itself in the granulocytic series. Palliative treatment can control the clinical manifestations; curative treatment, however, must eradicate the malignant stem cell, and hence cure of the disease requires marrow ablative therapy.

Malignant lymphomas in similar fashion manifest themselves clinically as the accumulation of cells at one particular stage of development. While some of these diseases are clearly curable (e.g., diffuse large cell lymphoma), others are merely palliated with current therapies (e.g., nodular poorly differentiated lymphocytic). By analogy with CML, the question arises as to whether the clinically apparent tumor is merely the tip of the iceberg representing the expansion of one cell compartment downstream from the malignantly transformed cell. Perhaps only palliation is possible in some of the malignant lymphomas because our therapy is directed against the wrong target cell.

Cytogenetics, Molecular Genetics, and Oncogenes

In the 1970s it was clearly established that the malignant cells of many neoplasms have chromosome abnormalities and that in many cases there are consistent abnormalities. At the time, only those patterns that represent gross defects by today's standards could be identified, and only leukemia cells could readily be analyzed. With improvements in banding and cell culture techniques it is now possible to study in detail the chromosomal patterns of leukemias, lymphomas, and several solid tumors. While a variety of abnormalities have been described, we shall discuss in detail those found in Burkitt's lymphoma.

In Burkitt's lymphoma, which is an aggressive neoplasm of B cells that affects mainly children, three reciprocal translocations have been detected. Manalov and Manalova first described the 14q+ chromosome in Burkitt's lymphoma cells from five biopsy specimens (9). Subsequently, it was shown that in most cases the donor chromosome was 8q (8q-) and that the specific bands involved in the translocation were 8q24 and 14q32 (10). In approximately ten percent of cases a variant translocation has been found (11,12). In each of the two variants, chromosome 8 is involved in the reciprocal translocation; either chromosome 2 or chromosome 22 donates chromosome material to chromosome 8. Hence it appears that a translocation involving chromosome 8 is the critical chromosomal change associated with Burkitt's lymphoma.

Because the majority of Burkitt's cells manifest the consistent breakpoint [14q32], investigators speculated on the significance of this breakpoint. With the techniques of genetic probing, the genes coding for immunoglobulin heavy chain have recently been mapped to band 32 on chromosome 14 (13). In Burkitt's cells, the 14q+ retains the genes coding for the constant region of the heavy chain, whereas genes coding for all or a portion of the variable region translocate to the 8q- chromosome. Is a similar immunoglobulin region involved in the variant translocations? In the less common variant translocation, the portion of the short arm of chromosome 2 which contains the gene coding for the kappa light chain translocates in a reciprocal fashion to chromosome 8 (14). In similar fashion, the segment of the long arm of chromosome 22 which contains the genetic code for the lambda light chain translocates in a reciprocal manner to chromosome 8 (15). Hence in all three translocations noted in Burkitt's lymphoma, an immunoglobulin gene translocates to chromosome 8. The obvious question then is whether there is some transforming potential on chromosome 8 which in some way favors the development of the malignant process.

The answer to this question has recently been elucidated. It is now appreciated that there are DNA sequences in the human genome which are analogous to known retrovirus oncogenes. The normal role of these cellular oncogenes (c-onc) is unknown, but they are presumed to play a role in normal cell differentiation. It is thought that in neoplastic transformation, these oncogenes are being either overexpressed, expressed at inappropriate times, or incorrectly expressed.

One of these oncogenes (c-myc) shares striking homology with v-myc, the transforming sequence of the avian retrovirus, mc-29. The c-myc gene has been localized to chromosome 8 band 24, which is the breakpoint on chromosome 8 consistently noted in Burkitt's lymphoma (16). There is now substantial evidence that chromosomal translocations involving the c-myc oncogene and one of the immunoglobulin genes are very common in human lymphoid malignancies. It is also known that the insertion of retroviral sequences in the vicinity of c-myc can greatly enhance its expression leading to oncogenic transformation. It is speculated that the juxtaposition of a human immunoglobulin gene in the vicinity of c-myc might also trigger malignant transformation (17). How this occurs is unknown, but obviously the answer to this question is the subject of intense investigation.

Hence, through elegant cytogenetic and molecular genetic analyses, we now know that in Burkitt's lymphoma there are certain consistent chromosome translocations, that each of these translocations involves an immunoglobulin gene, and that each translocation results in the juxtaposition of an immunoglobulin gene within the c-myc oncogene. The obvious questions then are [1] Do all malignant lymphomas demonstrate consistent chromosomal abnormalities? [2] What genes are involved in these translocations? and [3] Do all translocations result in the activation of an oncogene. While the answers to these questions are at present unknown, our knowledge of the cytogenetic abnormalities in malignant lymphoma cells has increased dramatically in the past several years.

Yunis et al. examined biopsy material from 44 patients with NHL; 27 of these patients had received no prior therapy (18). Forty-two of the samples were successfully analyzed, and all were found to have clonal chromosome abnormalities. Sixteen of 19 patients with follicular lymphoma had t[14;18]. Three of 3 patients with the aggressive histology small noncleaved non-Burkitt's lymphoma and 2 of 3 patients with immunoblastic sarcoma demonstrated t[8;14]; as noted above, this is the abnormality seen in the majority of patients with Burkitt's lymphoma. It appears then that the extra chromosome material on chromosome 14 may be derived from different chromosomes with the donor being chromosome 8 in aggressive lymphomas and chromosome 18 in indolent lymphomas. The breakpoint on chromosome 14 appears, however, to be fairly consistently located at band q32.

In a larger series of 94 patients, Bloomfield et al. examined cytogenetic abnormalities in lymph node biopsy specimens (19). Ninety-one of 94 samples showed chromosomal abnormalities. The authors observed seven recurring translocations in their series, and all except one of these involved chromosome 14 at band 32. As in Yunis's series, the t[14;18] was observed frequently in patients with follicular lymphoma. Overall the 14q+ chromosome was found in 66 % of 90 cases of non-Burkitt's lymphoma; no abnormality, however, was restricted invariably to a given histology or immunologic phenotype.

As noted above, immunoglobulin heavy chain genes are in the region of band 32 on chromosome 14. In Burkitt's lymphoma, the translocated material moves to the vicinity of the c-_myc_ oncogene. Does a similar situation arise in the t[14;18] and other translocations? Cellular oncogenes have been mapped to several human chromosomes and in many cases have been shown to be involved in specific chromosome translocations noted in various malignant processes. While it appears that such translocations result in oncogene activation, this activation is probably only one step which must occur in the course of malignant transformation. Subsequent progression may be further promoted by additional genetic and cytogenetic changes within the malignant clone, and these changes may in turn activate other oncogenes.

Retroviruses

Its has been known for several years that retroviruses can induce a variety of neoplasms in animals, and it has been postulated that they might also be capable of inducing neoplasia in man. It was not until 1970 that the viral specific DNA polymerase (reverse transcriptase) was discovered (20). This aided greatly in the biochemical probe of human cells for the presence of these retroviruses. Subsequently the purification of T cell growth factor (TCGF, interleukin-2) enabled malignant T cells to be cultured more effectively in vitro (21). A number of such cultured neoplastic T cells released a type C retrovirus which we now refer to as human T cell leukemia virus (HTLV). Two members of the HTL class of viruses have now been extensively studied and provide very useful clinical models of how these infectious agents are instrumental in the development of neoplasia.

HLTV-I is a human retrovirus associated with a subset of adult T cell malignancies as evidenced by viral specific natural antibodies (22). The virus has been shown to be exogenous, acquired by infection rather than transmitted in the germ line. While the category of T cell lymphoma includes a wide variety of diseases, HTLV is most commonly associated with lymphosarcoma cell leukemia, peripheral T cell lymphoma, and Japanese adult T cell leukemia/lymphoma. Epidemiologic studies indicate that human T cell leukemia/lymphoma clusters in areas where the HTL virus is endemic. The malignant cell which harbors the virus is most frequently a helper T cell (OKT4$^+$). In vitro, infected cells are polyclonal; however, the cells which are immortalized after 4 to 6 weeks in culture are clonal. It appears that by infecting a cell which occupies a central place in the hematopoietic and immune systems, the virus may exert widespread effects on cells beyond the actual target cell of infection. For example, preliminary data suggest that the virus may be indirectly responsible for the development of B cell malignancies. In their recent review, Wong-Staal and Gallo refer to work done by their collaborators suggesting that the clonal expansion of B cells in some patients with CLL may, in fact, represent a monoclonal response to antigen expressed on normal cells infected with HTLV-I (22).

While HTLV-I is capable of transforming and immortalizing cells in vitro, the effect of a second HTL virus, HTLV-III, which is the etiologic agent of the acquired immunodeficiency syndrome (AIDS), is quite different (22). Clinically AIDS is a disease manifested by severe immunosuppression resulting in a susceptibility to opportunistic infections and a propensity to the development of malignancy. The virus preferentially infects helper T cell ($OKT4^+$). Transiently the virus replicates; however, infected cells are gradually depleted. The resultant depletion of this major member of the immune system presumably then is responsible for the overwhelming infections which are frequently found in AIDS patients. Whether this depletion of helper T cells also contributes to the development of lymphoid malignancies in these patients is not clear. As in HTLV-I infection, the development of lymphoma may represent the clonal expression of B cells against antigens on the virally infected cells.

Hence it is now clear, as had been suspected for years, that in some instances viruses may be the cause of malignant transformation in humans. Though closely related, the two viruses act in quite different ways; one causes immortalization of infected cells whereas the other causes depletion. The outcome is, however, similar in that in both instances there are severe effects on the immune system and malignant proliferation of lymphoid tissue.

CURRENT THERAPEUTIC APPROACHES

As mentioned previously, the Working Formulation for Clinical Usage separates the non-Hodgkin's lymphomas into three basic categories: favorable, intermediate, and unfavorable histologies (5). Ironically from the standpoint of curative potential, the favorable histology lymphomas are at present unfavorable while the aggressive histology lymphomas are among the most curable of neoplasms. A brief summary of current treatment strategies for each of these subgroups follows.

Chemotherapy

Favorable Histology. This group of lymphomas usually occurs in middle-aged patients, and in the majority of cases, the disease is advanced (stage III and IV) at presentation. Approximately ten percent of these patients will have truly localized disease and can be shown to be pathologic stage (PS) I or II. This small group of patients may be cured with radiotherapy alone, and hence many authors recommend involved field radiotherapy (IRFT) for clinical stage I and II disease.

Considerable controversy exists regarding the optimum therapy of advanced disease. While both total nodal irradiation and aggressive combination chemotherapy have been used with excellent response rates, relapses are frequent; the risk of relapse remains high for as long as ten years. Smith et al. have recently provided an explanation of this clinical observation and have shown that when these diseases are in apparent complete remission, patients have persistence of a circulating monoclonal population of B lymphocytes (23). This is very suggestive of the fact that the disease is not truly eradicated even when clinically detectable disease is absent.

Portlock et al. have reported the results of a randomized trial comparing combination chemotherapy, combination chemotherapy with irradiation, and single alkylating agent therapy (24). There was no statistical difference in either response rate or survival among the three groups. Furthermore, in a retrospective analysis Portlock and Rosenberg reported on a group of patients who for various reasons had received no initial therapy for their lymphoma (25). When untreated patients were compared with patients randomized to one of the three treatment arms noted above, the authors found no difference in actuarial survival. Thus it may be beneficial to give no initial therapy to some patients with favorable histology. However, these patients should be followed closely for progression or complications of their disease.

Some researchers have reported very favorable results for some patients who received combination chemotherapy, particularly those patients with nodular mixed cell histology (26). Anderson et al. reported a 77 % complete remission (CR) rate using combination chemotherapy in 37 patients with nodular mixed histology (26). Of patients achieving a CR, 79 % remained disease free at 90+ months, as compared to a median survival of 13 months in patients who did not achieve a CR. There are conflicting results regarding therapy in this subset of patients, however, Glick et al. reported on a series of 51 patients with nodular mixed histology who were randomized to receive either one of two combination chemotherapy regimens (cyclophosphamide, vincristine, prednisone, and procarbazine or BCNU, cyclophosphamide, vincristine, and prednisone) or simply chlorambucil and prednisone (27). The authors found no significant differences in CR rate, response duration, or overall survival.

No consistent survival benefit appears to result from aggressive therapy of these favorable histology lymphomas. While complete remission can be achieved in some instances, late relapses make cure very unlikely in this group of diseases. It must be noted that most patients with these lymphomas will ultimately die of their disease. Hence, while a nonaggressive approach may be considered state of the art therapy in some of these lymphomas, it clearly is not an ideal approach. Randomized trials are being conducted at several centers which are addressing the question of the curative potential of modern regimens currently employed in the management of aggressive histology lymphoma.

Intermediate Histology. Included in this category are the histologic subtypes of diffuse poorly differentiated lymphocytic (DPDL), diffuse mixed cell (DM), nodular histiocytic (NH), and certain subclasses of diffuse histiocytic lymphoma (DHL). Therapy of DHL is treated in detail below, since a distinction regarding subclasses is usually not made clinically because all diffuse histiocytic subclasses are known to be clinically aggressive tumors.

The controversy regarding the therapy of other lymphomas in this category centers not on whether initial therapy is needed but on which is the best initial therapy to employ. While aggressive in their clinical behavior, these lymphomas have not been associated with prolonged CR. Currently aggressive chemotherapy identical to that used in the treatment of diffuse histiocytic lymphoma is being tested in the intermediate grade lymphomas. The results thus far are promising in terms of CR rate. Longer follow-up is needed to determine if these CRs are durable.

Unfavorable Histology. This category consists of a heterogeneous group of rapidly progressive lymphomas which are uniformly fatal in weeks to months if untreated. Included in this category are certain subclasses of DHL, lymphoblastic lymphoma, and undifferentiated lymphomas of both the Burkitt's and non-Burkitt's variety. As a group, though very aggressive, these lymphomas are highly curable with chemotherapy. We shall discuss in detail the evolution of therapeutic approaches in DHL, since this is by far the most common of these diseases seen in the adult population.

Fifteen years ago, DHL was generally a fatal disease regardless of therapy. Patients were usually treated with single drugs; complete responses were seen in approximately five percent of patients. Treated patients, overall, hat a median survival of nine months. Since that time, tremendous progress has occured in the treatment of these diseases so that today cure is the rule rather than the exception. These advances have come about through the use of effective combination chemotherapy and through a better understanding of the biology of these diseases.

Over the years a variety of combination therapy programs have been employed. Coleman suggests that these may be grouped into first, second, and third generation chemotherapy programs (28). First generation programs include CHOP (29) (cyclophosphamide, doxorubicin, vincristine, prednisone), C-MOPP (30) (cyclophosphamide, vincristine, prednisone, procarbazine), BACOP (31) (bleomycin, cyclophosphamide, doxorubicin, vincristine, prednisone) and COMLA (32) (cyclophosphamide, vincristine, methotrexate with leucovorin rescue, cytarabine). With the exception of COMLA, these regimens consisted of four or five drugs usually given in monthly cycles. Complete remission rates with these regimens ranged from 40 % to 60 % with approximately 35 % to 45 % of responding patients being cured. These regimens represented an obvious improvement over single agent therapy, but the majority of patients died of their disease.

With a better appreciation of the rapid proliferation capacity of these lymphomas, newer combination regimens were designed which introduced non-myelosuppressive but cytotoxic drugs midcycle in order to control the rapid proliferation. M-BACOD (high dose methotrexate with leucovorin rescue, bleomycin, doxorubicin, cyclophosphamide, vincristine, dexamethasone), an example of such a program, represented a considerable improvement over first generation regimens with a 77 % CR rate and a 57 % cure rate (33).

Because of the work of Goldie et al. in the late 1970s, oncologists began to appreciate the fact that resistant cells were in all probability present in the tumor population before the initiation of chemotherapy (34). The theoretical likelihood of overcoming this problem would be enhanced with the early use of multiple non-cross-resistant drugs. One of the first clinical applications of this hypothesis came from Santoro et al. who showed that the use of the non-cross-resistant drug regimens MOPP (mechlorethamine, vincristine, prednisone, procarbazine) and ABVD (doxorubicin, bleomycin, vinblastine, dacarbazine) produced superior results in terms of CR rates and disease-free survival compared with the use of MOPP alone in previously untreated patients with advanced Hodgkin's disease (35). Fisher et al. subsequently reported the use of an alternating-non-cross resistant regimen, ProMACE/MOPP (procarbazine, methotrexate with leucovorin rescue, doxorubicin, etoposide/mechlorethamine, vincristine, procarbazine, prednisone) in the treatment of patients with advanced diffuse aggressive lymphomas (36). This approach is unique because the number of cycles of each of the two regimens given depends upon the individual patient's tempo of response. A theoretical disadvantage of this approach is that by withholding the second regimen (MOPP) until the tempo of response has plateaued, one may be selecting out cells which are relatively drug resistant. With this program, 73 % of patients achieved a CR, and 60 % of the responding patients have had durable CR.

As our ability to cure certain malignancies has improved and as long-term survivals have been observed over time, it is obvious that there are significant long-term complications of intensive chemotherapy. Because of this and because of the intensity of initial therapy seems to be the essential factor in cure, the newest chemotherapy regimens employ short-term highly intensive therapy. Klimo and Connors recently reported the results obtained using the MACOP-B (methotrexate with leucovorin rescue, doxorubicin, cyclophosphamide, vincristine, prednisone, bleomycin) regimen (37). The total duration of therapy with this regimen is 12 weeks. Myelosuppressive agents are given on alternate weeks with nonmyelosuppressive agents. Eighty-four percent of patients treated achieved a CR, and with a median follow-up of 23 months, the relapse-free survival of responders is 90 %.

Thus, a once fatal disease is now highly curable. Keys to the successful management of this disease have included [1] effective non-cross-resistant drug combinations, [2] intensive therapy, and [3] excellent supportive care.

Biologic Response Modifiers. Biologic response modifiers include a variety of agents and therapeutic approaches, all of which have in common alteration of the biological response in the host-tumor interaction. Use of biologic response modifiers as alternatives to or adjuncts to traditional chemotherapy has been the subject of intense study. While there are a variety of agents in this therapeutic category, we will focus our attention on clinical trials involving the use of monoclonal antibodies and interferon.

Monoclonal Antibody Therapy. In 1975 Milstein and Kohler described a technique resulting in the fusion of murine B lymphocytes and murine myeloma cells leading to the production of hybridoma cell lines secreting antibodies of predefined specificity (38). Several investigators have subsequently attempted with varying success to use monoclonal antibodies in the therapy of lymphoma.

In 1980 Nadler et al. developed a monoclonal antibody, Ab89, which reacted with the malignant cells of a minority of patients with DPDL and B cell chronic lymphocytic leukemia (CLL); the antibody did not react with normal hematopoietic tissue (39). Ab89 was infused into a patient with a very large tumor burden; following infusion there was a transient decrease in the tumor cells in the circulation although the number returned to pretreatment levels in 24 hours. The authors made the significant observation that the patient's plasma contained large amounts of circulating antigen that blocked the binding of Ab89 to tumor cells in vivo.

The following year, 1981, Miller and Levy reported on their use of a monoclonal antibody in the treatment of two patients, one with T cell leukemia and the other with T cell lymphoma (40). The patient with lymphoma initially showed a good response with clearance of circulating Sezary cells and regression of tumor infiltrates in the skin and lymph nodes. However, the patient did not achieve a CR, and after seven weeks the tumor began to proliferate despite continued antibody infusion. A biopsy specimen showed that the "resistant" tumor cells reacted with the monoclonal antibody despite in vivo proliferation of the tumor.

In 1982, Miller et al. described a more specific approach in the use of monoclonal antibody therapy (41). A patient with B cell DPDL had become resistant to cytotoxic drugs and interferon. An anti-idiotype monoclonal antibody was developed which was specific for the idiotype on the surface membrane of the patient's malignant cells. Circulating IgM idiotype was detected prior to therapy, but with continued monoclonal antibody therapy the idiotype level fell and then disappeared. A complete clinical response was eventually achieved. Such therapy is not always so successful, and it has recently been shown that in certain instances the neoplastic process is not monoclonal, i.e., the tumor population includes more than one clone of malignant cells with different idiotypic determinants (42).

These preliminary studies illustrate that monoclonal antibodies can be therapeutically useful. Moreover the production of an idiotype antibody represents the goal of therapy which has "specific toxicity" because the antibody is generated to a specific protein which is unique to the tumor cell and theoretically should have no effect on normal cells. There are, however, several significant problems that currently limit the efficiency of monoclonal antibody therapy (43).

In several studies, circulating antigen which effectively blocks antibody binding in vivo or in vitro has been demonstrated. The presence of blocking factors does not appear to be a predictable phenomenon, although the presence of such factors may correlate with the total number of tumor cells. Hence, it may be possible to overcome this problem if monoclonal antibodies are used when the tumor burden is small.

A second observed problem has been antigenic modulation. Apparently the binding of the monoclonal antibody to the cell surface causes the cell membrane to alter so that the antigen forms microaggregates which are not susceptible to antibody attack. Another theory is that the binding of antibody to the cell surface antigen causes the antigen-antibody complex to be internalized within the cell. This latter mechanism, though posing a problem when antibody alone is used therapeutically, may ultimately be of great therapeutic benifit, for it may enable a specific cell toxin to be introduced into the cell.

The third and perhaps most significant problem with the use of monoclonal antibodies is that these antibodies are not inherently toxic, and the mere binding of antibody to the cell surface does not in most instances affect cell growth. The cytotoxic effects of the antibody are mediated by normally occuring effector mechanisms such as complement. In the clinical situation, however, of a patient with lymphoma who has become refractory to chemotherapy, the tumor burden is usually great and the patient's immune system is in many instances suppressed, both as a result of the disease and as a result of prior treatment. Possible ways to overcome this problem are [1] the development of new monoclonal antibodies, such as IgM antibodies, which are more efficient in the activation of complement; [2] use of monoclonal antibodies when the body burden of tumor is small, e.g., as an adjunct to cytotoxic therapy; and [3] the development of antibody-drug or antibody-toxin conjugates to kill specific tumor cells if the antibody were of such specificity that it reacted only with tumor cells.

<u>Interferons</u>. The interferons are small biologically active compounds which as a class appear to have some growth regulating capacity in that their antiproliferation effects are measurable both in vitro and in animal model systems. These proteins are known to have profound effects on the immune system with varying effects noted as dose range varies. It is still unclear in cancer therapy whether interferons work primarily through an antiproliferative effect or through alterations of the immune

response. What is clear is that they do have antitumor activity which has been most reproducibly seen in the lymphomas. Unfortunately responses to interferon have been few and too often short lived.

In a multi-institutional trial of recombinant leukocyte A interferon (Hoffman-La Roche), Sherwin et al. noted significant antitumor activity in favorable histology NHL and mycosis fungoides with the median duration of response being 5+ months in the former and 3+ months in the later (44). A similar study by Ozer et al. employing recombinant DNA α_2-interferon (Schering) noted a partial response in 4 of 10 patients with NPDL and 2 of 10 patients with DHL (45). All patients except one had been previously treated and had experienced progression of their disease while on conventional therapy. Interestingly, 2 patients whose disease had progressed on chemotherapy and then progressed on interferon subsequently responded to chemotherapy to which they had previously been resistant, suggesting that interferon may enhance the cytotoxic activity of chemotherapy agents (46).

One disease in which interferon has shown considerable activity is hairy cell leukemia. Golomb et al. recently reported on a series of 38 evaluable patients who were treated for at least three months with interferon (47). Eighty-four percent of the patients showed improvement in one of the measured hematologic parameters; 58 % achieved a partial remission of their disease. These results are better than those observed with interferon therapy of other lymphoid malignancies. The reason interferon is effective in this B cell disorder and relatively ineffective in others is not clear at present.

As with monoclonal antibodies, current trials with interferon have been restricted mainly to patients with disease which has become refractory to conventional therapy. Perhaps it will be possible in the future to identify subgroups of patients who will benefit from interferon therapy early in the course of their disease, possibly as an adjunct to cytotoxic therapy.

A VIEW TO THE FUTURE

Chemotherapy/Bone Marrow Transplantation

One obvious area of needed research is the development of new drugs which are not only more effective but less toxic. While a variety of new drugs have been introduced in the last decade, progress in this area has been slow. The most obvious problem with current chemotherapy agents is the very narrow and, in some instances, nonexistent therapeutic index.

In an attempt to overcome the problem of toxicity, many investigators are exploring the possibility of using bone marrow transplantation as a salvage maneuver to rescue the patient from the hematologic toxicity of very intensive, high dose therapy. Because allogeneic transplantation is available only to patients with a suitable donor, attention has turned to the use of autologous marrow transplantation. With this technique, marrow is harvested from the patient during a period of complete remission and reinfused at a later time to rescue the patient from lethal doses of chemotherapy.

While autologous transplantation is an attractive approach, a major drawback is that a marrow that appears free of tumor when examined by light microscopy may in fact harbor malignant cells. Reinfusion of such marrow could contribute to relapse. With monoclonal antibody technology, it is now becoming possible to identify malignant cells in a marrow which may appear histologically normal. Furthermore, by conjugating antibody with a toxin, we are approaching the day when we can purge the marrow of malignant cells for safe reinfusion into the patient.

Several authors have now shown that autologous bone marrow transplant is at least as effective as current salvage chemotherapy in selected heavily pretreated patients. We anticipate that such an approach may find its way into earlier and perhaps even initial therapeutic approaches for some of the non-Hodgkin's lymphomas.

Active Specific Immunotherapy

The development of lymphoma is a significant problem in the feline species. In 1964 Jarett et al. identified the feline leukemia virus (FeLV) as the causative agent of lymphoma in the cat (48). Subsequently investigators developed an immunofluorescent antibody test for FeLV. Testing of several members of the feline species led to the recognition that FeLV is a contagious virus for cats. Spread of disease could be effectively controlled by identifying infected cats and removing them from the environment of healthy animals. The logical next step was to develop a vaccine which could be administered prophylactically. Our veterinary colleagues have thus provided us with a model of etiology, epidemiology, detection, and prevention of a lethal disease which may in the future be very relevant to tumor control in humans.

The recognition that a specific retrovirus, HTL, is the causative agent of a specific class of leukemia/lymphoma and that this virus is endemic in certain areas of the world suggests that it might be possible to develop a vaccine to the virus which could be administered to the population at risk. Intense investigations are currently under way in the development of a vaccine against the AIDS virus, HTLV-III. With this disease there are clearly defined groups at risk, and the incidence of infection is rapidly increasing. Because of minor antigenic diversity within the population of the HTLV-III viruses, such a vaccine is at present not available. It would seem, however, that soon it may be available, and hence, as with previous lethal viral diseases of humans, control may soon be possible.

While up to this point a clear-cut relationship between a given virus and malignancy exists only for the HTL virus, further research in tumor virology may identify other virus-tumor relationships. If epidemiologic studies enable us to identify the populations at risk for these malignancies, the potential exists for the control of these diseases in much the same way that viral illnesses of the 1950s and 1960s are now controlled by large scale immunization programs.

Adoptive Immunotherapy

Interleukin-2 (IL-2), a T cell derived cytokine, was originally identified as a result of its nonspecific enhancing effect on murine thymocyte production. Interleukin-2 can mediate the growth of T cells in vitro, and thus it is possible to expand T lymphocytes from tumor-bearing hosts in vitro. If it were possible to identify cells with anti-tumor reactivity or if such cells could be generated from the immune system of the tumor-bearing host, these cells could theoretically be cloned in vitro. With this technique, a hyperimmune lymphoid population with the desired antitumor reactivity could be generated. Rosenberg has reported on the successful use of this approach in an animal system and has demonstrated that the adoptive transfer of sensitive syngeneic lymphocytes can mediate the regression of established transplantable tumors (49).

There are, however, three major challenges to the use of adoptive immunotherapy in humans. It is first necessary to identify the appropriate cell type for use in adoptive transfer. Second, suitable human tumor cell preparations are currently not available. Finally, effective means of blocking suppressor systems which would inhibit the transfused activated cells need to be explored. While there are considerable barriers to our current use of this approach, it is possible that adoptive immunotherapy may have a place in our therapeutic armamentarium to lymphoma in the future.

Genetic Manipulation

We now know that the majority of non-Hodgkin's lymphomas have clonal chromosomal abnormalities. Further, it appears that at least in some instances, these chromosomal rearrangements lead to the inappropriate expression of cellular oncogenes. Current research is seeking to define the cellular products of these oncogenes and to delineate their function in normal cells. If malignant transformation results from an abnormal expression of these genes, might it not be possible to block this expression or to convert the function of the altered gene to that which is typical of its function in the normal cell? The field of DNA analysis is young, but with currently available technology significant information is being added to a rapidly expanding data base. The therapeutic implications are speculative at the present time; however, one we understand the intricate coding system within the cell, it seems that the logical next step will be to manipulate that system to our advantage in therapy.

CONCLUSION

Clearly significant advances have occured in the past fifteen years in our understanding and management of the diverse group of diseases known as the non-Hodgkin's lymphomas. Currently we are in the midst of an exciting information explosion in our understanding of the pathophysiology of these diseases. While the development of more effective conventional therapy must continue, it is hoped that within the next several years more effective and less toxic means will be discovered as a result of a clearer understanding of the pathologic mechanisms underlying the lymphomas.

REFERENCES

1. H. Fox, Remarks on the presentation of microscopical preparations made from some original tissue described by Thomas Hodgkin, 1832. Annals of Medical History, 8:370 (1926).
2. H. Rappaport, Tumors of the hemopoietic system, in: "Atlas of Tumor Pathology", Armed Forces Institute of Pathology, Washington, D.C. (1966).
3. R. J. Lukes, and R. D. Collins, New approaches to the classification of the lymphomata, Br. J. Cancer, 31 (Suppl.2):1 (1975).
4. R. Gerard-Marchant, I. Hamlin, K. Lennert, F. Rilke, A. C. Stansfield, and J. A. Van Unik, Classification of non-Hodgkin's lymphomas, Lancet 2:406-409 (1974) [Letter to the editor].
5. Non-Hodgkin's Lymphoma Pathologic Classification Project, National Cancer Institute sponsored study of classifications of non-Hodgkin's lymphomas: summary and description of a working formulation for clinical usage, Cancer 49:2112 (1982).

6. A. Arnold, J. Cossman, A. Bakhshi, E. Jaffe, T. A. Waldmann, and S. J. Korsmeyer, Immunoglobulin-gene rearrangements as unique clonal markers in human lymphoid neoplasms, N. Engl. J. Med., 309:1593 (1983).
7. T. A. Waldmann, S. J. Korsmeyer, A. Bakhshi, A. Arnold, and J. R. Kirsch, Molecular genetic analysis of human lymphoid neoplasms, Ann. Intern. Med., 102:497 (1985).
8. I. T. Magrath, Lymphocyte differentiation: an essential basis for the comprehension of lymphoid neoplasia, JNCI, 67:501 (1981).
9. G. Manalov, and Y. Manalova, Marker band in one chromosome 14 from Burkitt lymphoma, Nature, 237:33 (1972).
10. Y. Manalova, G. Manalov, J. Kieler, A. Lenan, and G. Klein, Genesis of the 14q+ marker in Burkitt's lymphoma, Hereditas, 90:5 (1979).
11. H. Van den Berghe, C. Parloir, S. Gosseye, V. Englebienne, G. Carnu, and G. Sokal, Variant translocation in Burkitt lymphoma, Cancer Genet. Cytogenet., 1:9 (1979).
12. R. Berger, A. Bernheim, H. J. Web, G. Flandren, M. T. Daniel, J. C. Brovet, and N. Colbert, A new translation in Burkitt's tumor cells, Hum. Genet., 53:111 (1979).
13. J. R. Kirsch, C. C. Morton, K. Nakahara, and P. Leder, Human immunoglobulin heavy chain genes map to a region of translocations in malignant B lymphocytes, Science, 216:303 (1982).
14. S. Malcolm, P. Barton, C. Murphy, M. A. Ferguson-Smith, D. L. Bentley and T. H. Rabbits, Localization at human immunoglobulin kappa light chain variable region genes to the short arm of chromosome 2 by in situ hybridizations, Proc. Nat. Acad. Sci. USA, 79:4957 (1982).
15. J. Erickson, J. Martinus, and C. M. Croce, Assignment of the genes for human lambda immunoglobulin chains to chromosome 22, Nature, 294:173 (1981).
16. R. Dalla-Favera, M. Bregni, J. Erikson, D. Pattereson, R. C. Gallo, and C. M. Croce, Human c-myc oncogene is located on the region of chromosome 8 that is translocated in Burkitt lymphoma cells, Proc. Nat. Acad. Sci. USA, 19:7824 (1982).
17. G. Klein, Specific chromosomal translocations and the genesis of beta-cell derived tumor in mice and men, Cell, 32:311 (1983).
18. J. J. Yunis, M. M. Oken, M. E. Kaplan, K. M. Ensrud, R. R. Howe, and A. Theologidas, Distinctive chromosomal abnormalities in histologic subtypes of non-Hodgkin's lymphomas, N. Engl. J. Med., 307:1231 (1982).
19. C. D. Bloomfield, D. C. Arthur, E. G. Frizzera, B. A. Peterson, and K. J. Gajl-Peczalska, No random chromosome abnormalities in lymphoma, Cancer Res., 43:2975 (1983).
20. H. Temin, and S. Mizutani, RNA-dependent DNA polymerase in virions of Rous sarcoma virus, Nature, 226:1211 (1970).
21. D. A. Morgan, F. W. Ruscetti, and R. C. Gallo, Selective in vitro growing of T-lymphocytes from normal human bone marrow, Science, 193:1007 (1983).
22. F. Wong-Staal, and R. C. Gallo, The family of human T-lymphotrophic leukemia virus: HTLV-I as the cause of adult T cell leukemia and HTLV-III as the cause of acquired immunodeficiency syndrome, Blood, 65:253 (1985).
23. B. R. Smith, D. S. Weinberg, N. J. Robert, M. Towle, E. Luther, G. S. Pinkus, and K. A. Ault, Circulating monoclonal B lymphocytes in non-Hodgkin's lymphoma, N. Engl. J. Med.,311:1476 (1984).
24. C. S. Portlock, S. A. Rosenberg, E. Glatstein, and H. S. Kaplan, Treatment of advanced non-Hodgkin's lymphoma with favorable histologies: preliminary results of a prospective trial, Blood, 47:747 (1976).
25. C. S. Portlock, and S. A. Rosenberg, No initial therapy for stage III and IV non-Hodgkin's lymphomas of favorable histologic types.

Ann. Intern. Med., 90:10 (1979).
26. T. Anderson, R. A. Bender, R. I. Fisher, V. T. DeVita, B. A. Chabner, C. W. Bernard, L. Norton, and R. C. Young, Combination chemotherapy in non-Hodgkin's lymphoma: results of long-term follow-up. Cancer Treat. Rep., 6:1057 (1977).
27. J. H. Glick, J. M. Barnes, E. Z. Ezdinli, C. W. Berard, E. L. Orlow, J. M. Bennett, Nodular mixed lymphoma: results of a randomized trial failing to confirm prolonged disease-free survival with COPP chemotherapy, Blood, 58:920 (1981).
28. M. Coleman, Chemotherapy for large cell lymphoma: Optimism and caution, Ann. Intern. Med., 103:140 (1985) [editorial].
29. J. O. Armitage, F. R. Dick, M. P. Corder, S. C. Garneau, C. E. Platz, and D. J. Slymen, Predicting therapeutic outcome in patients with diffuse histiocytic lymphoma treated with cyclophosphamide, adriamycin, vincristine and prednisone (CHOP), Cancer, 50:1695 (1982).
30. V. T. DeVita, G. P. Canellos, B. Chabner, P. Schein, S. P. Hubbard, and R. C. Young, Advanced diffuse histiocytic lymphoma, a potentially curable disease: results with combination chemotherapy, Lancet, 1:248 (1975).
31. P. S. Schein, V. T. DeVita, S. P. Hubbard, B. A. Chabner, G. P. Canellos, and C. W. Berard, Bleomycin, Adriamycin, cyclophosphamide, vincristine and prednisone (BACOP) combination chemotherapy in the treatment of advanced diffuse histiocytic lymphoma, Ann. Intern. Med., 85:417 (1976).
32. D. L. Sweet, H. M. Golomb, J. E. Ultmann, J. B. Miller, R. S. Stein, E. P. Lester, U. Mintz, J. D. Bitran, R. A. Streuli, K. Daly, and N. O. Roth, Cyclophosphamide, vincristine, methotrexate with leucovorin rescue and cytarabine (COMLA) combination sequential chemotherapy for advanced diffuse histiocytic lymphoma, Ann. Intern. Med., 92:785 (1980).
33. A. T. Skarin, G. P. Canellos, D. S. Rosenthal, D. C. Case, J. M. MacIntyre, G. S. Pinkus, W. C. Moloney, and E. Frei, Improved prognosis of diffuse histiocytic lymphoma by use of high dose methotrexate alternating with standard agents, J. Clin. Oncol., 1:91 (1983).
34. J. H. Goldi, A. J. Goldman, and G. A. Gudauskas, Rationale for the use of alternating non-cross-resistant chemotherapy, Cancer Treat. Rep., 66:39 (1982).
35. A. Santoro, G. Bonnadonna, and V. Bonfante, Alternating drug combinations in the treatment of advanced Hodgkin's disease, N. Engl. J. Med., 306:770 (1982).
36. R. I. Fisher, V. T. DeVita, S. M. Hubbard, D. L. Long, R. Wesley, B. Chabner, and R. C. Young, Diffuse aggressive lymphomas, increased survival after alternating flexible sequences of ProMace and MOPP chemotherapy, Ann. Intern. Med., 98:304 (1983).
37. P. Klimo, and J. M. Connors, MACOP-B chemotherapy for the treatment of diffuse large cell lymphoma, Ann. Intern. Med., 102:596(1985).
38. G. Kohler, and C. Milstein, Continous cultures of fused cells secreting antibody of predefined specificiticy, Nature, 256:495 (1975).
39. L. M. Nadler, P. Stashenko, R. Hardy, W. D. Kaplan, L. N. Button, D. W. Kufe, K. H. Antman, and S. F. Schlossman, Serotherapy of a patient with a monoclonal antibody directed against a human lymphoma associated antigen, Cancer Res., 40:3147 (1980).
40. R. A. Miller, and R. Levy, Response of cutaneous T cell lymphoma therapy with hybridoma monoclonal antibody, Lancet, 2:226 (1981).
41. R. A. Miller, D. G. Maloney, R. Warnke, and R. Levy, Treatment of B cell lymphoma with monoclonal anti-idiotype antibody, N. Engl. J. Med., 306:517 (1982).
42. M. Raffeld, L. Neckers, D. L. Longo, and J. Cossman, Spontaneous

alteration of idiotype in a monoclonal B-cell lymphoma: escape from detection by anti-idiotype, N. Engl. J. Med., 312:1653 (1985).
43. J. Ritz, and S. F. Schlossman, Utilization of monoclonal antibodies in the treatment of leukemia and lymphoma, Blood, 59:1 (1982).
44. S. Sherwin, K. Foon, P. Bunn, D. Longo, and R. Oldham, Recombinant leukocyte A interferon in the treatment of non-Hodgkin's lymphoma, chronic lymphocytic leukemia, and mycosis fungoides, Proc. Amer. Soc. Hematol., 62:764 (1983).
45. H. Ozer, R. Leavett, V. Ratanatharathorn, J. E. Ultmann, C. Portlock, D. L. Kisner, R. Ferraresi, S. A. Rudnick, and E. M. Bonnem, Experience in the use of DNA α_2-interferon in the treatment of malignant lymphomas, Proc. Amer. Soc. Hematol., 62:761 (1983).
46. R. Ferraresi, S. A. Rudnick, E. M. Bonnem, J. Ochs, C. Karenes, V. Ratanatharathorn, and S. S. Costanzi, Enhanced response to chemotherapy after treatment with DNA α_2 human interferon, Proc. Amer. Soc. Hematol., 62:752 (1983).
47. H. Golomb, A. Fefer, M. Ratain, J. Thompson, H. Ozer, C. Portlock, R. Spiegel, and J. Brady, A Recombinant α_2 interferon for the treatment of hairy cell leukemia, Proc. Amer. Clin. Oncology, 4:874 (1985).
48. W. F. H. Jarrett, E. M. Crawford, W. B. Martin, and F. David, A virus like particle associated with leukemia (lymphosarcoma), Nature, 202:567 (1964).
49. S. A. Rosenberg, Adoptive immunotherapy of cancer: accomplishments and prospects, Cancer Treat. Rep., 68:233 (1984).

Participants

Dr. Anders, A.
Institut für Genetik der
Universität Gießen
Heinrich-Buff-Ring 58-62
D-6300 Gießen
FRG

Prof. Dr. Anders, F.
Institut für Genetik der
Universität Gießen
Heinrich-Buff-Ring 58-62
D-6300 Gießen
FRG

Dr. Aoki, T.
Department of Internal Medicine
Shirakuen Hospital
1-27 Nishi-Ariake-cho
Niigata 951-21
Japan

Prof. Baldwin, R. W.
Cancer Research Campaign Labor.
University of Nottingham
University Park
Nottingham NG7 2RD
England

Prof. Dr. Bertino, J.
Department of Pharmacology and
Medicine
Yale University School
of Medicine
333 Cedar Street
New Haven, Conn. 06510
USA

Prof. Dr. Bonadonna, G.
Istituto Nazionale per lo Studio
E La Cura Dei Tumori
Via Venezian 1
I-20133 Milano
Italy

Dr. Borrinans, B.
Institut voor Molekulaire
Biologie
Vrije Universiteit Brussel
Paardenstraat 65
B-1640 St. Genesius Rode
Belgium

Dr. Bruck, Claudine
University of Brussels
Department of Molecular Biology
Labor. of Biological Chemistry
67, Rue des Chevaux
B-1640 Rhode St. Genese
Belgium

Cand. med. Carella, Alessandra
Department of Dermatology
University Medical School
Theodor-Stern-Kai 7
D-6000 Frankfurt 70
FRG

Prof. Dr. Chandra, P.
Laboratory of Molecular Biology
Center of Biological Chemistry
University Medical School
Theodor-Stern-Kai 7
D-6000 Frankfurt 70
FRG

Dr. De Beatselier, P.
Institut voor Molekulaire
Biologie
Vrije Universiteit Brussel
Paardenstraat 65
B-1640 St. Genesius Rode
Belgium

Prof. Dr. De Palo, G.
Istituto Nazionale per lo Studio
E La Cura Dei Tumori
Via Venezian 1
I-20133 Milano
Italy

Dr. De Villiers, E. M.
German Cancer Research Center
Im Neuenheimer Feld 280
D-6900 Heidelberg 1
FRG

Dr. Demirhan, I.
Laboratory of Molecular Biology
Center of Biological Chemistry
Theodor-Stern-Kai 7
D-6000 Frankfurt 70
FRG

Dipl.-Ing. Doerk, K.-H.
Department of Medical Technology
Siemens AG
Rödelheimer Landstr. 17-19
D-6000 Frankfurt 90
FRG

Dzwonkowski, Maria
Laboratory of Molecular Biology
Center of Biological Chemistry
Theodor-Stern-Kai 7
D-6000 Frankfurt 70
FRG

Dr. Eisenberg, Lea
Department of Cell Biology
The Weizmann Institute
of Sciences
Rehovot 76100
Israel

Prof. Epstein, M. A.
Nuffield Department of
Clinical Medicine
University of Oxford
John Radcliffe Hospital
Headington
Oxford OX3 9DU
England

Prof. Feldman, M.
Department of Cell Biology
The Weizmann Institute
of Science
P.O.B. 26
Rehovot 76100
Israel

Dr. Ferreira, J. S. M.
Clinica Oncologica VIII
Instituto Portugues de Oncologia
Francisco Gentil
Palhava-Lisboa
Portugal

Dr. Gallo, R. C.
Tumor Cell Biology
National Cancer Institute
Building 37, 6C 09
Bethesda, Maryland 20205
USA

Dr. Gaynor, E.
Cancer Research Center
The University of Chicago
5841 South Maryland Avenue
Chicago, Ill., 60637
USA

Geist, Lydia
Secretary-NATO Conference
Feldbergstr. 15
D-6000 Frankfurt-Niederhöchstadt
FRG

Dipl.-Biol. Gerber, Th.
Laboratory of Molecular Biology
Center of Biological Chemistry
Theodor-Stern-Kai 7
D-6000 Frankfurt 70
FRG

Prof. Dr. Ghione, M.
Istituto di Microbiologia
Universita degli Studi
Via Mangiagalli 31
I-20133 Milano
Italy

Prof. Goldstein, A.
Department of Biochemistry
George Washington University
School of Medicine
2300 Eye Street NW
Washington, DC. 20037
USA

Prof. Dr. Gros, F.
Institut Pasteur
Department Biologie Moleculaire
25,rue du Dr. Roux
F-75015 Paris
France

Dr. Hareuveni, M.
Department of Microbiology
University of Tel Aviv
Ramat Aviv
Tel Aviv 69 978
Israel

Prof. Hnilica, L. S.
Department of Biochemistry
University School of Medicine
Station 17
Vanderbilt University
Nashville, Tenn. 37232
USA

Prof. Dr. Hobbs, J. R.
Dept. of Chemical Immunology
Westminster Hospital
Page Street Wing
London SW 1P 2AP
England

Prof. Dr. Hohorst, H. J.
Laboratory of Cell Biology
Center of Biological Chemistry
Theodor-Stern-Kai 7
D-6000 Frankfurt 70
FRG

Dr. Kaneko, Y.
Pharmaceutical Department
Ajinomoto Company
1-5-8 Kyobashi
Chuoka, Tokyo 104
Japan

Dr. Kaul, S.
Laboratory of Therapeutical
Biochemistry
Center of Biological Chemistry
University Medical School
Theodor-Stern-Kai 7
D-6000 Frankfurt 70
FRG

Prof. Dr. Keydar, I.
Department of Microbiology
University of Tel Aviv
Ramat Aviv
Tel Aviv 69 978
Israel

Prof. Dr. Klein, Eva
Karolinska Institute
Department of Tumor Biology
S-10401 Stockholm 60
Sweden

Prof. Dr. Koch, M. A.
Division of Virology
Robert-Koch-Institut
Nordufer 20
D-1000 Berlin 65
FRG

Prof. Koprowski, H.
The Wistar Institute of Anatomy
and Biology
Philadelphia, Pennsylvania 19104
USA

Prof. Dr. Kornhuber, B.
Klinikum der
J. W. Goethe-Universität
Zentrum der Kinderheilkunde
Abt. für Pädiatrische
Hämatologie und Onkologie
Theodor-Stern-Kai 7
D-6000 Frankfurt 70
FRG

Dr. Kottaridis, S. D.
Hellenic Anticancer Institute
171, Alexandra Avenue
Athens 603
Greece

Cand. Biol. Krug, K. P.
Institut für Biochemie der
J. W. Goethe-Universität
Haus 75A
Theodor-Stern-Kai 7
D-6000 Frankfurt 70
FRG

Dr. Kragh Larsen, Annette
Institut Gustav-Roussy
Unite de Biochimie et
Enzymologie
Rue Camille Desmoulins
F-94800 Villejuif
France

Dr. Lawless, G. B.
Specialty Diagnostics Division
E. I. du Pont de Nemours
Biomedical Products Department
Wilmington, Delaware 19898
USA

Dr. Lederer, T. W. P.
Institute for Experimental
Surgery
Technical University Munich
Ismaninger Straße 22
D-8000 München 80
FRG

Luibrand, T.
Hoechst AG
D-6000 Frankfurt
FRG

Dr. Lübke, K.
Schering AG
Pharma Forschung
P.O.B. 65 0311
D-1000 Berlin 65
FRG

Matteus, Renate
Gottfried-Schaider-Str. 26
D-6050 Offenbach/Main
FRG

Dr. McCaffrey, R.
Boston University Medical Center
75 East Newton Street
Boston, Mass. 02118
USA

Dr. Musil, H.
Großgartenweg 15
D-7811 Sulzburg-Laufen
FRG

Dr. Musil, J.
Medical Department
Serono Pharmaceutical GmbH
Merzhauser Str. 134
D-7800 Freiburg
FRG

Prof. Dr. Pompidou, A.
Hopital Saint-Vincent de Paul
74, Avenue Denfert Rochereau
F-75014 Paris
France

Dr. Rankin, Elaine M.
Medical Oncology Clinic
Guy's Hospital
London SE1 9RT
England

Dr. Rossi, L.
Du Pont de Nemours Italia
Via Volta 16
Cologno Monzese
I-20093 Milano
Italy

Dr. Sarangadharan, M. G.
Litton Bionetics, Inc.
Department of Cell Biology
5516 Nicholson Lane
Kensington, Maryland 20895
USA

Dr. Sarin, P.
Tumor Cell Biology
National Cancer Institute
Building 37, 6B 09 15
Bethesda, Maryland 20205
USA

Dr. Schioppacassi, G.
Istituto Farmacologico Serono
Via San Pietro All'Orto 17
I-20121 Milano
Italy

Dr. Schlatterer, B.
Umweltbundesamt
Bismarckplatz 1
D-1000 Berlin 33
FRG

Dr. Sedlacek, H. H.
Abt. f. Experimentelle Medizin
Forschungslaboratorie
Behringwerke AG
Postfach 1140
D-3550 Marburg
FRG

Dr. Shrieve, D. C.
Department of Radiation Oncology
Radiation Oncology Research
Laboratory CED-200
University of California
San Francisco, Cal. 94143
USA

Dr. Siedentopf, Dörthe
Bergstraße 7
D-6057 Dietzenbach
FRG

Prof. Dr. Siedentopf, H. J.
Zentrum der Frauenheilkunde
University Medical School
Theodor-Stern-Kai 7
D-6000 Frankfurt 70
FRG

Dr. Stehlin, Dominique
Laboratoire d'Oncologie
Moleculaire
Institut Pasteur
1, Rue Calmette
F-59109 Lille Cedex
France

Dr. Ting, R. C. Y.
Biotech Research Laboratories
1600 East Gude Drive
Rockville, Maryland 20850
USA

Dr. Tritton, T. R.
Department of Pharmacology
The University of Vermont
Given Medical Building
Burlington, Vermont 05405
USA

Prof. Dr. Turano, A.
Istituto di Microbiologia
Spedali Civili
Universita di Brescia
I-25100 Brescia
Italy

Dr. Tzehoval, Ester
Department of Cell Biology
The Weizmann Institute
of Sciences
Rehovot 76100
Israel

Prof. Ultmann, J. E.
Cancer Research Center
The University of Chicago
5841 South Maryland Avenue
Chicago, Ill. 60637
USA

Prof. Undheim, K.
Department of Chemistry
University of Oslo
Blindern, Oslo
Norway

Dr. Vanderlinden, B.
Rue De La Bruyere 41
B-1320 Genval
Belgium

Dr. Voegeli, R.
Cancer Research Laboratory
ASTA-Werke AG
DEGUSSA Pharmagruppe
D-4800 Bielefeld 14
FRG

Dr. Vogel, Angelika
Laboratory of Molecular Biology
Center of Biological Chemistry
University Medical School
Theodor-Stern-Kai 7
D-6000 Frankfurt 70
FRG

Prof. Witz, I. P.
University of Tel Aviv
P.O.B. 39040
Ramat Aviv
Tel Aviv 69 978
Israel

Dr. Wong-Staal, F.
Laboratory of Tumor Cell Biology
National Cancer Institute
National Institutes of Health
Building 37, 6 C 09
Bethesda, Maryland 20205
USA

Prof. Dr. Zagury, D.
Universite Pierre et Marie Curie
4 Place Jussien
F-75005 Paris
France

Prof. Dr. Zur Hausen, H.
German Cancer Research Center
Im Neuenheimer Feld 280
D-6900 Heidelberg 1
FRG

Index

Adriamycin
  action on membranes, 196
  cytotoxicity, 195
  intracellular activity, 196
  mechanism of action, 197
  and protein phosphorylation, 199
  receptor binding site, 198
Adenosine deaminase
  inhibitors of, 209
  in leukemia 203, 208
  properties of, 208
Acquired Immune Deficiency Syndrome (AIDS)
  associated neoplasms in, 304
  associated virus, see HTLV-III
  IL-2 production in, 359
  immunologic disorders in, 304
  immunosuppression in, 288
  infections in, 304
  neurologic disorders in, 305
  therapeutic strategies for, 305 - 328
  virus etiology 288, 303

Burkit's lymphoma
  and EBV, 263
  chromosomal translocation in, 263
Bovine Leukosis Virus (BLV)
  and asymptomatic infection, 280
  cloned env gene of, 283
  envelope glycoprotein, 282
  and enzootic bovine leukemia, 280
  induced leukemogenesis, 279
  induced tumors, 281
  and persistent lymphocytosis, 280
  proviral sequences, 281
  similarity to HTLV-I, 281, 295
Bone marrow transplantation 387

Carcinoma
  mouse mammary, 123
  nasopharyngeal and EBV, 263
Cell differentiation and onc genes, 7
Cellular oncogenes, see Oncogenes

Cyclophosphamide
  antitumor efficacy of, 243, 244
  DNA cross-linking by, 243
  membrane alteration by, 243
Cyclosporin
  effect on HTLV-replication, 338
  mode of action, 337
  structure of, 336
Cytotoxicity
  membrane-mediated, 195
  of adriamycin, 195
  of cis-diaminedichloro-platinum (II), 223 - 233
  of elliptinium, 235 - 241

DNA-polymerases, 215, 309 - 311, 317, 331
DNA-Protein crosslinking
  and cytotoxicity, 225
  immunological detection, 227
  in-vivo determination, 228, 229
  and Novikoff hepatoma, 226
  of platinum complex, 223 - 227
  and sarcoma growth, 223
Drug action
  at cell surface, 195, 196
  intracellular level, 196
Drug-conjugation to monoclonal antibodies, see Monoclonal antibodies
Drug-resistance, 183 - 193
  impaired transport and, 185
  in leukemia, 183
  mechanism of, 184
  to methotrexate, 183, 184, 186
  in patients, 186
  in resistant cells, 187
  role of gene amplification in, 185

Elliptinium
  antitumor activity of, 235, 236
  clinical trials with, 238 - 239
  cytotoxicity of, 235, 236
  derivatives, 236

Elliptinium (continued)
  metabolism of, 238
  mode of action of, 236, 237
  pharmacology of, 237
  structure of, 235
Epstein-Barr (EB) virus
  antibodies to membrane antigens of, 268
  associated cancers, 263, 264
  induced lymphoma, 266
  membrane antigen, 265, 267, 268
  vaccine against, 264, 265

Fibroblast growth factor, 9, 76
  induction of Fc-receptor, 76
Foscarnet
  inhibition of HTLV-III RT, 334
  inhibition of HTLV-III replication, 334
  mode of action, 333
  structure of, 333

Genes
  cellular oncogenes, see Oncogenes
  differentiation, 26, 28
  viral oncogenes, see Oncogenes
Glutathione
  and cell growth, 249, 250
  in cell protection, 249, 251, 252
  and cellular radiosensitivity, 250
  depletion, 250-252

Human T-Lymphotropic Retrovirus
  HTLV-I
    in ATL patients 287, 288
    gene structure, 295
    and human disease, 287
    mechanism of transformation by, 295
    properties of, 293
    reverse transcriptase, 311
    serological analysis of, 310, 311
    similarity to BLV, 281, 295
    and tat-mRNA, 295
  HTLV-II
    and hairy-cell leukemia, 287
    reverse transcriptase, 311
    serological analysis of, 310, 311
  HTLV-III
    and AIDS, 288
    antigen expression, 297, 344-353
    associated disorders, 289, 304, 305
    cDNA, 296
    clinical impact of, 289, 304, 305

Human T-Lymphotropic Retrovirus (continued)
  HTLV-III (continued)
    cytopathic activity of, 297, 357-361
    env gene, 298
    env gene products, 346, 351, 352
    gag gene products, 345, 346
    gene localization of, 307, 343, 344
    heterogeneity of isolates of, 298
    inhibition of replication of, 319-321, 329-342
    mode of transmission of, 289, 357
    molecular biology of, 293, 296
    pol gene products 345, 346
    proteins encoded by, 343-356
    replication in cell culture, 329-342
    replicative cycle of, 306-308, 330
    reverse transcriptase, 307-318, 334
    and tat function, 296

Immune reactivities
  in chemical carcinogenesis, 74
  in early cancer, 71-80
  in precancer, 71-80
  in primary carcinogenesis, 72
Interferons
  and MHC-expression, 85
  and metastasis, 89
  in NH-lymphoma treatment, 386
Interleukin-2
  cloning of gene, 169
  in endometrial carcinoma, 72, 73
  expression in human cells, 171, 172
  gene expression, 169, 172
  gene localization, 176
  gene transcription, 173
  in HTLV-infected cells, 169, 175
  induction of, 170, 174
  monoclonal antibodies to, 167
  in NH lymphoma treatment, 388
  nuclear transcripts of, 173
  production of, 166
  properties of, 168
  specific mRNA, 170

Lymph-Adenopathy Virus (LAV), see HTLV-III

Mafosfamide
  hydrolysis of, 245
  in immunotherapy, 245
  pharmacokinetics, 245

Major histocompatibility complex (MHC)
  effect of interferon on expression of, 85
  expression and metastases, 84
  and metastatic growth, 82
  metastatic potentials and, 83, 88
Membrane
  adriamycin action on, 196, 197
  alteration by drugs, 196, 243
  cyclophosphamide action on, 243
Methotrexate
  Analogs 188
  Binding affinity of, 189
  resistance, 183-193
  transport of, 185
Monoclonal antibody (MA)
  and adriamycin conjugates, 98
  to breast cancer, 91
  to CEA, 91
  and cis-platinum conjugates, 98
  and drug conjugates, 91-99
  to EBV-membrane antigen, 271
  to human breast cancer, 124
  to human MTV, 123, 127, 132
  to HTLV-III reverse transcriptase, 310-316, 347-349
  imaging by, 92, 93, 112-114
  and interferon conjugates, 98
  Interleukin-2, 167
  to idiotypic antibodies, 101-122
  in lymphoma treatment, 106, 115
  localization in colorectal cancer, 93
  to melanoma, 91
  and methotrexate conjugates, 95, 97
  targeting of biological response modifiers to, 91
  therapy for NH lymphoma, 385
  tumor localization by, 91

Nerve growth factor, 8
Non-Hodgkin's Lymphoma
  active immunotherapy of, 388
  adoptive immunotherapy of, 388
  classification of, 377
  cytogenetics of, 379
  pathophysiology of, 379
  role of retroviruses in, 381
  therapeutic approaches for, 382-387
  treatment by genetic manipulation, 389

Oncogenes
  activation, 4, 5
  amplification, 22, 31
  and cell differentiation, 7
  cellular oncogenes, 2, 8, 16, 18

Oncogenes (continued)
  conservation of sequences, 8
  cooperation between, 6
  in development, 15, 17, 35
  distribution of, 34
  in eukaryotes, 34
  in evolution, 15, 32
  history of, 15
  location, 3
  in melanoma, 29
  and metastasis, 57-70
  myc, 3, 5
  in neoplasia, 15, 21, 24
  organ specificity of, 18
  phylogeny, 1, 6
  products of, 3, 7
  protooncogenes, 1, 4, 8
  ras oncogenes, 4
  regulation of activity of, 24, 26
  restriction analysis of, 22, 31, 33
  transduction, 3, 8
  tumorigenicity, 3
  and tyrosine kinase, 21, 23, 64
  viral oncogenes, 3, 7, 17, 18
  in Xiphophorus, 16
Oxazaphosphorines, 243-247 (see also cyclophosphamide)

Papilloma viruses
  in anogenital cancer, 257
  in cervical cancer, 258
  classification, 255
  DNA chromosomal localization, 258
  DNA integration pattern, 258
  DNA sequencing, 256
  and genital infections, 256
  in human cancer, 255-261
  infective cycle, 255
  and non-genital cancer, 259
  in perianal cancer, 258
Penicillamine
  inhibition of HTLV-III replication by, 319-321
  structure of, 319
Platelet-derived growth factor, 9
Protein kinases
  amino-acid specificity of, 64-66
  in carcinogen-treated animals, 20
  effect of adriamycin, 199
  effect of c-AMP on, 65
  effect of c-GMP on, 65
  in embryo, 19
  in growth, 20
  in metastasis, 57-70
  in neoplasms, 21, 23
  organ specificity, 18
  in plasma membrane, 61, 62
  $pp60^{v-src}$, 2, 8, 19
  in Xiphophorus, 19-21, 23, 29

Protein phosphorylation, see
    Protein kinases
Proto-oncogenes, see
    Oncogenes

Retroviruses
  AEV, 7
  BLV, see Bovine leukosis Virus
  cellular onc genes, see
      Oncogenes
  HuMTV
    anti-HuMTV activity, 130, 134
    electron microscopy, 127
    induction, 126
  MMTV, 123
  MV, 133
  prototypes, 18
  RSV, 7
  SSV, 133
  STLV-I, 293, 294
  STLV-II, 293, 294
  T-lymphotropic viruses, see
      human T-lymphotropic
      viruses
  transforming, 7
Reverse transcriptase
  heterogeneity of, 314
  of HTLV-III, 310-316
  inhibitors, 314-318, 339
  in Xiphophorus, 24, 25

Suramin
  clinical trials of, 335
  inhibition of HTLV-III RT, 318

T-cell
  hybridomas, 43, 53
  invasion of, 53
Terminal transferase
  characteristics of, 207
  characterization of CML, 207
  effect of uracil analogs on,
      214, 216
  enrichment of activity, 206
  in hematological diseases, 204
  in human leukemia, 203, 204
  inhibitors of, 213-221
  physiological function of, 219
  properties of, 203, 213
  TdT^+ cell lines, 216
  TdT^- cell lines, 216, 217
  therapeutic regulation, 207
  in tissues, 205
Thymic factors
  and autoimmune disease, 152
  biochemical properties, 138
  biological activities, 140, 141
  and cancer, 153-155, 363-375
  clinical aspects of, 363-375
  contraindication to therapy of,
      373

Thymic factors (continued)
  and helper T-cell activity,
      143
  immune response of, 138
  and immunological diseases,
      151, 363-375
  and induction of suppressor
      T-cell, 144
  and infectious diseases, 151,152
  lymphocyte maturation by, 366
  mode of action of, 365
  and neuroendocrine system, 145
  pharacokinetics of, 146-149
  phylogeny, 149
  in primary immunodeficiency, 367
  production, 137,149
  and proliferation, 143
  in secondary T-cell
      differentiation, 368
  and T-cell differentiation,
      141, 142
  therapeutic properties, 364, 373
  thymopoietin, 139, 364
  thymosins, 138-141, 364
  thymostimulin, 370
Thymic hormones, see Thymic
    factors
Tumor Metastases
  and carcinoma, 67
  by cell fusion, 41, 45, 52
  and collagenase activity, 58,
      59, 88
  and EGF binding, 64
  generation of, 48, 51
  role of H-2 gene, 81-83
  immunomodulation of, 81-90
  invasive capacity of, 45, 47, 53
  of lymphoid tumor cells, 42, 45
  mechanisms of, 50
  MHC-expression and, 84
  multistep nature of, 87
  role of oncogenes in, 67, 69
  organotropism and, 45
  plasminogen activator in, 60
  properties of, 57
  and protein kinases, 57-70
  spontaneous, 58
  tumoric burden, 44
  tumoric potential, 45, 47,
      53, 58, 82
  tumoric variants, 48, 50

Viral Induction, 126
Viral Oncogenes, see Oncogenes